Free boundary problems:
theory and applications

CHAPMAN & HALL/CRC
Research Notes in Mathematics Series

Submission of proposals for consideration
Suggestions for publication, in the form of outlines and representative samples, are invited by the Editorial Board for assessment. Intending authors should approach one of the main editors or another member of the Editorial Board, citing the relevant AMS subject classifications. Alternatively, outlines may be sent directly to the publisher's offices. Refereeing is by members of the board and other mathematical authorities in the topic concerned, throughout the world.

Preparation of accepted manuscripts
On acceptance of a proposal, the publisher will supply full instructions for the preparation of manuscripts in a form suitable for direct photo-lithographic reproduction. Specially printed grid sheets can be provided. Word processor output, subject to the publisher's approval, is also acceptable.

Illustrations should be prepared by the authors, ready for direct reproduction without further improvement. The use of hand-drawn symbols should be avoided wherever possible, in order to obtain maximum clarity of the text.

The publisher will be pleased to give guidance necessary during the preparation of a typescript and will be happy to answer any queries.

Important note
In order to avoid later retyping, intending authors are strongly urged not to begin final preparation of a typescript before receiving the publisher's guidelines. In this way we hope to preserve the uniform appearance of the series.

CRC Press UK
Chapman & Hall/CRC Statistics and Mathematics
Pocock House
235 Southwark Bridge Road
London SE1 6LY
Tel: 0171 407 7335

I. Athanasopoulos
University of Crete, Greece

G. Makrakis
IACM, Foundation for Research and Technology, Greece

J.F. Rodrigues
CMAF-Universidade de Lisboa, Portugal

(Editors)

Free boundary problems:
theory and applications

CRC Press
Taylor & Francis Group
Boca Raton London New York

CRC Press is an imprint of the
Taylor & Francis Group, an informa business

A CHAPMAN & HALL BOOK

Published 1999 by Chapman & Hall/CRC Press
Taylor & Francis Group
6000 Broken Sound Parkway NW, Suite 300
Boca Raton, FL 33487-2742

© 1999 by Taylor & Francis Group, LLC
CRC Press is an imprint of Taylor & Francis Group, an Informa business

No claim to original U.S. Government works

ISBN 13: 978-1-58-488018-9 (pbk)

Visit the Taylor & Francis Web site at
http://www.taylorandfrancis.com

and the CRC Press Web site at
http://www.crcpress.com

Library of Congress Cataloging-in-Publication Data

Free boundary problems : theory and applications / I. Athanasopoulos,
    G. Makrakis, J.F. Rodrigues, editors.
            p. cm. -- (Chapman & Hall/CRC research notes in mathematics
    series)
        Includes bibliographical references.
        ISBN 1-58488-018-X (alk. paper)
        1. Boundary value problems--Congresses. I. Athanasopoulos, I.
    II. Makrakis, G. III. Rodrigues, José-Francisco. IV. Series.
    TA347.B69F743 1999
    515'.35—dc21   99-22939
                                                                           CIP

Library of Congress Card Number 99-22939

Contents

Preface

This volume contains the proceedings of the International Interdisciplinary Congress on "Free Boundary Problems, Theory and Applications", held in Crete (Greece), from 8 to 14 June, 1997.

The Scientific Committee was composed by H. Berestycki, L. Caffarelli, F. Durst, L.G. Leal, N. Markatos, A. Quarteroni, J.F. Rodrigues and J.L. Vazquez.

The Organizing Committee consisted of I. Athanasopoulos, S. Boudouvis, G. Kossioris, C. Makridakis and A. Tertikas.

Free Boundaries: an interdisciplinary theme

As an active topic in mathematical modelling during the last three decades, free boundaries problems hold a strategic position in pure and applied sciences. The occurrence of free boundaries and interfaces, i.e., material or geometrical frontiers between regimes with different physical properties not "a priori" prescribed, arises in an enormous number of situations in nature and in technology.

As a consequence of a valuable experience, "a regular series of conferences with a distinctive style" was established in the eighties. An interdisciplinary tradition between theory and applications was shaped in conferences held in Montecatini, Italy (1981), Maubuisson, France (1984), Irsee, Germany (1987), Montreal, Canada (1990), Toledo, Spain (1993), Zakopane, Poland (1995), and continued in Crete, Greece (1997).

The last three conferences have had the support of the European Science Foundation's scientific programme "Mathematical Treatment of Free Boundaries Problems" (FBP Programme), which has reduced the previous tri-annual period to bi-annual.

Scope of the Crete Congress

The FBP'97 congress consisted of Plenary Lectures, Focus and Discussion Sessions, Contributed Talks and Poster Presentations from more than one hundred scientists working in mathematics, material sciences, chemical engineering, biology, physics and other disciplines, among a total of 158 participants from 30 countries.

The main topics included free boundary problems in fluid and solid mechanics, combustion, financial mathematics, the theory of filtration and glaciology. Material science modeling, as well as recent mathematical developments and numerical analysis advances were presented in talks of more specific topics, such as singularities of interfaces, cusp cavitation and fracture, capillary fluid dynamics of film coating, dynamics of surfaces growth and phase transition kinetics and also phase fields models.

These Proceedings

Since it was not possible to include all contributions of these proceedings, the selection of the articles and their distribution in different chapters creating only one volume were done by the editors.

The balanced rearrangement of topics, excluding the invited lectures that constitute the first chapter of this book, led us to the following four other chapters:
- Mathematical Developments of Free Boundary Problems
- Free Boundary Problems in Fluid Mechanics
- Phase Change in Material Science
- Computational Methods and Numerical Analysis

Acknowledgments

In addition to the task of organizing such a congress where a substantial number of participants were at least partially supported directly by the Organizing Committee, there was the task of finding supporters (sponsors). Above all we must thank the European Science Foundation not only for the quantitative importance of the financial support allocated to us but also for the encouragement and support provided via the Steering Committee of the FBP Programme.

We are most thankful to the University of Crete, National Technical University of Athens, Technical Chamber of Greece, FORTH, Silicon Graphics Computer Systems, Ergobank, Hellenic General Secretariat, Hellenic Ministry of Culture, U.S. Air Force Office of Aerospace Research and Development, and European Research Office of U.S. Army. We thank also the following companies for their contribution: Instant Lottery S.A., Agricultural Bank of Greece, Interbank, Metaxa S.A., Aspropyrgos Hellenic Refineries, Psiktiki Ellados S.A., Plastika Kritis S.A., and TEE/TAK.

Finally, we wish to thank Mrs. Maridée Morales who brought the manuscripts into the final TeX-form as well as our publishers for their helpful assistance during the preparation of these proceedings.

Heraklion and Lisbon, November 1998
I. Athanasopoulos, G. Makrakis and J.F. Rodrigues

Part 1.

Plenary Lectures

L. BADEA, R. E. EWING, AND J. WANG[*]

A Study of Free Boundary Problems of Fluid Flow in Porous Media by Mixed Methods

Abstract

In this article the flow of fluids in porous media is studied as a free or moving boundary problem by the mixed method. In particular, a new weak formulation for the problem of seepage of fluids through a porous media is discussed and analyzed mathematically and numerically. The new formulation is in a mixed form and is suitable for the use of mixed finite element methods in the numerical approximation. It is proved that the weak formulation and its finite element discretization have a solution which can be approximated by a sequence of regularized problems.

1 Introduction

In this paper, we are concerned with the free or moving boundary value problem in the study of the flow of fluids through a heterogeneous porous media. Such problems are important in many branches of science and engineering. For example, in the areas of soil science, agricultural engineering, and groundwater hydrology, the movement of fluids and their dissolved components in both saturated and unsaturated soils is an important environmental consideration. In petroleum engineering, improved recovery of oil and gas is based on simulation of multiphase and multicomponent fluid transport in deep rocks. In both application areas, mass transfer across phase boundaries is an important consideration which can be discussed in the context of free or moving boundary problems.

Another important application of the free or moving boundary problem is water seepage through a dam, or rain water creeping through an unsaturated zone. The underlying physics of the petroleum and seepage problems are very similar. For comparison, assume that there are two fluids flowing simultaneously in the porous medium. In unsaturated flow, these fluids are water and air; while in the petroleum problem, the fluids are assumed to be water and oil. Relevant material properties, including the capillary pressure and relative permeability are assumed to be known.

Free boundary problems are also seen in other areas of the petroleum industry such as basin simulation. The research of basin simulation is important because it attempts to determine when and where organic matter was put down and how hydrocarbons may have been produced. This research requires treating the top layer of rock as a boundary which varies in time. Due to the deposition and erosion of materials, the

---

[*]The research of Ewing and Wang is partially supported by the NSF grant #DMS-9706985.

shifting of land masses, and natural processes which change solid organic matter into hydrocarbons, the precise definition of the rock strata and the surface of the basin, at any instant in time is not known *a priori*. In the description and modelling of the basin evolution, free boundaries are necessary.

For simplicity of presentation, we restrict ourselves to the 2-dimensional flow of fluids through a dam in which the free or moving boundary divides the saturated soil from the unsaturated part. This phenomena can be characterized by using the Richards equation [22]; similar phenomena in the petroleum engineering are governed by the Buckley-Leverett equation. The method of variational inequalities has been extensively studied for this problem in the past two decades. See Baiocchi [2], Oden and Kikuchi [20], and Chipot [11] for details. The existing results are based on Baiocchi or similar transformations which are based on some contour integrals. While the approach is powerful for analysis, it has some serious drawbacks in practical simulations. First, the hydraulic conductivity of the dam from the application must be in some special form which is not satisfied for the usual industrial problems. Second, there must be no holes in the dam since otherwise the contour integral will depend upon the choice of curves. Nevertheless, the method of variational inequalities has been a successful one in this application.

There was also a regularization method proposed by Brezis, Kinderlehrer, and Stampacchia [3] for the underlying free boundary value problem. The method of Brezis et al. does not have any restrictions on the hydraulic conductivity and the connectedness of the domain. This method has been proved to be powerful and general in the study of the existence and uniqueness of solutions.

The main objective of this paper is to propose a new weak formulation for the problem in a mixed form. The approach taken herein is to rewrite the governing equations in the form of a system of first order equations in which the flux is treated as an independent variable. It is shown that the new weak problem has a unique solution by using the Shauder fixed point theorem and a regularization technique. Moreover, the weak solution is approximated by using the mixed finite element method, which has the advantage of preserving the mass locally on each element. In addition, most of the techniques developed earlier by Ewing and colleagues [9, 10, 14, 15, 16, 17] for the mixed method have a strong potential for solving free or moving boundary value problems.

The paper is organized as follows. In §2, we shall describe the model problem in simulation. In §3, we derive a weak formulation for the steady-state problem in mixed form. In §4, we prove an existence result on weak solutions. In §5, we shall approximate the weak solution by a mixed finite element method.

2 Statement of the problem

Let $D \subset \mathbf{R}^2$ be the domain of the dam on a horizontal impervious foundation, through which the water is filtered so as to produce a two-dimensional flow field. Let us take the specific weight γ of the water as unity. The problem is to find the pressure of the fluid in D and that portion of D which is wet; the wet part of the domain D is called the saturated zone and shall be denoted by Ω.

Experimentally, the velocity $\mathbf{u}$ of the fluid in the saturated zone Ω is given by Darcy's law:

$$\mathbf{u} = -k\nabla(p + y), \qquad \text{in } \Omega, \tag{2.1}$$

where k is the hydraulic conductivity of the dam and the vertical direction is given by the y-axis. Using the mass conservation law, one arrives at the following Richards equation [22]:

$$\frac{\partial \theta}{\partial t} + \nabla \cdot \mathbf{u} = 0, \qquad \text{in } \Omega, \tag{2.2}$$

where θ represents the volumetric water content.

The geometry of our problem is illustrated in Figure 1. The cross-section D of the dam D is the region enclosed by $OACB$ with $A \in Ox$ and $B \in Oy$. The x-axis is an impermeable foundation, and the fluid lies on the left-hand side of y-axis with level H. Assume that point B has coordinates $B = (0, H)$, and on the right-hand side of AC there is accumulated fluid with fixed level h. The free boundary is the interface of the saturated region from the unsaturated part of the dam. Notice that the free boundary is changing with time.

Assume now that the atmospheric pressure is chosen to be 0 and neglect the effects of capillarity and evaporation. The pressure has to satisfy the following boundary conditions:

$$\begin{aligned}
\mathbf{u} \cdot \mathbf{n} &= 0 & \text{on } OA, \\
p &= H - y & \text{on } OB, \\
p &= h - y & \text{on } AF, \\
p &= 0 & \text{on } BCF,
\end{aligned} \tag{2.3}$$

where $\mathbf{n}$ denotes the unit outward normal direction to D. Denote by $\mathcal{C}(t)$ the free boundary at time t. The fact that no fluid can flow through the free boundary leads to two more conditions

$$\mathbf{u} \cdot \mathbf{n} = 0, \qquad p = 0, \qquad \text{on } \mathcal{C}(t), \tag{2.4}$$

where $\mathbf{n}$ indicates the unit outward normal vector to the saturated region Ω.

The problem is to find a pressure $p = p(x, y)$ and a region Ω such that (2.1), (2.2), (2.3), and (2.4) hold.

5

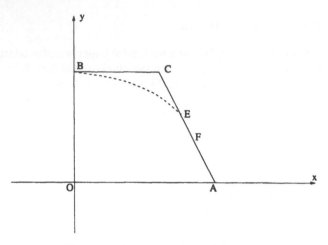

Figure 1: A cross-section of the dam.

3   Weak formulation in mixed form

In this section, we introduce a weak formulation for the model problem (2.1) and (2.2) in mixed form by following the idea of H. Brezis, K. Kinderlehrer, and G. Stampacchia [3]. For simplicity, we shall consider only the steady-state problem in which the time derivative in the Richards equation (2.2) disappears.

Assume that we have found a pair (p, Ω) which is a solution of (2.1), (2.2), (2.3), and (2.4). Let us extend the pressure p by 0 in the unsaturated region $D\backslash\Omega$. Also, we extend the velocity $\mathbf{u}$ by the vector $\mathbf{0}$ in the same unsaturated region. Since $\mathbf{u}\cdot\mathbf{n} = 0$ on the free boundary $\mathcal{C}$, the extended velocity is a vector-valued function in the Hilbert space

$$H(div) = \{\mathbf{v} \in \left[L^2(D)\right]^2, \nabla\cdot\mathbf{v} \in L^2(D)\},$$

which is equipped with the following norm:

$$\|\mathbf{v}\|_{H(div)} = \left(\|\mathbf{v}\|_0^2 + \|\nabla\cdot\mathbf{v}\|_0^2\right)^{1/2}.$$

Here $\|\cdot\|_0$ denotes the standard L^2 norm for either vector-valued or real-valued functions, as appropriate. Thus, it follows from (2.2) and the fact that the flow is of steady state that

$$\nabla\cdot\mathbf{u} = 0, \qquad \text{in } D. \tag{3.1}$$

The Darcy's law (2.1) is equivalent to

$$k^{-1}\mathbf{u} + \nabla p + \nabla y = 0, \qquad \text{in } \Omega. \tag{3.2}$$

Since p was extended by 0 in the unsaturated region and the pressure is positive in the saturated region Ω, then the Darcy's law (3.2) can be rewritten as follows:

$$k^{-1}\mathbf{u} + \nabla p + \nabla y H(p) = 0, \qquad \text{in } D, \tag{3.3}$$

where here, and in the rest of this paper, we shall denote by $H = H(x)$, the Heaviside function which has value 1 for $x > 0$ and 0 for $x \leq 0$.

The hydraulic conductivity k in the saturated region is quite different from the one in the unsaturated zone. Thus, the practical problem has a variable or even discontinuous hydraulic conductivity. Consequently, the underlying flow problem should be solved by numerical schemes which are appliable to heterogeneous porous media.

Let us now derive a weak formulation for the model free boundary value problem. First, we test (3.3) against any vector-valued function $\mathbf{v} \in \left(C^1(\bar{D})\right)^2$, yielding

$$(k^{-1}\mathbf{u}, \mathbf{v}) + (\nabla p, \mathbf{v}) + (\nabla y H(p), \mathbf{v}) = 0. \tag{3.4}$$

Using Green's formula (integration by parts), one arrives at

$$(\nabla p, \mathbf{v}) = -(\nabla \cdot \mathbf{v}, p) + \langle p, \mathbf{v} \cdot \mathbf{n} \rangle_{\partial D}, \tag{3.5}$$

where $\langle \cdot, \cdot \rangle_{\partial D}$ indicates the L^2 inner product on $L^2(\partial D)$. Observe that the value of the pressure p is known on ∂D, except on OA. Denote by ϕ the trace of the pressure p on $ACBO$. It is clear that ϕ is given by

$$\phi = \begin{cases} H - y & \text{on } OB, \\ 0 & \text{on } BCF, \\ h - y & \text{on } FA. \end{cases}$$

By choosing $\mathbf{v}$ such that $\mathbf{v} \cdot \mathbf{n} = 0$ on OA, one arrives at the following

$$(\nabla p, \mathbf{v}) = -(\nabla \cdot \mathbf{v}, p) + \langle \phi, \mathbf{v} \cdot \mathbf{n} \rangle_{ACBO}, \tag{3.6}$$

for all $\mathbf{v} \in H(div)$ such that $\mathbf{v} \cdot \mathbf{n} = 0$ on OA. Substituting the above into (3.4) yields

$$(k^{-1}\mathbf{u}, \mathbf{v}) - (\nabla \cdot \mathbf{v}, p) + (\nabla y H(p), \mathbf{v}) = -\langle \phi, \mathbf{v} \cdot \mathbf{n} \rangle_{ACBO}. \tag{3.7}$$

Next, by testing (3.1) against any $w \in L^2(D)$ we obtain

$$(\nabla \cdot \mathbf{u}, w) = 0. \tag{3.8}$$

The weak formulation seeks $p \in L^2(D)$, $\chi \in L^\infty(D)$, and $\mathbf{u} \in H(div)$ such that

1. $p \geq 0$ a.e. in D.

2. $0 \leq \chi \leq 1$ a.e. in D, $\chi = 1$ a.e. on the set where $p > 0$.

3. $\mathbf{u} \cdot \mathbf{n} = 0$ on OA.

4. The following equations

$$(k^{-1}\mathbf{u}, \mathbf{v}) - (\nabla \cdot \mathbf{v}, p) + (\nabla y \chi, \mathbf{v}) = -\langle \phi, \mathbf{v} \cdot \mathbf{n} \rangle_{ACBO},$$
$$(\nabla \cdot \mathbf{u}, w) = 0, \tag{3.9}$$

are fulfilled for any $\mathbf{v} \in H(div), \mathbf{v} \cdot \mathbf{n} = 0$ *on* OA *and* $w \in L^2(D)$.

Actually, the first equation in (3.9) should be replaced by (3.7) and the function χ should be related to the pressure field as $\chi = H(p)$ a.e. in D. A detailed discussion in this direction can be found from [3] and [11].

4 Existence of a solution

In this section, we show that the weak formulation has a solution. To this end, let H_ϵ be a regularization of the Heaviside function H given by

$$H_\epsilon(x) = \begin{cases} 1 & \text{if } x \geq \epsilon \\ x/\epsilon & \text{if } 0 \leq x \leq \epsilon \\ 0 & \text{if } x \leq 0. \end{cases}$$

Correspondingly, we consider the following problem which is a regularization of (3.7) and (3.8). Find $p_\epsilon \in L^2(D)$ and $\mathbf{u}_\epsilon \in H(div)$ with $\mathbf{u}_\epsilon \cdot \mathbf{n} = 0$ on OA such that

$$(k^{-1}\mathbf{u}_\epsilon, \mathbf{v}) - (\nabla \cdot \mathbf{v}, p_\epsilon) + (\nabla y H_\epsilon(p_\epsilon), \mathbf{v}) = -\langle \phi, \mathbf{v} \cdot \mathbf{n} \rangle_{ACBO},$$
$$(\nabla \cdot \mathbf{u}_\epsilon, w) = 0, \tag{4.1}$$

for all $\mathbf{v} \in H(div), \mathbf{v} \cdot \mathbf{n} = 0$ on OA, and $w \in L^2(D)$.

Theorem 4.1 *The regularized problem (4.1) has a solution.*

Proof. Consider the mapping T_ϵ which for any $q \in L^2(D)$, associates $\tilde{p}_\epsilon = T_\epsilon(q)$, the solution of the following problem:

$$(k^{-1}\tilde{\mathbf{u}}_\epsilon, \mathbf{v}) - (\nabla \cdot \mathbf{v}, \tilde{p}_\epsilon) + (\nabla y H_\epsilon(q), \mathbf{v}) = -\langle \phi, \mathbf{v} \cdot \mathbf{n} \rangle_{ACBO},$$
$$(\nabla \cdot \tilde{\mathbf{u}}_\epsilon, w) = 0, \tag{4.2}$$

for all $\mathbf{v} \in H(div), \mathbf{v} \cdot \mathbf{n} = 0$ on OA, and $w \in L^2(D)$. Here the solution $\tilde{\mathbf{u}}_\epsilon$ should satisfy $\tilde{\mathbf{u}}_\epsilon \cdot \mathbf{n} = 0$ on OA. The *inf-sup* condition of Brezzi and Babuška [4, 1] implies that the problem (4.2) has a unique solution. Moreover, there is a constant $C = C(\phi)$ such that

$$\|\tilde{\mathbf{u}}_\epsilon\|_{H(div)} + \|\tilde{p}_\epsilon\|_0 \leq C(\phi). \tag{4.3}$$

Notice that the first equation in (4.2) implies

$$k^{-1}\tilde{\mathbf{u}}_\epsilon + \nabla \tilde{p}_\epsilon + \nabla y H_\epsilon(q) = 0.$$

Thus,

$$\nabla \tilde{p}_\epsilon = -k^{-1}\tilde{u}_\epsilon - \nabla y H_\epsilon(q)$$

is a function in $[L^2(D)]^2$. This together with the estimate (4.3) shows that $T_\epsilon(q) = \tilde{p}_\epsilon \in H^1(D)$ and

$$\|T_\epsilon(q)\|_1 \leq C(\phi). \tag{4.4}$$

The mapping T_ϵ is then bounded from $L^2(D)$ to $H^1(D)$. From the compactness of the identity map from $H^1(D)$ to $L^2(D)$, it follows that the map T_ϵ is completely continuous from $H^1(D)$ to $H^1(D)$. The estimate (4.4) shows that T_ϵ maps the ball of radius $C(\phi)$ in $H^1(D)$ into itself. Thus, the Schauder fixed point theorem can be applied to yield the existence of a function $p_\epsilon \in H^1(D)$ such that $T_\epsilon(p_\epsilon) = p_\epsilon$. The fixed point p_ϵ, together with the corresponding u_ϵ, comprises a solution of (4.1). In fact, it can be shown that the fixed point p_ϵ takes the boundary value ϕ on the segment $ACBO$. □

The solution of the problem (4.1) can be regarded as an approximation of (3.7) and (3.8). It is then natural to look for solutions of (3.9) as the limit of p_ϵ, u_ϵ, and $H_\epsilon(p_\epsilon)$.

Theorem 4.2 *There exists a triple (p, u, χ) which is a solution of the weak formulation (3.9).*

Proof. For any $\epsilon > 0$, let (p_ϵ, u_ϵ) be the solution of (4.1). We know from the proof of Theorem 4.1 that there exists a constant $C(\phi)$ such that

$$\|p_\epsilon\|_1 + \|u_\epsilon\|_{H(div)} \leq C(\phi).$$

Thus, one can find a sequence $\{\epsilon_n\}_{n=1}^\infty$ with $\epsilon_n \to 0$ as $n \to \infty$ such that

$$\begin{aligned}
p_{\epsilon_n} &\longrightarrow p && \text{weakly in } H^1(D),\\
p_{\epsilon_n} &\longrightarrow p && \text{strongly in } L^2(D),\\
u_{\epsilon_n} &\longrightarrow u && \text{weakly in } H(div),\\
H_{\epsilon_n}(p_{\epsilon_n}) &\longrightarrow \chi && \text{weakly in } L^2(D).
\end{aligned}$$

Here we have used the fact that $\{H_{\epsilon_n}(p_{\epsilon_n})\}_n$ is bounded in $L^\infty(D)$ and, therefore, in $L^2(D)$.

Since χ is the limit of $H_{\epsilon_n}(p_{\epsilon_n})$, then the function χ satisfies the property that $\chi = 1$ a.e. on the set where $p > 0$. Furthermore, one has $0 \leq \chi \leq 1$ a.e. on D. Finally, it is easily seen that (p, u, χ) satisfies (3.9). □

Next, we claim that the solution of (4.1) satisfies $p_\epsilon \geq 0$ a.e. in D. In fact, the solution $p_\epsilon \in H_1(D)$ can also be characterized as the solution of the following problem: Find $p_\epsilon \in H^1(D)$ such that $p_\epsilon = \phi$ on $ACBO$ and

$$(k\nabla p_\epsilon, \nabla q) + (k\nabla y H_\epsilon(p_\epsilon), \nabla q) = 0, \tag{4.5}$$

9

for all $q \in H^1(D), q = 0$ *on* $ACBO$. Let $p_\epsilon^- = \min(0, p_\epsilon)$. It is clear that $p_\epsilon^- \in H^1(D)$ and $p_\epsilon^- = 0$ on the segment $ACBO$. By choosing $q = p_\epsilon^-$ in (4.5) we arrive at

$$(k\nabla p_\epsilon, \nabla p_\epsilon^-) + (k\nabla y H_\epsilon(p_\epsilon), \nabla p_\epsilon^-) = 0. \tag{4.6}$$

Observe that the following holds true for p_ϵ^-:

$$\nabla p_\epsilon^- = \begin{cases} 0, & p_\epsilon \geq 0, \\ \nabla p_\epsilon, & p_\epsilon < 0. \end{cases}$$

A proof of the above can be found from [18]. Thus, using the fact that $H_\epsilon(x) = 0$ for $x \leq 0$ we obtain from (4.6)

$$(k\nabla p_\epsilon^-, \nabla p_\epsilon^-) = 0,$$

which leads to $p_\epsilon^- = 0$ a.e. in D. It follows that $p_\epsilon \geq 0$ a.e. in D. Consequently, the weak solution obtained in the proof of Theorem 4.2 is non-negative for the pressure field.

5   Mixed finite element approximations

The weak formulation (3.9) allows us to approximate the underlying free boundary value problem by using mixed finite element methods. To this end, let $\mathcal{T}_h$ be a finite element partition of the domain D into triangles or quadrilaterals. The pressure p will be approximated in a finite element space, denoted by W_h, consisting of piecewise polynomials on $\mathcal{T}_h$. The velocity field $\mathbf{u}$ shall be approximated from a finite element space denoted by $\mathbf{V}_h$ which makes a stable pair with W_h. Details on the construction of $\mathbf{V}_h$ and W_h can be found from [21, 23, 8, 12, 7, 6, 5, 13].

For simplicity, we shall consider the Raviart-Thomas elements of lowest order on triangles. The pressure space consists of piecewise constants and the velocity space is defined by

$$\mathbf{V}_h = \{\mathbf{v} : \ \mathbf{v} \in H(div), \mathbf{v} \cdot \mathbf{n} = 0 \text{ on } OA, \mathbf{v}|_K \in V(K), \forall K \in \mathcal{T}_h\},$$

where $V(K)$ is the set of vector-valued polynomials of form $(a + cx, b + cy)$ on the triangle K.

Our mixed finite element method seeks a triple $(p_h, \mathbf{u}_h, \chi_h)$ from $W_h \times \mathbf{V}_h \times W_h$ satisfying

1. $p_h \geq 0$ in D,

2. $0 \leq \chi_h \leq 1$ in D, $\chi_h = 1$ on the elements where $p_h > 0$,

3. the following equations

$$\begin{aligned} (k^{-1}\mathbf{u}_h, \mathbf{v}_h) - (\nabla \cdot \mathbf{v}_h, p_h) + (\nabla y \chi_h, \mathbf{v}_h) &= -\langle \phi, \mathbf{v}_h \cdot \mathbf{n}\rangle_{ACBO}, \\ (\nabla \cdot \mathbf{u}_h, w_h) &= 0, \end{aligned} \tag{5.1}$$

are fulfilled for any $\mathbf{v}_h \in \mathbf{V}_h$ and $w_h \in W_h$.

Theorem 5.1 *The discrete problem (5.1) has a solution in the mixed finite element spaces.*

A proof for Theorem 5.1 can be given along the same line as that of Theorem 4.2. The details are thus omitted.

It is not clear whether or not the approximations p_h and χ_h from (5.1) are related by the following identity:

$$\chi_h = H(p_h). \tag{5.2}$$

A study should be conducted on the validity of (5.2).

Once the finite element approximations $(p_h, \mathbf{u}_h, \chi_h)$ are obtained from solving (5.1), one may choose a small, but positive real number δ (e.g., $\delta = 10^{-12}$) and approximate the saturated region Ω by the following set:

$$\Omega_h = \{K : \; p_h \geq \delta \text{ on } K\}.$$

Correspondingly, the unsaturated region is approximated by $D\backslash\Omega_h$. The free boundary is then approximated by the part of the boundary of Ω_h which is located in the interior of the domain D.

The mixed finite element method allows a natural use of locally refined grids in practical simulations. Therefore, most of the techniques which have been developed by Ewing and colleagues [14, 9, 10, 15, 16, 17] are applicable to the free boundary value problem considered in this article.

There are a couple of other open problems regarding the numerical scheme (5.1). First, the error between the true solution $(p, \mathbf{u}, \chi)$ and its approximation $(p_h, \mathbf{u}_h, \chi_h)$ should be established. Second, the uniqueness of solutions to (5.1) should be studied. Finally, some iterative methods should be discussed in order to solve the discrete problem (5.1) efficiently. Those aspects shall be discussed in our future research in this area.

Acknowledgment: We are grateful to Dr. Suzanne L. Weekes for a helpful discussion on this subject.

References

[1] I. BABUŠKA, The finite element method with Lagrangian multiplier, *Numer. Math.*, 20 (1973), 179–192.

[2] C. BAIOCCHI, Annali Mat. Pura et Appli., 92, 107 (1972).

[3] H. BREZIS, D. KINDERLEHRER AND G. STAMPACCHIA, *Sur une nouvelle formulation du problème de l'écoulement à travers une digue*, C.R. Acad. Sc. Paris, t. 287, Série A (1978), pp. 711-714.

[4] F. BREZZI, On the existence, uniqueness, and approximation of saddle point problems arising from Lagrangian multipliers, *RAIRO, Anal. Numér.*, 2 (1974), 129-151.

[5] F. BREZZI, J. DOUGLAS, R. DURÁN, AND L. MARINI, Mixed finite elements for second order elliptic problems in three variables, *Numer. Math.*, 52 (1987), 237-250.

[6] F. BREZZI, J. DOUGLAS, M. FORTIN, AND L. MARINI, Efficient rectangular mixed finite elements in two and three spaces variables, *RAIRO, Modélisation Mathématique Analyse Numérique*, 21 (1987), 581-604.

[7] F. BREZZI, J. DOUGLAS, AND L. MARINI, Two families of mixed finite elements for second order elliptic problems, *Numer. Math.*, **47** (1985), 217-235.

[8] F. BREZZI AND M. FORTIN, "Mixed and Hybrid Finite Element Methods", Springer-Verlag, New York, 1991.

[9] Z. CHEN, R. E. EWING, AND R. D. LAZAROV, Domain decomposition algorithms for mixed methods for second order elliptic problems, *Math. Comp.* **65(214)** (1996), 467–490.

[10] Z. CHEN, R. E. EWING, R. LAZAROV, S. MALIASSOV, AND Y. KUZNETSOV, Multilevel preconditioners for mixed methods for second order elliptic problems, *Numer. Lin. Alg. Appl.*, **3** (1996), 427–453.

[11] M. CHIPOT, *Variational Inequalities and Flow in Porous Media*, Springer-Verlag, New York, 1984.

[12] J. DOUGLAS AND J. ROBERTS, Global estimate for mixed finite elements methods for second order elliptic equations, *Math. Comput.*, 44 (1985), 39-52.

[13] J. DOUGLAS AND J. WANG, A new family of spaces in mixed finite element methods for rectangular elements, *Comp. Appl. Math.* 12 (1993), 183-197.

[14] R. E. EWING, R. LAZAROV, AND J. WANG, Superconvergence of the velocity along the Gauss lines in mixed finite element methods, *SIAM J. Numer. Anal.*, 28 (1991), 1015-1029.

[15] R. E. EWING AND J. WANG, Analysis of mixed finite element methods on locally refined grids, *Numerische Mathematik*, **63** (1992), 183–194.

[16] R. E. EWING AND J. WANG, Analysis of the Schwarz algorithm for mixed finite element methods, *R.A.I.R.O. Modélisation Mathématique Analyse Numérique*, **26** (1992), 739–756.

[17] R. E. EWING AND J. WANG, Analysis of multilevel decomposition iterative methods for mixed finite element methods, *R.A.I.R.O. Mathematical Modeling and Numerical Analysis*, **28(4)** (1994), 377–398.

[18] D. GILBARG AND N. S. TRUDINGER, "Elliptic Partial Differential Equations of Second Order" Springer-Verlag, New York, 1983.

[19] P. L. LIONS, *On the Schwarz alternating method I*, in R. Glowinski et al., eds., First International Symposium on Domain Decomposition Methods for Partial Differential Equations, Philadelphia, 1988, SIAM, pp. 2-42.

[20] J. T. ODEN AND N. KIKUCHI, Theory of variational inequalities with applications to problem of flow through porous media, *Internat. J. Engng. Sci.*, 18 (1980), pp. 1173-1284.

[21] P. RAVIART AND J. M. THOMAS, A mixed finite element method for 2nd order elliptic problems, In the *"Mathematics Aspects of Finite Element methods"*, *Lecture Notes on Mathematics*, Vol. 606, 292-315, 1977.

[22] L. A. RICHARDS, Capillary conduction of liquids through porous mediums, *Physics*, 1 (1931), pp. 318-333.

[23] J. M. THOMAS, Sur L'Analyse Numerique des Methodes D'elements Finis Hybrides et Mixtes, Ph.D. Thesis, Université Pierre et Marie Curie, 1977.

Lori Badea
Institute of Mathematics
Romanian Academy of Sciences, Romania

Richard E. Ewing
Institute for Scientific Computation
Texas A&M University
College Station, TX, 77843, USA

Junping Wang
Department of Mathematics
University of Wyoming
Laramie, WY 82071, USA

A. CĒBERS AND I. DRIĶIS

Labyrinthine Pattern Formation in Magnetic Liquids

1 Introduction

In different physical systems due to the competing attractive and long-range repulsive forces, the formation of labyrinthine patterns is observed [1]. This allows us to believe that the phenomenon of the labyrinthine pattern formation is insensitive to the concrete details of the properties of the physical systems. Interest in the investigation of the peculiarities of the labyrinthine pattern formation in magnetic fluids arose in early 1980s, when it was shown that the magnetic fluid in the plane layer under the action of the normal magnetic field forms labyrinthine patterns [2, 3]. Further different properties of the magnetic fluid labyrinthine pattern formation have been found [4, 5, 6, 7, 8, 9]. The crucial issue concerning the understanding of the labyrinthine pattern formation is the concept of the renormalization by the long-range magnetic interaction forces of the surface tension of the magnetic liquid in the plane layer. For the first time the proof that effective surface tension of the equilibrium stripe structure is exactly equal to zero was given in [10, 11]. Later that basic chevron instability was predicted and confirmed experimentally [12] as undulation instability in magnetic fluid and in the amphiphile monolayer foams [13, 14] described.

Here we are giving the general background of the description of the labyrinthine pattern formation in magnetic liquids. In the first part the equation of the motion of the magnetic fluid under the action of the magnetic field forces in the Hele-Shaw approximation is derived and general properties of the labyrinthine pattern formation established. Two different approaches to calculation of the magnetic field forces are discussed. Furthermore, in the second part boundary integral equation technique for the simulation of the labyrinthine pattern formation is briefly described and different numerical simulation results of the labyrinthine pattern formation are given. The concept of the surface tension renormalized by magnetic interactions is introduced in the third part. On that basis the chevron instability of the stripe pattern and undulation instability of polarizable foams are described.

2 Dynamics of magnetic liquid pattern formation

Let us derive the equation of motion of magnetic fluid in a plane layer in the Hele-Shaw approximation. Introducing the potential ψ of the self-magnetic field $\mathbf{H} = \mathbf{H_0} + \nabla\psi$ and neglecting its influence on the magnetization, the magnetic force on the volume element of the magnetic fluid arises from the surface forces acting on the fictious magnetic charges $\pm\mathbf{Mn}$ on the boundaries of the layer. Then for the volume density of magnetic forces in the layer of thickness h, we obtain accounting for symmetry

$$\mathbf{f_m} = \frac{-M\Delta S(-2\,\nabla\,\psi)}{\Delta Sh} = \frac{2M\,\nabla\,\psi}{h} \qquad (1)$$

Potential ψ is expressed as the Coulomb potential created by charges on the parts of the boundary of the layer wetted by magnetic fluid

$$\psi(\mathbf{r}) = -M\int \frac{dS'}{|\mathbf{r} - \mathbf{r'}|} + M\int \frac{dS'}{\sqrt{(\mathbf{r} - \mathbf{r'})^2 + h^2}} \qquad (2)$$

The equation for the mean velocity of fluid assuming Poiseuille profile

$$\mathbf{v} = \frac{3}{2}\langle\mathbf{v}\rangle\left(1 - \left(\frac{2z}{h}\right)^2\right)$$

then is obtained from the balance of viscous, pressure and magnetic forces

$$\mathbf{f_m} - \nabla p + \frac{2\eta}{h}\frac{\partial \mathbf{v}}{\partial z}\bigg|_{z=h/2} = 0$$

which leads to the Darcy equation taking into account the self-magnetic field forces [15]

$$-\alpha\langle\mathbf{v}\rangle - \nabla p + \frac{2M\,\nabla\,\psi}{h} = 0; \qquad \mathrm{div}\langle\mathbf{v}\rangle = 0 \qquad (3)$$

Further brackets for mean velocity are omitted.

Pressure on the free boundary of magnetic liquid is determined by Laplace law

$$p|_\Sigma = \frac{\sigma}{R_c}, \qquad (4)$$

where R_c is curvature of the boundary.

Energy of the magnetic fluid volume can be expressed in two equivalent ways. From one point of view, the energy can be expressed as Coulomb interaction energy of magnetic charges which account for symmetry between upper and lower boundaries of the plane layer can be written as follows:

$$E_m = \frac{1}{2}\int -2M\psi\,dS' = -\int M\psi\,dS'$$

From another point of view, the problem of the calculation of magnetic fields created by the fictious magnetic charges on the upper and lower boundaries of the plane layer can be easily reduced to the calculation of the magnetic field created by azimuthal currents along the free boundary of the volume. Namely, the problem for Laplace equation at boundary conditions

15

$$\psi_i = \psi_e, \qquad H_{en} = H_{in} + 4\pi Mn$$

on top and bottom surfaces of magnetic fluid volume by transformation $\mathbf{H_i} = -4\pi M\Theta$ $+\mathbf{H'_i}$ (Θ - Heaviside function of the magnetic fluid region) is reduced to the problem at boundary condition on free interface [16, 17]

$$\mathrm{rot}\mathbf{H'} = 0; \quad \mathrm{div}\mathbf{H'} = 0; \quad H'_{iz} - H_{ez} = \frac{4\pi}{c}cM$$

This shows that magnetic field $\mathbf{H'}$ is produced by azimuthal current with linear density $\mathbf{i} = cM\mathbf{t}$. As a result the magnetic energy can be expressed as follows:

$$E_m = \frac{1}{8\pi} \int \mathbf{H}^2 \, dV = 2\pi M^2 V - \frac{1}{2c} \int \mathbf{i} A \, dS,$$

where $\mathbf{H'} = \mathrm{rot}\,\mathbf{A}$. After substitution the expression for the vector potential

$$\mathbf{A(r)} = \frac{1}{c} \int \frac{cM\mathbf{t}\, dl\, dz}{\sqrt{(\mathbf{r} - \mathbf{r'})^2 + (z - z')^2}}$$

the magnetic field energy in "current" formulation can be expressed as follows [7, 16, 17, 18]:

$$E_m = 2\pi M^2 V - \frac{M^2}{2} \int d\mathbf{l}\, dz \int \frac{d\mathbf{l'}\, dz'}{\sqrt{(\mathbf{r} - \mathbf{r'})^2 + (z - z')^2}} \tag{5}$$

For calculation of the dynamics of the labyrinthine pattern formation the concept of magnetic pressure is useful. It can be introduced as follows:

$$\delta E_m = -\oint p_m h \delta \xi_n \, dl$$

where $\delta\xi$ is Lagrange displacement. According to the energy in "charge" formulation,

$$p_m = \frac{2M\psi}{h} \tag{6}$$

Variation of the energy in "current" formulation gives

$$p_m = -2\pi M^2 - \frac{2M^2}{h} \int \frac{(x - x')y'_l - (y - y')x'_l}{|\mathbf{r} - \mathbf{r'}|^2} \left(\sqrt{(\mathbf{r} - \mathbf{r'})^2 + h^2} - \sqrt{(\mathbf{r} - \mathbf{r'})^2} \right) dl'$$

To prove the identity of the last expression with (6), the following relation is necessary [9]:

16

$$-\int \frac{(x-x')y_l' - (y-y')x_l'}{|\mathbf{r}-\mathbf{r}'|^2}\sqrt{(\mathbf{r}-\mathbf{r}')^2 + h^2}\, dl' \;=\; \int \frac{\partial}{\partial n'} \ln|\mathbf{r}-\mathbf{r}'|\sqrt{(\mathbf{r}-\mathbf{r}')^2 + h^2}\, dl'$$

$$= \int \frac{dS'}{\sqrt{(\mathbf{r}-\mathbf{r}')^2 + h^2}}$$

for all $\mathbf{r}$ outside the region occupied by magnetic fluid. Taking the limit of the last expression at $\mathbf{r}$ tending to the point on contour from outside and applying the theorem for the limiting value of the double layer potential we obtain

$$\int \frac{dS'}{\sqrt{(\mathbf{r}-\mathbf{r}')^2 + h^2}} = -\pi h + P\int \frac{\partial}{\partial n'} \ln|\mathbf{r}-\mathbf{r}'|\sqrt{(\mathbf{r}-\mathbf{r}')^2 + h^2}\, dl'$$

where P is the principal value by Cauchy. This proves the relation

$$p_m = \frac{2M^2}{h}\int \frac{dS'}{\sqrt{(\mathbf{r}-\mathbf{r}')^2 + h^2}} - \frac{2M^2}{h}\int \frac{dS'}{|\mathbf{r}-\mathbf{r}'|}$$

The equivalence of "charge" and "current" formulation is useful for understanding the analogies which exist between labyrinthine patterns of magnetic liquids and structures observed in superconductors of I kind [19].

For discussion of the properties of magnetic liquid labyrinthine pattern formation it is useful to note that dynamics in the framework of the present model is completely dissipative. This means that total energy $E = E_v + E_m$ including surface $E_s = \sigma h \oint dl$ and magnetic contributions E_m is diminishing at the rate determined by viscous dissipation

$$\frac{d}{dt}(E_s + E_m) = -\alpha \int \mathbf{v}^2\, dS$$

Properties of the labyrinthine pattern formation in the present model are dependent on only two parameters magnetic Bond number $2M^2 h/\sigma$ characterizing the ratio of magnetic and surface forces, and dimensionless parameter $2R/h$ characterizing the dimensions of the volume of magnetic liquid.

It should be mentioned that the model of dipolar interactions considered above can be also applied to the description of domain shape instabilities in polarized amphiphile monolayers [16, 17, 18, 20]. In that case the thickness h becomes of the order of magnitude of molecular length. Different characteristics of shape instabilities of amphiphile monolayers by introducing molecular "cutoff" at the calculation of the dipolar interaction energy are obtained in papers [21, 22, 23].

Shape dynamics of amphiphile monolayer domains due to coupling with the hydrodynamic motion of the substrate is more complicate as described by equation (3). Neglecting the dissipation in the monolayer domain interface dynamics was considered

17

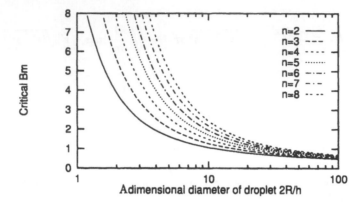

Figure 1: The critical values of the magnetic Bond number depending on the adimensional diameter of droplet for the instability with respect to n-lobe mode, calculated according to relation (7).

in [24]. Increments of the growth of symmetry destroying small domain perturbations accounting for subphase dynamics have been calculated in [25, 26].

As already has been mentioned above the main interest connecting the dynamics of the magnetic liquid in thin plane layers consists of the development of labyrinthine patterns. The physical reason for the instabilities leading to the formation of labyrinthine patterns is quite simple and can be explained as follows. Under the magnetization of magnetic fluid in the external field parallelly oriented dipoles repulse each other. Since for dipoles nearer to the periphery of the circular droplet, the average distance to other dipoles is larger than for dipole in the center, the energy of this repulsive interaction is decreasing approaching to the boundary. That means that force exists directed to the interface and at equilibrium is balanced by a corresponding pressure gradient. If the magnetic fluid volume undergoes some deformation from an initially symmetric configuration to a nonsymmetric one, that means that some amount of magnetic liquid is transferred further from the center of droplet. Thus repulsive magnetic interaction energy decreases at the deformation of magnetic fluid volume to nonsymmetric configuration. If that decrease will take over the corresponding increase of the surface energy of the free boundary, then instability takes place and the magnetic fluid volume undergoes the transition to a new nonsymmetric figure of equilibrium. The critical Bond number $2M^2h/\sigma$ depending on an adimensional diameter of the droplet $p = 2R/h$ at which the instability with respect to the n-lobe mode of free interface deformation takes place $r = R_0(1 + a_n \cos n\varphi)$ [2, 7, 27] can be expressed as follows:

$$Bm_n^c = \frac{2(n^2 - 1)}{p^2 \left[\psi\left(n + \frac{1}{2}\right) - \psi\left(-\frac{1}{2}\right) + Q_{n-\frac{1}{2}}\left(\frac{p^2+2}{p^2}\right) - Q_{\frac{1}{2}}\left(\frac{p^2+2}{p^2}\right) \right]}, \tag{7}$$

where $\psi(x)$ is the digamma function and $Q_n(x)$ the Legandre function of II kind.

At large R and n relation (7) transforms to the expression of the critical Bond number for the straight edge of magnetic fluid in the plane layer [3]. Values of the critical Bond number calculated according to the relation (7) are shown in Figure 1.

When the magnetic Bond number is greater than the critical value for the corresponding mode development, instability takes place and magnetic fluid volume undergoes transition to the new equilibrium configuration. We say that at corresponding values of magnetic Bond numbers bifurcation to new families of figures of equilibrium takes place. The precise shape of the figure of equilibrium belonging to a definite family can be found in principle by the numerical simulation of the dynamics of instability development.

3   Algorithm of numerical simulation

For the first time efficient numerical technique based on the boundary integral equation method for the numerical simulation of free surface flows in Hele-Shaw cells was proposed in [28]. It is based on the solution of the boundary integral equation of II kind. The numerical method based on the solution of boundary integral equations of I kind was ellaborated in [29, 30]. In [9] numerical simulation results of magnetic free surface flows obtained by conformal mapping method [31] are given.

From equation of motion (3) it follows that magnetic Hele-Shaw flow in the given approximation $M = $ const is potential - rot $\mathbf{v} = 0$. More complex phenomena, for example, magnetic Saffman-Taylor instability in the miscible case when magnetic ponderomotive force due to the magnetization dependent on concentration is nonpotential are considered in [32] and will not be considered here. Thus introducing the stream function

$$v_x = \frac{\partial \varphi}{\partial y}; \qquad v_y = -\frac{\partial \varphi}{\partial x} \tag{8}$$

condition of the potential flow shows that it satisfies Laplace equation

$$\Delta \varphi = 0$$

Since on the free boundary the pressure is known according to Laplace law (4) then the equation of motion allows expression of the tangential to the boundary component of the velocity of flow as follows:

$$\mathbf{v}_t|_\Sigma = -\frac{1}{\alpha} \frac{d}{dl} \left(\frac{\sigma}{R_c} - \frac{2M}{h} \psi \right)$$

19

Thus magnetic free boundary Hele-Shaw flow in the case of a magnetic droplet in plane layer can be found as a solution of the internal Neuman problem for the Laplace equation. Consideration of a more general case with viscosity contrasted between inner and outer regions is straightforward.

$$\left.\frac{\partial\varphi}{\partial n}\right|_{-} = -\frac{1}{\alpha}\frac{d}{dl}\left(\frac{\sigma}{R_c} - \frac{2M}{h}\psi\right); \qquad \triangle\varphi = 0 \tag{9}$$

Solution of the Neuman problem by representing the stream function as a single layer potential

$$\varphi = \frac{1}{2\pi}\int f(l')\ln|\mathbf{r} - \mathbf{r}'|\,dl'$$

can be reduced to the solution of boundary integral equation of II kind [28]

$$\frac{1}{\alpha}\frac{d}{dl}\left(\frac{\sigma}{R_c} - \frac{2M}{h}\psi\right) = -\frac{1}{2}f + \frac{1}{2\pi}P\int f(l')\frac{\partial}{\partial n}\ln|\mathbf{r} - \mathbf{r}'|\,dl' \tag{10}$$

If the solution of the boundary integral equation is found, the dynamics of the interface is calculated according to the stream function (8).

At numerical realization the unknown function f is represented with a finite number of pyramidal basis functions and the boundary integral equation is discretized by Galerkin method [28]. The obtained linear set of equations is solved by GM-RES method [33]. The free boundary from known coordinates of N marker points is represented by interpolation with cubic splines. From those interpolated values the curvature and magnetic pressure are calculated. Since at the evolution of the magnetic fluid shapes the length of the interface and its complexity are changing considerably, the new marker points should be introduced during the calculations [34].

In [35] the free interface is represented by dependence of the tangent angle on contour length and different gauges for tangential to boundary velocity are discussed.

4   Numerical simulation results of the labyrinthine pattern formation.

In Figure 2 the dynamics of establishment of equilibrium dumbbell and three lobe shapes at $Bm = 8$ and $p = 1$, $r = R_0(1 + \varepsilon_n\cos n\phi)$, $\varepsilon_n = 0.1$ is shown. We draw attention that the value of the magnetic Bond number is greater as a critical value for the instability with respect to 2-lobe and 3-lobe modes which means that at least 2 families of the figures of equilibrium exist at a given magnetic Bond number. That, of course, is consistent with numerical simulation results shown in Figure 2. It is interesting to note here that according to the finite element calculations [36] accounting for gravity effects and wetting angle, the transition circle-ellipse-dumbbell is subcritical at large layer thickness but supercritical for small. Analytical investigations of

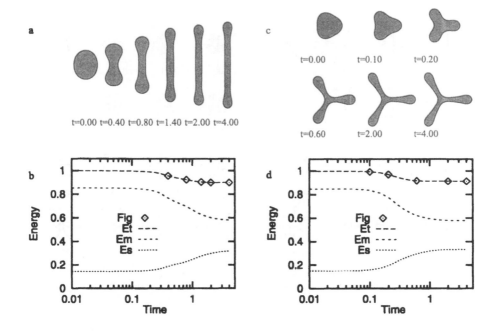

Figure 2: Results of numerical experiments at $Bm = 8$ and $p = 1$. The dynamics of establishment of equilibrium dumbbell shape (a) and three-lobe shape (c) and time evolution of characteristic energies in (b,d) (squares correspond to figures shown in a,c).

the energy of the family of the ovals of Cassini show that transition from circular to dumbbell shape is of first order [37]. It is of interest also to draw attention to the competition of the magnetic and surface energies during the establishment of the figures of the equilibrium. As is shown in Figure 2b and Figure 2d during the establishment of the figures of the equilibrium, magnetic energy of the droplet is decreasing and surface energy increasing. The total energy in accordance with dissipative character of the dynamics of labyrinthine pattern formation is, of course, decreasing.

It should be emphasized that figures of equilibrium shown in Figure 2 are quite similar to those observed in experiments for magnetic liquids [2] as well as domains in amphiphile monolayers [38].

Considering the dynamics of the development of 2-lobe and 3-lobe figures of equilibrium shown in Figure 2 the question arises if it will be possible to observe at corresponding initial conditions the formation of the figures of equilibrium with n-fold symmetry. The results of the numerical simulation show that it is impossible due

21

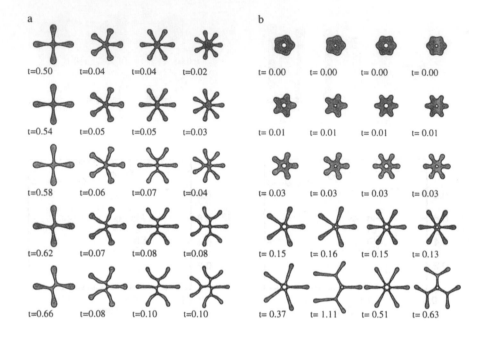

a

t=0.50 t=0.04 t=0.04 t=0.02

t=0.54 t=0.05 t=0.05 t=0.03

t=0.58 t=0.06 t=0.07 t=0.04

t=0.62 t=0.07 t=0.08 t=0.08

t=0.66 t=0.08 t=0.10 t=0.10

b

t= 0.00 t= 0.00 t= 0.00 t= 0.00

t= 0.01 t= 0.01 t= 0.01 t= 0.01

t= 0.03 t= 0.03 t= 0.03 t= 0.03

t= 0.15 t= 0.16 t= 0.15 t= 0.13

t= 0.37 t= 1.11 t= 0.51 t= 0.63

Figure 3: a) Splitting of vertexes arising in development of n-fold deformation mode. Part of numerical experiment results from initial conditions with $n = 4, 5, 6, 7$. In left column $Bm = 30$, $p = 1$ and $Bm = 30$, $p = 2$ in all other colums. b) Stabilization of vertices by trapped bubble for $Bm = 18$ and $p = 2$.

to vertex splitting instability [34]; namely, n-fold vertex arising at the development of n-fold deformation mode is unstable and by cascade of splitting events evolves to several 3-fold vertices.

How it happens for several initial perturbation modes is shown in Figure 3a. Those vertex splitting events lead to the breaking of n-fold symmetry. It is interesting to note here that the general proof of the vertex splitting instability for our system at the present moment is lacking. It should be emphasized that this issue does not seem to be trivial at all as is illustrated by the numerical simulation results of the development of the n-lobe deformation mode for the annulus of the magnetic fluid shown in Figure 3b. Those results show that if the radius of the inner bubble is large enough n-fold vertex can be stabilized. Even more interesting to mention is that this conclusion at the present moment has experimental confirmation in amphiphile monolayer stabilization of n-fold ($n > 3$) vertex by inner bubble is observed [39]. In cellular patterns in magnetic garnet films, in addition to threefold vertices pentagonal

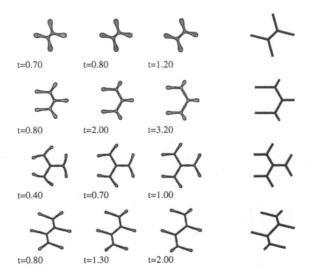

Figure 4: Formation of Steiner trees. Part of shape evolution for different initial conditions with $n = 4, 5, 6$. In the upper row $Bm = 30$, $p = 1$, in the second row $Bm = 15$, $p = 2$ and $Bm = 25$, $p = 2$ in other rows. On the right side Steiner trees for regular polygons are shown.

domain structures which contain trapped bubbles are observed [40].

By vertex splitting events at intermediate stages the configurations are formed which are very close to Steiner trees which are characteristic for 2D soap films expanded between a fixed number of pins [41]. Comparison between the configuration of the magnetic fluid volume in plane layer at intermediate stage of the instability development and Steiner trees calculated theoretically for positions of pins on the vertices of regular polygons are shown in Figure 4. We can see an astonishing coincidence although in the case of the magnetic fluid droplet there do not exist external fixed pins; the role of pins is played by the magnetic interaction forces.

Due to the dynamics of the magnetic fluid, volume turns out to be more complicate. After establishment at intermediate stages the configurations corresponding to Steiner trees persist for quite a long period of time as is shown in Figure 5. During that time adjustments of the thickness of the stripes take place. After the end of the intermediate stage the rearrangement of the Steiner tree to a much simpler configuration as is illustrated in Figure 5 takes place. At this rearrangement as we can see from energy curves on Figure 5b the decrease of surface energy and increase of the magnetic energy occur opposite to the initial stage of structure formation driven by the decrease

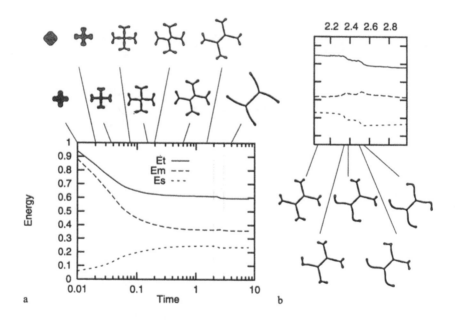

Figure 5: Numerical simulations of a magnetic fluid droplet's long-term evolution for $Bm = 30$, $p = 2$. a) Transient shapes and characteristic energies in semilogarithmic plot, b) enlarged view of time interval $2\dots3$.

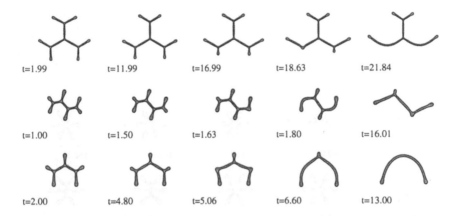

Figure 6: Part of shape evolution at rearrangement of Steiner trees. Bond numbers are $Bm = 30$ (upper row), $Bm = 15$ (middle and lower row) and $p = 2$. The last figures are not equilibrium shapes.

of the magnetic energy. Total energy at the rearrangement of the Steiner tree, of course, decreases. It should be mentioned that a conclusion about rearrangements of Steiner trees to simpler configurations is consistent with numerical simulation results by contour dynamics [8] where it is shown that from final configurations with zero, one and two vertices the smallest energy has configurations without vertices. With respect to the structure formation events shown in Figure 5, it is interesting to note that the curves of the energy changes do not show any peculiarities at vertex splitting events. Thus it occurs as symmetry destroying second order transition. The scenario of the Steiner tree formation at intermediate stages and its rearrangement at later stages that still persist at other initial conditions has been checked. In Figure 6 it is illustrated that the formation of a Steiner tree at the intermediate stage with its subsequent rearrangement to a simpler configuration takes place for different initial conditions. Another issue worthy of note concerns the influence of the viscosity contrast on the formation of the intermediate Steiner tree phase and its subsequent rearrangement. A comparison of the dynamics of a bubble in magnetic liquid and magnetic liquid droplet for the same initial conditions and the values of physical parameters is given in Figure 7. It is possible to see the delay in the case of a bubble of the vertex splitting events. As well, the Steiner tree configuration formed in the case of the magnetic fluid droplet is much longer living.

Concluding the issue concerning vertex splitting events, it is interesting to claim that the vertex splitting events can be responsible for the nonlinear suppression of the development of the overextension instability and formation of the alternating finger

25

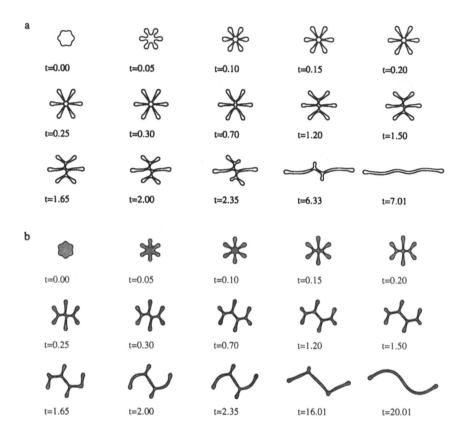

Figure 7: Comparison between dynamics of a bubble in magnetic fluid (a) and magnetic fluid droplet (b). Initial shape and parameters ($Bm = 15, p = 2$) are the same.

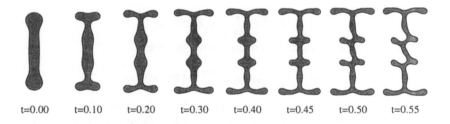

t=0.00   t=0.10   t=0.20   t=0.30   t=0.40   t=0.45   t=0.50   t=0.55

Figure 8: Development of overextension instability and splitting of 4-fold vertices. Numerical simulations of magnetic fluid droplet evolution for $Bm = 1.5$, $p = 20$.

patterns of the magnetic fluid stripes. How vertex splitting can at nonlinear stages suppress the development of overextensions is illustrated by numerical simulation results in Figure 8. It is possible to observe that development of overextensions at nonlinear stages is suppressed by vertex splitting and as a result the pattern of alternating fingers arises. Such kinds of patterns are observed also in experiments for different systems (magnetic fluid stripes, stripe domains in ferromagnetic films) [42, 43]. This phenomenon of overextension suppression by vertex splitting could be the reason why magnetic fluid stripe rupture under the action of external field is usually not observed in experiments.

5   Effective surface tension of the free boundaries under the action of long-range magnetic forces.

It turns out that long-range magnetic interactions renormalize the surface tension of free boundaries in such a way that effective surface tension of the equilibrium structure turns out to be exactly equal to zero. A simple illustration of this fact can be given by the example of magnetic fluid stripe. This example also has interest itself since properties of polarized stripes allow understanding of the behavior of the magnetic fluid foams, foams in amphiphile monolayers, etc. Let us consider the magnetic stripe along y-axis. At small bending deformation the equations of its boundaries will be $x = \pm d + \zeta(y)$. The length of free boundary can be expressed as

$$l = \int \sqrt{1 + \zeta_y'^2}\, dy = \text{const} + \frac{1}{2} \int \zeta_y'^2\, dy$$

So accounting for two free boundaries and neglecting end effects the surface energy of the stripe can be expressed as follows:

$$E_{surface} = 2\sigma l h = \text{const} + \sigma h \int \zeta_y'^2\, dy$$

27

Representing the free boundary displacement by Fourier integral

$$\zeta(y) = \frac{1}{2\pi} \int \zeta(k) e^{iky} \, dk$$

free surface energy of stripe can be expressed as follows:

$$E_{surface} = \text{const} + \frac{\sigma h}{2\pi} \int k^2 |\zeta(k)|^2 \, dk$$

For a magnetized stripe it is necessary also to account for magnetic interaction energy. Then considering the long wavelength surface deformations $k \to 0$ the total energy of stripe introducing effective surface tension σ_e can be expressed as follows:

$$E = E_{surface} + E_{magnetic} = \frac{\sigma_e h}{2\pi} \int k^2 |\zeta(k)|^2 \, dk$$

It turns out that if the stripe has equilibrium length L when $\partial E / \partial L = 0$, then effective surface tension σ_e is exactly equal to zero. Let us illustrate that fact.

The total energy of a magnetic stripe with length L can be expressed as follows:

$$E = 2\sigma h L + L M^2 h^2 \, e\left(\frac{2d}{h}\right), \tag{11}$$

where the second term represents the magnetic interaction energy and function $e(x)$ has the following expression:

$$e(x) = x^2 \ln\left(1 + \frac{1}{x^2}\right) - \ln(1 + x^2) + 4x \arctan x$$

Minimizing (11) with respect to stripe length at a fixed surface area of stripe $S = 2dL$, the equation for equilibrium width of the stripe is as follows:

$$Bm\left(\frac{4d^2}{h^2} \ln\left(1 + \frac{h^2}{4d^2}\right) + \ln\left(1 + \frac{4d^2}{h^2}\right)\right) = 4 \tag{12}$$

From equation (12) we obtain that the equilibrium width of the stripe is decreasing with the magnetic Bond number. Energy of a slightly deformed stripe can be expressed as follows:

$$E = 2\sigma h L + L M^2 h^2 \, e\left(\frac{2d}{h}\right) + \frac{1}{2\pi} \int |\zeta(k)|^2 \left\{\sigma h k^2 - \right. \tag{13}$$

$$\left. - 4M^2 \left(\ln\frac{kh}{2} - \ln\sqrt{1 + \frac{h^2}{4d^2}} + \gamma + K_0(kh) + K_0(2dk) - K_0\left(k\sqrt{4d^2 + h^2}\right)\right)\right\} \, dk,$$

where γ is the Euler constant, K_0 is the McDonald function. Considering the long-wavelength limit we obtain that the coefficient at the k^2 term for the stripe of equilibrium width is exactly equal to zero. That is the proof that effective surface tension of

28

the stripe with equilibrium width is exactly equal to zero. From that we immediately can conclude that if magnetic Bond number is larger, the value at which the width of stripe corresponds to equilibrium σ_e is negative. It means that the stripe must undulate. In Figure 9a that phenomenon is illustrated by numerical simulation results of the behavior of a magnetic stripe at an abrupt increase of the magnetic Bond number. For undulation to take place in the case of the stripe with finite length L, critical magnetic Bond number can be determined from the relation ($k = 2\pi/L$)

$$Bm_c = \frac{(kh)^2}{2\left(\ln \frac{kh}{2} - \ln \sqrt{1 + \frac{h^2}{4d^2}} + \gamma + K_0(kh) + K_0(2dk) - K_0\left(k\sqrt{4d^2 + h^2}\right)\right)} \tag{14}$$

The conclusions about the undulations of the stripes allow understanding of some peculiarities of the behavior of polarized foams. Representing the magnetic foam as a set of stripes with different lengths we see that at the increase of the magnetic Bond number if the width of stripe is fixed due to fixed positions of vertices, at first the longest stripes must undulate, and after that the shorter ones. That is what is observed in experiment [14] and in the numerical simulation of magnetic fluid foam shown in Figure 9b. One can see that as the magnetic Bond number increases subsequently more short stripes start to undulate. It should be noted that the behavior of the magnetic foam illustrated in Figure 9b corresponds to the predictions of our model not only qualitatively but also quantitatively (relation (13)). This supports the possibility of applying the model of the single stripes to the description of the magnetic foam at a large stripe length to width ratio. The approach of the effective surface tension turns out to be productive also for the understanding of some properties of foams in amphiphile monolayers, namely, in [13] it was observed that at enough fast temperature increases stripes of amphiphile monolayers foam undulate. The theoretical understanding of that phenomenon can be obtained from the point that as the temperature increases the width of the stripes becomes larger at equilibrium one due to the fixed positions of vertices in the foam. Theoretical relation for the wavelength of developing stripe undulations of the amphiphile monolayer foam can be obtained from the relation (12) and (13) in the limit $2d/h \rightarrow \infty$ since in amphiphile monolayers h is small (h has the order of magnitude of the dipole length). From the relation (12) for critical magnetic Bond number we have

$$Bm_c = \frac{2}{\ln \frac{2d}{h}} \tag{15}$$

Wavelength of the developing undulation deformations can be found from relation (13). In the limit $2d/h \ll 1$ and $kh \rightarrow 0$ we have from the relation (13)

$$kh = 2\exp\left(-\frac{2}{Bm_c}\right)$$

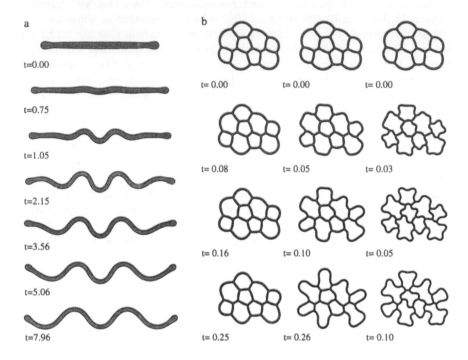

Figure 9: a) Bending instability for single stripe. $Bm=1.4$, $p=20$. The last is not an equilibrium shape. b) Results of numerical experiments. Magnetic foam structure at time moments increasing downward for $p=20$ and magnetic Bond number $Bm = 2.5$ (left column), $Bm = 3.0$ (middle column), $Bm = 3.5$ (right column). The average wall thickness $\langle 2d \rangle = 1.2h$.

As a result of substitution Bm_c by relation (15) we obtain the following relation for the dependence of the wavelength of developing undulation deformations on the width of the stripe ($D = 2d$ - width of stripe)

$$\lambda = \pi D \tag{16}$$

The relation (16) gives the prediction which is in reasonable aggreement with experimental data – for stripe width $D = 3\mu m$, the calculated wavelength $\lambda = 9\mu m$ is in range of the wavelengths $10 \div 25\mu m$ observed for amphiphile monolayers in experiment [13]. It is interesting to note that the estimate for the critical Bond number obtained according to relation (15) $Bm_c \approx 0.250$ ($D = 3\mu m$; $h = 10\text{Å}$) is quite near to the estimate $Bm = 0.163$ obtained in [45], although for different amphiphile, from the analysis of domain shape fluctuation data of paper [46].

Patterns similar to those obtained numerically and shown in Figure 9 are also observed for foam of gaseous domains in an amphiphile monolayer under its expansion [47].

Another system where the concept of the magnetic interaction renormalized effective surface tension turns out to be quite fruitful concerns infinite system of magnetic liquid stripes and, in fact, also labyrinthine patterns.

Considering the periodic system of magnetic fluid stripes in the plane layer, it is possible to show in the same way as has been done above for the case of the single stripe that effective surface tension of the equilibrium system of magnetic stripes is exactly equal to zero. The period of the equilibrium stripe pattern l satisfies the equation

$$\frac{d}{dl}\tilde{e}(l) = 0,$$

where $\tilde{e}$ is the surface density of the total energy of stripe pattern. The expression for the surface density of the total energy of stripe pattern [48] (φ_m is volume fraction of magnetic liquid in the layer)

$$\tilde{e} = \frac{2\sigma h}{l} + 2\pi M^2 h \varphi_m (1 - \varphi_m) - \frac{2M^2 h^2}{l} \int_0^1 (1-t) \ln\left(1 + \frac{\sin^2 \pi \varphi_m}{\sinh^2 \frac{\pi h t}{l}}\right) dt$$

For a slightly deformed stripe pattern it is possible to obtain the following expression for the energy [10, 11]:

$$\tilde{e} = \tilde{e}^0 + \frac{1}{2\pi} \int |\zeta(k)|^2 \left(F(k) - F(0)\right) dk,$$

where

$$F(k) = \sum_{n=1}^{\infty} \frac{16\pi M^2}{l^2 \sqrt{(2\pi n/l)^2 + k^2}} \left(1 - \exp\left(-h\sqrt{(2\pi n/l)^2 + k^2}\right)\right) \sin^2 \frac{2\pi d n}{l}$$

which in the limit of long-wavelength deformation (N - number of stripes, L_y - extension of the pattern along the stripes) gives

$$E_{cur} = \frac{N l L_y K_c}{2\pi} \int k^4 |\zeta(k)|^2 dk = \frac{1}{2} \int K_c \zeta_{yy}^2 dx dy$$

We see that the first nonvanishing contribution to the energy of the stripe pattern at its long wavelength undulation deformations is arising due to the curvature elasticity. The term corresponding to effective surface tension is exactly equal to zero. A similar conclusion for the stripe pattern in an amphiphile monolayer is obtained

31

in [44]. The calculation gives the following expression for the curvature elasticity constant:

$$K_c = \frac{3M^2 l^3}{8\pi^4} \sum_{k=1}^{\infty} \frac{\sin^2 \pi k \varphi_m}{k^5} \left(1 - \left(1 + \frac{2\pi kh}{l} + \left(\frac{2\pi kh}{l}\right)^2 /3\right) \exp\left(-\frac{2\pi kh}{l}\right)\right)$$

Another contribution to the energy of the deformed stripe pattern arises from the compressional elasticity and can be expressed as follows (l_0 - equilibrium pattern period)

$$E_{com} = N l L_y \frac{1}{2} \frac{d^2 \tilde{e}}{dl^2}(l - l_0)^2 = \int \frac{1}{2} B \left(\frac{l - l_0}{l_0}\right)^2 dx dy,$$

where for the compressional elasticity constant the following expression is valid

$$B = \frac{4\pi M^2 h^2}{l^2} \int_0^1 \frac{t^2 \coth \frac{\pi h t}{l} \sin^2 \pi \varphi_m \, dt}{\sin^2 \pi \varphi_m + \sinh^2 \frac{\pi h t}{l}}$$

Expressions of elasticity constants at sinusoidal perturbations of magnetization of garnet films ("weak segregation" limit) were obtained in [49].

Accounting for the compressional deformation arising due to nonhomogeneous displacement of the stripes and their undulation, finally, the following expression for the energy of a slightly deformed stripe pattern can be obtained ($\alpha = (l - l_0)/l$)

$$E = \frac{1}{2} \int \left\{ B(\alpha + \zeta_x - 1/2\zeta_y^2)^2 + K_c \zeta_{yy}^2 \right\} dx dy \tag{17}$$

We see that due to the canceling of the surface energy term the energy of the stripe pattern is expressed in the same way as for 2D smectic. The calculation described above allows expression of the elasticity constant of 2D magnetic smectic through material properties of magnetic fluid – its magnetization and surface tension as well as the thickness of the plane layer.

The established smectic-like behavior of the magnetic liquid stripe patterns immediately allows drawing several interesting physical conclusions. Since the equilibrium period of the smectic pattern is decreasing with the increase of the magnetic field strength [10, 11, 12], then switching the field strength from value H_1 to $H_2 > H_1$, the extension α of the stripe pattern will be

$$\alpha = \frac{l_0(H_1) - l_0(H_2)}{l_0(H_2)}$$

For the new field strength value the old period will be conserved. Thus in complete analogy with the behavior of real smectic at their mechanical extension [50] undulation instability of the straight stripe pattern at the critical extension will arise. The critical

32

value of the extension for the stripe pattern between fixed boundaries at distance L from each other $(\zeta(L) = \zeta(0) = 0)$ is

$$\alpha_c = 2\pi \frac{\lambda}{L},$$

where $\lambda = \sqrt{K_c/B}$ – penetration length of smectic. The period of arising undulation deformations will be expressed as follows:

$$l_{ch} = 2\sqrt{\pi \lambda L}$$

Those physical conclusions are in complete accordance with experiment [12]. Measurements of the rising structure period allow obtaining the value of the penetration length. Undulation instabilities at thermal heating leading to the formation of chevron patterns in garnet films were investigated in [51, 52]. There the necessary equilibrium decrease is reached by temperature increase.

The next issue which is possible to describe on the basis of established smectic analogy concerns the deformation of the stripe pattern along dislocation. Again results of calculation are in good agreement with experiment [12].

Under further field increase the development of the undulation deformation leads to the formation of chevron patterns [12]. As magnetic field strength is increased, the angle between stripes is decreased. At some critical value, the field strength pattern becomes unstable and hairpins are formed [14]. The critical angle of the hairpin formation can be obtained as a result of comparing the energy of the domain wall between the two regions of zero curvature of the chevron pattern per unit of length equal to [53]

$$u = \frac{2B}{3} \lambda \tan^{3/2} \beta,$$

where $\pi - 2\beta$ is the angle between stripes in the regions of zero curvature, to the energies of two domain walls with the angle between stripes $\pi/4 - \beta/2$ and the angle with an initial domain wall $\pi/4 + \beta/2$ (L_x - length of domain wall)

$$2\frac{2}{3} B \frac{\lambda \tan^{3/2}(\pi/4 - \beta/2) L_x}{\sin(\pi/4 - \beta/2)}$$

The condition of the equality of those two energies gives the condition when splitting of the domain wall into two becomes energetically advantageous. By the solution of the equation

$$\tan^{3/2} \beta = 2 \frac{\tan^{3/2}(\pi/4 - \beta/2)}{\sin(\pi/4 - \beta/2)}$$

the value of $\beta \sim 50^0$ for the critical angle can be obtained. This value is in good accordance with existing experimental data [14].

6 Conclusions

Results given above allow the drawing of some conclusions:

1. Complex labyrinthine patterns are formed in plane layers of magnetic fluids under the action of a normal field.

2. The figures of equilibrium with n-fold ($n \geq 4$) symmetry do not exist due to vertex splitting instability. At the development of labyrinthine instability from initial stages with n-fold symmetry by a cascade of vertex splitting events, the configurations very close to Steiner trees at intermediate stages are formed. At later stages transition to simpler configurations driven by surface tension forces occurs.

3. Vertex splitting events can be suppressed by a small gas bubble trapped by the vertex.

4. Transitory states at labyrinthine instability development depend on viscosity contrast – in the case of a bubble in the infinite layer of the magnetic liquid, vertex splitting is delayed but the life time of the intermediate Steiner tree phase in the case of the bubble is less than for the droplet.

5. Vertex splitting suppresses the development of overextension instability of the stripes and leads to the formation of alternating finger patterns of magnetic liquid stripes. This can explain the absence of the rupture of stripes in experiments.

6. Magnetic interactions renormalize effective surface tension – for magnetic stripe of equilibrium width's effective surface tension is exactly equal to zero, the same is valid for the system of straight magnetic stripes at equilibrium distance between them. Due to that system of magnetic liquid, stripes behave as a 2D smectic liquid crystal.

7. Due to the renormalization of effective surface tension – undulation instability of magnetic foam as magnetic field strength increase occurs. For a system of magnetic fluid stripes, undulation instability as magnetic field increase takes place and leads to the formation of chevron patterns. Formation of hairpins can be understood on the basis of magnetic smectic analogy.

References

[1] M. SEUL, D. ANDELMAN. Domaines shapes and patterns: The phenomenology of modulated phases. *Science*, 267:476–483, 1995.

[2] A. CEBERS, M. M. MAIOROV. Magnetostatic instabilities in plane layers of magnetizable liquids. *Magnitnaya Gidrodinamika (in Russ.)*, 1:27–35, 1980 // *English translation: Magnetohydrodynamics* 16:21–27, 1980.

[3] A. CEBERS, M. M. MAIOROV. Structures of the interface of a bubble and a magnetic fluid in a field. *Magnitnaya Gidrodinamika (in Russ.)*, 3:15–20, 1980.

[4] R. E. ROSENSWEIG, M. ZAHN, R. SHUMOVICH. Labyrinthine instability in magnetic and dielectric fluids. *Journal of Magnetism and Magnetic Materials*, 39:127–132, 1983.

[5] J.-C. BACRI, R. PERZYNSKI, D. SALIN. Les liquides magnétiques. *Recherche*, 18:1150–1159, 1987.

[6] A. G. BOUDOUVIS, J. L. PUCHALLA, L. E. SCRIVEN. Shape instabilities of captive ferrofluid drops in magnetic field: Routes to labyrinthine pattern formation. *Journal of Colloid and Interface Science*, 124:688–690, 1988.

[7] S. A. LANGER, R. E. GOLDSTEIN, D. P. JACKSON. Dynamics of labyrinthine pattern formation in magnetic fluids. *Phys. Rev. A*, 46:4894–4904, 1992.

[8] A. J. DICKSTEIN, S. ERRAMILLI, R. E. GOLDSTEIN, D. P. JACKSON, S. A. LANGER. Labyrinthine pattern formation in magnetic fluids. *Science*, 260:1010–1015, 1993.

[9] D. P. JACKSON, R. E. GOLDSTEIN, A. O. CEBERS. Hydrodynamics of fingering instabilities in dipolar fluids. *Phys. Rev. E*, 50:288–307, 1994.

[10] A. CEBERS. On elastic properties of stripe structures of magnetic fluid. *Magnitnaya Gidrodinamika (in Russ.)*, 30:179–187, 1994 // *English translation: Magnetohydrodynamics* 30:148–155, 1994.

[11] A. CEBERS. Liquid magnetic stripe patterns and undulation instabilities. *Journal of Magnetism and Magnetic Materials*, 149:93-96, 1995.

[12] C. FLAMENT, J.-C. BACRI, A. CEBERS, F. ELIAS, R. PERZYNSKI. Parallel stripes of ferrofluid as a macroscopic bidimensional smectic. *Europhysics Letters*, 34:225–230, 1996.

[13] K. J. STINE, CH. M. KNOBLER, R. C. DESAI. Buckling instability in monolayer network structures. *Phys. Rev. Lett.*, 65:1004-1007, 1989.

[14] F. ELIAS, C. FLAMENT, J.-C. BACRI, S. NEVEU. Macro-organized patterns in ferrofluid layer: Experimental studies. *J. Phys. I France*, 7:711–728, 1997.

[15] A. CEBERS. Dynamics of magnetostatic instabilities. *Magnitnaya Gidrodinamika (in Russ.)*, 2:3–15, 1981 // *English translation: Magnetohydrodynamics* 17:113–122, 1981.

[16] D. J. KELLER, J. P. KORB, H. M. MCCONNELL. Theory of shape transition in two-dimensional phospholipid domains. *Journal of Physical Chemistry*, 91:6417–6422, 1987.

[17] H. M. MCCONNELL, V. T. MOY. Shapes of finite two-dimensional lipid domains. *Journal of Physical Chemistry*, 92:4520–4525, 1988.

[18] A. CEBERS. Labyrinthine structures of magnetic liquids and their electrostatic analogues. *Magnitnaya Gidrodinamika (in Russ.)*, 2:13–20, 1989 // *English translation: Magnetohydrodynamics* 25:149–155, 1989.

[19] F. HAENSSLER, L. RINDERER. Statique et dynamique de l'etat intermediaire des superconducteurs du type I. *Helv. Phys. Acta*, 40:659–687, 1967.

[20] P. A. RICE, H. M. MCCONNELL. Critical shape transitions of monolayer lipid domains. *Proc. Natl. Acad. Sci. USA*, 86:6445–6448, 1989.

[21] H. M. MCCONNELL. Harmonic shape transitions in lipid monolayer domains. *Journal of Physical Chemistry*, 94:4728–4731, 1990.

[22] H. M. MCCONNELL, R. DE KOKER. Note on the theory of the sizes and shapes of lipid domain monolayers. *Journal of Physical Chemistry*, 96:7101–7103, 1992.

[23] J. M. DEUTCH, F. E. LOW. Theory of shape transitions of two-dimensional domains. *Journal of Physical Chemistry*, 96:7097–7101, 1992.

[24] D. K. LUBENSKY, R. E. GOLDSTEIN. Hydrodynamics of monolayer domains at the air-water interface. *Physics of Fluids*, 8:843–854, 1996.

[25] H. A. STONE, H. M. MCCONNELL. Lipid domain instabilities in monolayers overlying sublayers of finite depth. *Journal of Physical Chemistry*, 99:13505–13508, 1995.

[26] H. A. STONE, H. M. MCCONNELL. Hydrodynamics of quantized shape transitions of lipid domains. *Proc. Royal Society, London A*, 448:97–111, 1995.

[27] J. A. MIRANDA, M. WIDOM. Stability analysis of polarized domains. *Phys. Rev. E*, 55:3758–3762, 1997.

[28] A. CEBERS, A. ZEMITIS. Numerical simulations of MHD instability of the free surface of a squeezed drop of magnetic liquid I. *Magnitnaya Gidrodinamika (in Russ.)*, 4:15–26, 1983 // *English translation: Magnetohydrodynamics* 19:360–369, 1983.

[29] A. J. DeGregoria, L. W. Schwartz. A boundary-integral method for two-phase displacement in Hele-Shaw cells. *Journal of Fluid Mechanics*, 164:383–400, 1986.

[30] A. J. DeGregoria, L. W. Schwartz. Saffman-Taylor finger width at low interfacial tension. *Phys. Rev. Lett.*, 58:1742–1744, 1987.

[31] D. Bensimon, L. P. Kadanoff, S. Liang, B. I. Shraiman, C. Tang. Viscous flows in two dimensions. *Reviews of Modern Physics*, 58:977–989, 1986.

[32] A. Cebers. Stability of difusion fronts of magnetic particles in porous media (Hele-Shaw cell) under the action of external magnetic field. *Magnitnaya Gidrodinamika*, 33:67–74, 1997.

[33] Y. Saad, M. H. Schultz. GMRES: A generalized minimal residual algorithm for solving nonsymmetric linear systems. *SIAM J. Sci. Statist. Comput.*, 7:856–869, 1986.

[34] A. Cebers, I. Drikis. A numerical study of the evolution of quasi-two-dimensional magnetic fluid shapes. *Magnitnaya Gidrodinamika*, 32:11–21, 1996.

[35] T. Y. Hou, J. S. Lowengrub, M. J. Shelley. Removing of stiffness from interfacial flows with surface tension. *J. Comp. Phys.*, 114:312–338, 1994.

[36] A. G. Papathanasiou, A. G. Boudouvis. Three-dimensional magnetohydrostatic instabilities of ferromagnetic liquid bridges. *(to be published)*

[37] R. de Koker, H. M. McConnell. Circle to dogbone. Shapes and shape transitions of lipid monolayer domains. *Journal of Physical Chemistry*, 97:13419–13424, 1993.

[38] K. Y. C. Lee, H. M. McConnell. Quantized symmetry of liquid monolayer domains. *Journal of Physical Chemistry*, 97:9532–9539, 1993.

[39] S. Perkovic, H. M. McConnell. Cloverleaf monolayer domains. *Journal of Physical Chemistry*, 101:381–388, 1997.

[40] K. L. Babcock, R. M. Westerwelt. Elements of cellular domain patterns in magnetic garnet films. *Phys. Rev. A*, 40:2022–2037, 1989.

[41] C. Isenberg. *The Science of Soap Films and Soap Bubbles*. Dover Publ. Inc., New York.

[42] J.-C. Bacri, A. Cebers, C. Flament, S. Lacis, R. Melliti, R. Perzynski. Fingering phenomena at bending instability of a magnetic fluid stripe. *Prog. Colloid Polym. Sci.*, 98:30–34, 1995.

[43] M. SEUL, R. WOLFE. Evolution of disorder in magnetic fluid stripe domains II. Hairpins and labyrinth patterns versus branched and comb patterns formed by growing minority component. *Phys. Rev. A*, 46:7534-7547, 1992.

[44] R. DE KOKER, W. JIANG, H. M. MCCONNEL. Instabilities of the stripe phase in liquid monolayers. *Journal of Physical Chemistry*, 99:6251–6257, 1995.

[45] R. E. GOLDSTEIN, D. P. JACKSON. Domain shape relaxation and the spectrum of thermal fluctuations in Langmuir monolayers. *Journal of Physical Chemistry*, 98:8626–9636, 1994.

[46] M. SEUL. Domain wall fluctuations and instabilities in monomolecular films. *Physica A*, 168:198–209, 1980.

[47] M. YONEYAMA, A. FUJI, S. MAEDA, T. MURAYAMA. Light-induced bubble-stripe transitions of gaseous domains in porphyrin Langmuir monolayers. *Journal of Physical Chemistry*, 96:8982–8988, 1992.

[48] A. CEBERS. Transformations of concentration domain structures of liquid magnetics in plane layers I. Energetical approach. *Magnitnaya Gidrodinamika*, 31:61-68, 1995.

[49] D. SORNETTE. Undulation instability in stripe domain structures of "bubble" material. *Journal de Physique*, 48:151–163, 1987.

[50] P.-G. DE GENNES. *The Physics of Liquid Crystals*. Clarendon Press, Oxford, 1974.

[51] M. SEUL, R. WOLFE. Evolution of disorder in magnetic stripe domains I. Transverse instabilities and disclination unbinding in lamellar patterns. *Phys. Rev. A*, 46:7519–7533, 1992.

[52] M. SEUL, R. WOLFE. Evolution of disorder in two-dimensional stripe patterns: "smectic" instabilities and disclination unbinding. *Phys. Rev. Lett.*, 68:2460–2463, 1992.

[53] A. C. NEWELL, T. PASSOT, N. ERCOLANI, R. INDIK. Elementary and composite defects of striped patterns. *J. Phys. II France*, 5:1863-1882, 1995.

Andrejs Cēbers
Institute of Physics, University of Latvia
Salaspils-1, LV-2169, Latvia

Ivars Driķis
University of Latvia, Department of Physics
Division of Electrodynamics
Riga LV-1586, Latvia

A. FASANO

Some Two-Scale Processes Involving Parabolic Equations

Abstract

In a large number of cases a physical system is appropriately described by a system of two (or more) equations associated with phenomena occurring on different spatial scales. Here we consider some cases in which at least one of the equations is parabolic and the coupling between the processes at the two scales may be nonlocal in space and in time. Free boundaries can also be present. After a general discussion we present some specific problem in more detail, pointing out the questions remaining open.

1 Introduction

Looking at the work performed during the last decade by our group at the Mathematics Department U. Dini in Florence in cooperation with industries and examining closely related studies made by others, I was surprised to notice that two-scale diffusion processes were occurring with impressive frequency in a variety of different contexts. Although two-scale (or multiple scale) processes find a very natural framework in homogenization theory (see, e.g. [10]), here we shall deal primarily with problems which have been studied with different techniques, even if in some cases, homogenization has been used effectively [17] and looks promising for further developments.

The general situation we are talking about is that of a system possessing a microstructure with diffusion (or transport) processes taking place at both scales. The processes at the two scales are coupled and the coupling may be nonlocal in space and in time. The system at the large scale is generally described as a continuum whose physical properties depend on the local state at the microscale.

Our principal aim here is to show the amazing diversity of problems in this class. Therefore, we shall present a (necessarily concise) list of mathematical models with two-scale diffusion, spending some more time on some of them. It is remarkable that many of the problems illustrated are partially or completely open from the mathematical point of view. Moreover, there is still much to do about numerical methods. Finally, some of the mathematical models presented are susceptible to further, although quite nontrivial, refinement. For this reason I hope that this paper will be stimulating.

In the next section we deal briefly with the so-called espresso coffee problem, describing some recent developments. Still in the framework of filtration theory we shall illustrate a new model for the wetting of a porous medium with hydrophile granules (Sect. 3).

Then we turn our attention to polymers which provide a number of examples of two-scale processes involving parabolic equations. For lack of space we will confine ourselves to sketching flow problems of solutions of polymers in which the polymer molecules can be schematized as rods (Sect. 4). The leading equation for such types of flow is a Fokker-Planck equation on the unit sphere (Smoluchowski equation).

However, we must recall that there are many more examples of comparable complexity that are equally interesting from the physical and the mathematical point of view. A large class of two-scale problems groups various polymer crystallization models (see e.g. [3], [2] and the survey papers [4], [12]). In particular, Ziabicki's nucleation theory [31], [32], [33] couples macro-diffusion (heat) with the micro-evolution of crystalline germs obeying again a Fokker-Planck equation (see [5] for some mathematical result).

Still in the framework of polymer science, a perfect example of the two-scale diffusion process is the Ziegler-Natta polymerization process of gaseous monomers in a spherical agglomerate of catalytic particles. In the model described in [6], [15] the agglomerate is considered as a continuum in the macroscopic scale. The main variables at the large scale are the density of catalytic sites, the temperature, the monomer concentration and the velocity field in the expansion due to the growth of a polymer layer around the catalytic particles. In the small scale we work precisely in the growing polymer layer and we want to find its thickness as a function of time (and of the large scale radial coordinate), the temperature, the monomer concentration. The two diffusion problems are coupled in a multiple and even nonlocal way (heat release and monomer absorption due to polymerization, expansion velocity, the boundary conditions at the surface of the growing polymer layer).

Again in the field of polymeric materials, several examples of two-scale processes come from the manufacturing of composite materials: wetting fronts with heat transfer and polymerization [7], [8], compression and relaxation during injection [28] [9], etc.

A remarkable example of two-scale diffusion in the presence of phase change is freezing (or thawing) of emulsions or any system with a fine dispersion in a range of temperature such that the disperse substance undergoes phase change while the surrounding material does not (see [29]). When the small-scale diffusion is considered quasi-steady it is not difficult to derive a kinetic law for the evolution of the solid volume fraction (a macroscopic quantity), stating that its rate of change is a known function of the temperature and of the volume fraction itself. The structure of this function depends on how the phase change process at the microscale is modelled (plain Stefan, kinetic undercooling, surface tension effect), see [21]. However, if the heat capacity of the freezing component is not neglected, the problem is open.

Another impressive source of two-scale processes with lots of free boundaries is provided by the flow of materials with aggregating components. This is the case of coal-water slurries (see e.g. [16], [19]), of waxy crude oils (a preliminary mathematical study is [11]), the flow of coalescing dispersions (totally open), etc.

40

2 Some recent developments in the espresso-coffee problem

Much research has been performed in cooperation with the Italian company illy-caffé s.p.a. (based in Trieste) with the aim of describing the flow of water through ground coffee in an espresso coffee machine. First, the purely mechanical problem of the flow of cold water has been considered, thus eliminating most of the chemistry. A mathematical model (in one space dimension) accounting for the striking experimental results obtained at illy-caffé's laboratories has been produced (for a full description see [20], [14]). The model is based on the assumption that a very fine component of the porous medium can be removed by the flow when its concentration is above some threshold depending on the flow intensity (the more intense is the flow, the larger the population of removable particles). The particles are transported by the flow as they accumulate near the outflow surface giving rise to the formation of a compact layer of low hydraulic conductivity. The growth of the layer explains the decrease in time of the total discharge. Moreover, increasing the injection pressure increases the initial value of the discharge, thus disseminating into the flow a large number of fine particles which eventually produce a thicker highly resistant layer at the bottom. As a consequence, the increase of injection pressure may lead to a lower value of the asymptotic discharge, a fact which has been observed experimentally.

Using dimensionless and suitably normalized variables, the abovementioned model consists of the following equations:

$$\frac{\partial b}{\partial t} = -f(q)[b - \beta(q)]_+, \qquad 0 < x < s(t),\ t > 0 \tag{2.1}$$

$$\frac{\partial m}{\partial t} + \alpha q(t)\frac{\partial m}{\partial x} = -\frac{\partial b}{\partial t}, \qquad 0 < x < s(t),\ t > 0 \tag{2.2}$$

$$q(t) = \{s(t) + R_c(1 - s(t))\}^{-1}, \quad t > 0 \tag{2.3}$$

$$\dot{s}(t) = -\frac{\alpha q m}{M - (m + b)}\Big|_{x = s(t)}, \qquad t > 0 \tag{2.4}$$

where $q(t)$ is the flux intensity (volumetric velocity), with the implicit simplifying assumption that the liquid is incompressible and porosity remains constant, b is the concentration of removable particles still bound to the skeleton, $f(q)$ is a smooth increasing function vanishing for $q = 0$, $\beta(q)$ is a smooth non-negative decreasing function, $[\cdot]_+$ denotes the positive part, m is the concentration of moving particles, α a positive constant, R_c the hydraulic resistance of the compact layer (the resistance of the complementary domain is normalized to 1), M is the concentration of the fine particles in the compact layer. Equation (2.3) incorporates Darcy's law (with no gravity), the boundary condition for pressure (the inflow pressure exceeds the outflow pressure by 1), and the continuity of pressure and of volumetric velocity across the moving interface $x = s(t)$ bounding the compact layer. The removal process is described by (2.1), while (2.2) is the transport equation for the fine particles (here assumed to be identical). The mass balance (2.4) describes the growth of the compact layer.

The initial conditions are

$$s(0) = 1, \qquad b(0) = 1, \qquad m(x,0) = m_0 \geq 0, \tag{2.5}$$

and, in particular, the uniform initial distribution of bound fine particles implies that b depends on t only and consequently that the sum $b + m$ is constant along the characteristics of (2.7).

Equation (2.2) requires the boundary condition

$$m(0,t) = 0 \tag{2.6}$$

expressing that no particles are injected from outside.

Global existence and uniqueness along with several qualitative properties of the solutions have been proved in the quoted papers.

Strictly speaking the model just described is not in the class of problems with which we want to deal. However, we come closer to it when we consider the chemical aspects, in particular the dissolution of substances, affecting both the porosity and the conductivity of the medium.

Leaving aside the whole class of free boundary problems related to the presence of a wetting front (see [13], [23]), we still consider a saturated layer and we introduce several species of particles and substances with concentrations b_i (in the solid), m_i (in the liquid). For each species we may introduce a removal or dissolution kinetic similar to (2.1) with corresponding thresholds $\beta_i(q)$, possibly vanishing. As to the motion of each species, we can take a pure transport mechanism like (2.2), but for the dissolved substances we can also introduce a diffusion matrix.

The most important complication of the model, however, is represented by the change of porosity due to both chemical and mechanical effects. If $\varepsilon(x,t)$ denotes the porosity, instead of the simple equation (2.3) for the flux, we have to write

$$\frac{\partial \varepsilon}{\partial t} = -\sum_i \delta_i \frac{\partial b_i}{\partial t} - g(q)[\varepsilon - \varepsilon_*(q, \mathbf{b})] + h(q)[\varepsilon^*(\mathbf{q}) - \varepsilon], \tag{2.7}$$

$$\frac{\partial \varepsilon}{\partial t} + \frac{\partial q}{\partial x} = 0, \tag{2.8}$$

$$q = -k(\mathbf{b}, \mathbf{m}, \varepsilon)\frac{\partial p}{\partial x}, \tag{2.9}$$

where in (2.7) δ_i represents the volume increase coefficient associated to the loss of the i-th species, g is a smooth, non-negative, nondecreasing function vanishing for $q = 0$, $\varepsilon_*(q, \mathbf{b})$ is the equilibrium value of ε under the compression exerted by the flow, while the last term describes the medium relaxation towards an equilibrium value $\varepsilon^* \geq \varepsilon_*$ ($h \geq 0$ being a smooth function). Equation (2.8) is the volume balance, and (2.9) is Darcy's law. Initial and boundary conditions must be modified accordingly. In [18] a model of this type has been studied with two main simplifications: (i) the transport of all species is purely convective, (ii) all species are allowed to leave the system at the outflow surface. The latter condition implies that no compact layer is present and therefore, no free boundary. Nevertheless, the problem is still very complicated. The main difficulty arises from (2.7), (2.8), which provide an o.d.e. for $q(x,t)$ at each time t, whose integration would require the knowledge of $q(0,t)$, i.e., the fluid injection rate. Instead, only the pressure p is specified at $x = 0$, $x = 1$. The way to overcome this difficulty is to integrate (2.9). For instance, at $t = 0$ we compute $q(x,0,q_0^*)$ from (2.7), (2.8), using the known initial value for b_i and imposing $q(0,0,q_0^*) = q_0^*$. Then from (2.9) we get

$$\int_0^1 \frac{q(x,0,q_0^*)}{k(\mathbf{b}_0, \mathbf{m}_0, \varepsilon_0)} = 1, \tag{2.10}$$

where ε_0 is the initial value of ε, and the r.h.s. is the rescaled injection pressure. It can be proved that (2.10) determines q_0^* uniquely. For positive times this same operation must be performed but the coupling with removal and transport equations comes into play. Global existence and uniqueness have been proved.

In [22] we have considered the complete problem with the only simplification writing the diffusive current of each diffusing species in the form $j_i = -D_i \dfrac{\partial m_i}{\partial x_i} + m_i q_i$, i.e., neglecting interdiffusion.

In addition to diffusion, which modifies the character of the governing p.d.e.s, another substantial change is the replacement of the single species free boundary condition (2.6). When the compact layer is built up with the contribution of a number $k > 1$ of different species of particles, we must introduce a vector $(M_1, \ldots, M_k)$ of packing concentrations subjected to the constraint

$$\sum_{i=1}^k \lambda_i M_i = M, \tag{2.11}$$

λ_i being known multipliers and $M > 0$ being a known constant. Then we write the mass balance equations

$$|\dot{s}|[M_i - (b_i + m_i)] = \alpha_i m_i q, \quad x = s(t), i = 1, \ldots, k, \tag{2.12}$$

which we combine with (2.11) in order to get the free boundary advancement equation

$$|\dot{s}| \left[M - \sum_{i=1}^k \lambda_i (b_i + m_i) \right] = \sum_i \lambda_i \alpha_i m_i q. \tag{2.13}$$

Note that equations (2.12) are used to determine the packing concentrations M_i.

The two-scale character of the problem emerges in the fact that the macroscopic quantity k (the hydraulic conductivity) depends on the microscopic concentrations vector $\mathbf{b}$, whose evolution is in turn governed by the flow.

3 Wetting fronts in porous media with hydrophile granules

The practical problem we are referring to is the wetting of the familiar baby diapers. A diffusive model has been formulated and studied numerically in [30] and the relative mathematical investigation can be found in [25]. Here we want to follow a different approach, based on the natural law for filtration, i.e., Darcy's law, neglecting gravity.

Thus we consider a porous medium whose pores contain water-absorbing granules and we make the assumption that the granules increase to some maximum volume without displacing the porous skeleton (so there is no macroscopic swelling).

We denote by $V_g(x, y)$ the volume fraction occupied by the granules. The porosity ε decreases at the same rate as V_g increases:

$$\frac{\partial \varepsilon}{\partial t} = -\frac{\partial V_g}{\partial t}. \tag{3.1}$$

As usual, we suppose that the saturation S is a function of pressure p, such that $S \equiv 1$ for $p \geq p_S$ (saturation pressure), $p_S > 0$ ($p = 0$ being the atmospheric pressure). Moreover $0 < S(p) < 1$ for $p < p_S$ with $S'(p) > 0$, and $S(0) = S_0 = 0$ is the irreducible moisture content (once the medium has been wet).

The flow is governed by Darcy's law (no gravity)

$$\mathbf{q} = -k(S, \varepsilon)\nabla p \tag{3.2}$$

($\mathbf{q}$ volumetric velocity, k hydraulic conductivity, a bounded and strictly positive C^1 function). The local liquid mass balance is now

$$\frac{\partial(S\varepsilon)}{\partial t} + \operatorname{div} \mathbf{q} = -\frac{\partial V_g}{\partial t} \tag{3.3}$$

and the liquid intake of the granules obeys an equation of the form

$$\frac{\partial V_g}{\partial t} = h(V_{\max} - V_g)(S - S_0)\chi_p \tag{3.4}$$

where $h(\xi)$ is a smooth increasing function such that $h(0) = 0$, χ_p is the characteristic function of pressure ($\chi_p = 0$ in the dry region).

The structure of the function $h(\xi)$ determines in particular whether or not the volume $V_{\max}$ can be reached in a finite time.

Due to (3.1), equation (3.3) simplifies to

$$\varepsilon \frac{\partial S}{\partial t} + (S - 1)\frac{\partial \varepsilon}{\partial t} + \operatorname{div} \mathbf{q} = 0. \tag{3.5}$$

Note that in the saturated region the latter equation reduces to

$$\operatorname{div} \mathbf{q} = 0 \tag{3.6}$$

44

even if $\dfrac{\partial \varepsilon}{\partial t} \neq 0$, because for the local volume balance it is irrelevant whether the liquid is in the pore or in the granule.

At this point let us make a selection of $h(\xi)$; in order to be specific, e.g., we set

$$h(\xi) = c\xi, \qquad c > 0, \text{ constant,} \tag{3.7}$$

so that we can integrate (3.4) with the initial condition

$$V_g(x, 0) = V_0 < V_{\max}. \tag{3.8}$$

Setting $\xi = \dfrac{V_g}{V_{\max}}$, $\xi_0 = \dfrac{V_0}{V_{\max}} < 1$, and

$$\varphi(S) = \exp\left[-c \int_0^t (S(\mathbf{x}, \theta) - S_0)\chi_p \, d\theta\right] \tag{3.9}$$

we obtain

$$1 - \xi = (1 - \xi_0)\varphi(S) \tag{3.10}$$

so that ε can be expressed as a functional of S, since the sum $\varepsilon + V_g$ is constant by (3.1) and therefore, $\varepsilon - \varepsilon_{\min} = V_{\max}(1 - \xi)$:

$$\varepsilon = \varepsilon_{\min} + (V_{\max} - V_0)\varphi(S), \tag{3.11}$$

whence

$$\frac{\partial \varepsilon}{\partial t} = -c(V_{\max} - V_0)(S - S_0)\chi_p \varphi(S), \tag{3.12}$$

and we derive the equation for the wet region

$$[\varepsilon_{\min} + (V_{\max} - V_0)\varphi(S)]\frac{\partial S}{\partial t} + c(1 - S)(S - S_0)(V_{\max} - V_0)\chi_p \varphi(S) =$$
$$\operatorname{div}[K(S, \varepsilon)\nabla p], \tag{3.13}$$

where the relationship $S = S(p)$ must be used. Thus we have an equation which is parabolic in the unsaturated zone and elliptic in the saturated region. Concerning the function $S(p)$ for $0 \leq p \leq p_S$ we take it C^1, but we may allow S' to be nonzero for $p = p_S$. In this case (3.13) is uniformly parabolic in the unsaturated region. The saturation front Γ_S is characterized by the conditions

$$p = p_S, \qquad \left[\frac{\partial p}{\partial n}\right] = 0 \quad \text{on } \Gamma_S, \tag{3.14}$$

expressing the continuity of pressure and of volumetric velocity.

At the wetting front Γ_w (i.e., the interface with the dry region) we have $p = 0$, $S = S_0$, $\varepsilon = \varepsilon_0$ and therefore, the boundary conditions are

$$p = 0, \qquad v_n = -\frac{1}{\varepsilon_0 S_0} K(S_0, \varepsilon_0)\frac{\partial p}{\partial n} \quad \text{on } \Gamma_w, \tag{3.15}$$

45

v_n being the normal component of the interface velocity.

From the mathematical point of view we can say that (3.13) and (3.15) are an elliptic-parabolic Stefan problem, although the known results about the latter problem (see [1]) are not immediately applicable to the present case.

The conditions on the external boundary are also very delicate. As in [30] we may consider a cylindrical domain bounded between two planes: $z = 0$ (inflow) and $z = 1$. We impose that the face $z = 1$ and the lateral boundary are impermeable and that injection occurs through a portion Ω on the surface at $z = 0$ in that q_n (the normal component of $\mathbf{q}$) is prescribed as a function of time over Ω and is zero elsewhere.

However, it is unfeasible to allow any value of the injection rate through Ω. We expect that when the pressure reaches some limit p^* the inflow current must be compatible with such a constraint and the excess liquid speads over the surface $z = 0$ and may or may not penetrate according to whether the constraint $p \leq p^*$ is satisfied or not. In other words the boundary condition at $z = 0$ is of the Signorini type. Even though the model described so far is already rather complicated, it looks quite sensible to introduce a nontrivial modification consisting of the replacement of the condition $S = S_0$ at the wetting front Γ_w with a condition of the type

$$S|_{\Gamma_w} = \sigma(v_n), \qquad (3.16)$$

where σ is a smooth nondecreasing function with values in the interval $[S_0, 1]$. The second condition in (3.15) should be modified accordingly (still keeping $\varepsilon = \varepsilon_0$). In terms of the phase change terminology, we would say that we have adopted the kinetic undercooling scheme. In the extreme situation $\sigma \equiv 1$ there is no unsaturated region and the wetting front coincides with the saturation front. In that case, the flow model is a variant of the well-known Green-Ampt problem [24], modeling the penetration of water into a dry soil.

Even in the one-dimensional geometry the problem is open.

4 A two-scale problem in polymers flow

A remarkable example of a system in which a diffusive type evolution of the microstructure is coupled with the rheological behaviour of the system as a whole is given by the flow of polymers diluted in solvents. Here we report very briefly some results of [26] and we refer to the fascinating paper [27] for the description of the physical background and for the relevant bibliography.

The starting point of the model is that the polymer molecules can be associated with a vector $\mathbf{R}$ (the end-to-end vector) rather than being seen as random coils and that all the relevant macroscopic quantities can be expressed as ensemble averages of expressions involving $\mathbf{R}$, through a suitable distribution function evolving with time.

More specifically, we consider just the case in which the average length of the vector $\mathbf{R}$ can be taken constant, so that its unit vector $\mathbf{u}$ is the only microscopic variable. In other words, the polymeric chains are modelled by rods of constant length and at each point $\mathbf{x}$ and time t the microscopic state of the system is described by a probability function $f(\mathbf{u})$.

At this point the incompressible flow of such a system is described by

$$\operatorname{div} \mathbf{v} = 0 \tag{4.1}$$

and by the momentum balance equation

$$\rho \frac{d\mathbf{v}}{dt} = -\nabla p + \operatorname{div} \sigma, \tag{4.2}$$

with the usual meaning of the symbols, where we have to specify the stress tensor σ in terms of averaged quantities involving $\mathbf{u}$. To this end we need to study the evolution of the distribution function $f(\mathbf{u})$, which is governed by the so called Smoluchowski's equation (see the quoted papers for the details):

$$\frac{\partial f}{\partial t} = D\nabla_{\mathbf{u}} \cdot (\nabla_{\mathbf{u}} f + \frac{1}{kT} f\nabla_{\mathbf{u}} V(\mathbf{u})) - \nabla_{\mathbf{u}} \cdot (f\dot{\mathbf{u}}) \tag{4.3}$$

where $V(\mathbf{u})$ is a given potential, T is the absolute temperature, k the Boltzmann constant, and

$$\dot{\mathbf{u}} = \nabla_{\mathbf{v}} \cdot \mathbf{u} - (\mathbf{u} \cdot \nabla \mathbf{v} \cdot \mathbf{u})\mathbf{u}. \tag{4.4}$$

Now, defining the average $< g >$ of $g(\mathbf{u})$ as the integral of the product fg over the unit sphere, the constitutive equation of the system is written as follows:

$$\sigma = \sigma^v + \sigma^e, \tag{4.5}$$

$$\sigma^v = \frac{ckT}{D_0} \nabla \mathbf{v} :< \mathbf{uuuu} >, \tag{4.6}$$

$$\sigma^e = 2ckT(1 - 2U) < \mathbf{uu} > + \\ 2U < \mathbf{uuuu} >:< \mathbf{uu} >, \tag{4.7}$$

where c is the polymer concentration, D_0 the diffusion coefficient, and U a non-dimensional parameter.

In [27] the authors analyze in full detail the case in which a unidirectional linear shear field is prescribed and $\mathbf{u}$ varies in a plane, so that it is defined by the angle θ formed with the shear direction. Choosing

$$V(\mathbf{u}) = -2UkT < \mathbf{uu} >: \mathbf{uu} \tag{4.8}$$

equation (4.3) has a stationary solution of the form

$$f(\theta) = C \exp[-g(\theta)] \left\{ \int_0^\theta \exp[g(x)]\, dx + K \right\} \tag{4.9}$$

47

where

$$g(\theta) = -U[< \cos 2\theta > \cos 2\theta + < \sin 2\theta > \sin 2\theta] +$$
$$+G[\theta - \frac{1}{2}\sin 2\theta], \qquad (4.10)$$

G being the dimensionless shear rate.

The constant K must be determined by the periodicity condition $f(\pi) = f(0)$, while C is a normalization coefficient.

Rather surprisingly the stationary solution is shown to exist and be unique only in a certain range of the constants U, G, while for U sufficiently large a critical value $G(U)$ is found below which the solution disappears, leading to the conjecture that in such a situation the system switches to a tumbling behaviour.

References

[1] H.W. ALT AND S. LUCKHAUS. Quasilinear elliptic-parabolic differential equations. *Mathematische Zeitschrift* **183**, (1983), 311-341.

[2] D. ANDREUCCI, A. FASANO, M. PAOLINI, M. PRIMICERIO, AND C. VERDI. Numerical simulation of polymer crystallization. *Math. Models Meth. Appl. Sci.* **4**, (1994), 135-145.

[3] D. ANDREUCCI, A. FASANO, AND M. PRIMICERIO. On a mathematical model for the crystallization of polymers. *Proc. 4th ECMI Conference*, H.J. Wacker, W. Zulehner eds., Teubner and Kluwer, (1991), 3-16.

[4] D. ANDREUCCI, A. FASANO, M. PRIMICERIO, AND R. RICCI. Mathematical problems in polymers crystallization. *Survey Math. Ind.* **6**, (1996), 7-20.

[5] D. ANDREUCCI, A. FASANO, M. PRIMICERIO, R. RICCI, AND C. VERDI. Diffusion driven crystallyzation in polymers. In *Free Boundary Problems, Theory and Applications*, M. Niezgodka, P. Strzekeli eds., Pitman Tes. Notes Math., vol. **363**, pp. 359-367, Longman 1996.

[6] D. ANDREUCCI, A. FASANO, AND R. RICCI. On the growth of a polymer layer around a catalytic particle: a free boundary problem. NoDEA **4** (1997), 511-520.

[7] L. BILLI. Incompressible flows through porous media with temperature variable parameters. *Nonlinear Analysis* **31**, (1998), 363-384.

[8] L. BILLI. Non-isothermal flows in porous media with curing. NoDEA **8**, (1997), 623-637.

[9] L. BILLI AND A. FARINA. Unidirectional infiltration in deformable porous media: mathematical modelling and self-similar solutions. To appear.

[10] U. HORNUNG (ED.). *Homogenization and Porous Media.* Springer-Verlag, N.Y. Berlin, Heidelberg, 1996.

[11] A. FARINA AND A. FASANO. Flow characteristics of waxy crude oils in laboratory experimental loops. *Math. Comp. Modelling* **25**, (1997), 75-86.

[12] A. FASANO. Modelling the solidification of polymers: an example of an ECMI cooperation. *ICIAM 91,* R.E. O'Malley ed., (1991), 99–118.

[13] A. FASANO. The penetration of a wetting front through a porous medium accompanied by the dissolution of a substance. In *Math. Modelling of Flow through Porous Media,* A. Bourgeat et al. eds., World Scientific, (1995), 183–195.

[14] A. FASANO. Some non-standard one-dimensional filtration problems. *Bull. Fac. Ed. Chiba Univ.* **44**, (1996), 5–29.

[15] A. FASANO, D. ANDREUCCI, AND R. RICCI. Modello matematico di replica nel caso limite di distribuzione continua dei centri attivi. In *Simposio Montell e "Meccanismi di accrescimento di poliolefine su catalizzatori di Ziegler-Natta".* S. Mazzullo, G. Cecchin, eds., (Ferrara 1996), 154–173.

[16] A. FASANO, C. MANNI, AND M. PRIMICERIO. Modelling the dynamics of fluidizing agents in coal-water slurries. *Int. Symp. Nonlinear Problems in Engineering and Science,* Shutie Xiao, Xian-Cheng Hu eds., Science Press, Beijing, (1992), 64–71.

[17] A. FASANO, A. MIKELIČ, AND M. PRIMICERIO. Homogenization of flows through porous media with permeable grains. *Adv. Math. Sci. Appl.* **8** (1998), 1-31.

[18] A. FASANO AND M. PRIMICERIO. Flows through saturated mass exchanging porous media under high pressure gradients. In *Calculus of Variations, Applications and Computations.* Pont-à-Mousson 1994, C. Bandle et al. eds., Pitman Res. Notes Math. **326**, 109–129.

[19] A. FASANO AND M. PRIMICERIO. Modelling the rheology of a coal-water slurry. *Fourth ECMI Conference,* Hj. Wacker, W. Zulehner eds., Teubner, (1991), 269–274.

[20] A. FASANO AND M. PRIMICERIO. Mathematical models for filtration through porous media interacting with the flow. In *Nonlinear Mathematical Problems in Industry, I.* M. Kawarada, N. Kenmocki, N. Yanagilan eds., Math. Sci. & Appl. **1** Gakkotosho, Tokyo, (1993), 61–85.

[21] A. FASANO AND M. PRIMICERIO. An analysis of phase transition models. *EJAM* **7**, (1996) 1–12.

[22] A. FASANO AND F. TALAMUCCI. Formation of a composite compact layer in a porous medium due to the accumulation of the fine particles. To appear.

[23] A. FASANO AND P. TANI. Penetration of a wetting front in a porous medium with flux dependent hydraulic parameters. In *Nonlinear Problems in Applied Mathematics*, (K. Cooke et al. eds.), SIAM, (1995).

[24] W.H. GREEN AND G.A. AMPT. Studies on soil physics. The flow of air and water through soils. *J. Agric. Sci.* **4**, (1911), 1–24.

[25] P. MANNUCCI. Study of the mathematical model for adsorption and diffusion in ultra-napkins. *Le Matematiche* **50**, (1995), 3–14.

[26] P. MARRUCCI. Micro-rheological modelling. In *Rheological Fundamentals of Polymer Processing*. (J. Covas et al. eds.), Kluwer, (1996), 37–60.

[27] P. MARRUCCI AND F. MAFFETTONI. Description of the liquid-crystalline phase of rodlike polymers at high shear rates. *Micromolecules* **22**, (1989), 4076–4082.

[28] L. PREZIOSI. The theory of deformable porous media and its applications to composite material manufacturing. *Surveys Math. Ind.* **6**, (1996), 167–214.

[29] G. VALLET. Analyse mathématique des transferts thermiques dans des systèmes disperses subissant des transformations de phases. *Math. Mod. Num. Anal.* **27**, (1993), 895–923.

[30] J. WEICKERT. A mathematical model for diffusion and exchange phenomena in ultra napkins. *Math. Meth. Appl. Sci.* **16**, (1973), 759–777.

[31] A. ZIABICKI. Generalized theory of nucleation kinetics I, II. *J. Chem. Phys.*, **48**, (1968), 4368–4380.

[32] A. ZIABICKI. Theoretical analysis of oriented and non-isothermal crystallization, Part I and Part II. *Colloid & Polym. Sci.* **252**, (1974), 207– 221; 433–447.

[33] A. ZIABICKI AND G.C. ALFONSO. Memory effects in thermal crystallization, I, II. *Colloid & Polym. Sci.* **272**, (1994), 1027; **273** (1995), 317-323.

A. Fasano
Dipartimento di Matematica "U. Dini"
Univ. degli Studi di Firenze
Viale Morgagni 67/A, 50134 Firenze, Italy

B. Merriman, R. Caflisch and S. Osher

Level Set Methods, with an Application to Modeling the Growth of Thin Films

Abstract

The level set method was devised in 1987 by S. Osher and J.A. Sethian [OSe] as a versatile and useful tool for analyzing the motion of fronts. It has proven to be phenomenally successful as both a theoretical and computational device.

In this paper we review its properties, discuss the advances in level set technology since the original paper, highlight some of the application areas, and present a new application to the modeling of epitaxial growth of thin film semiconductor devices (see also [Ca, Me, CGMORVZ]).

1 Introduction

The idea behind the level set method is a simple one. Given an interface Γ in R^n of co-dimension one, bounding a (perhaps multiply connected) open region Ω, we wish to analyze and compute its subsequent motion under a velocity field $\vec{v}$. This velocity can depend on position, time, the geometry of the interface (its normal, mean curvature...) and the external physics. The idea, as devised in 1987 by S. Osher and J.A. Sethian, is merely to define a smooth (at least Lipschitz continuous) function $\varphi(x,t)$, that represents the interface as the set where $\varphi(x,t) = 0$. Here $x = (x_1, \ldots, x_n) \, \epsilon \, R^n$.

The level set function φ has the following properties:

$$\begin{aligned}
\varphi(x,t) &> 0 \text{ for } x \, \epsilon \, \Omega \\
\varphi(x,t) &< 0 \text{ for } x \, \not\epsilon \, \bar{\Omega} \\
\varphi(x,t) &= 0 \text{ for } x \, \epsilon \, \partial\Omega = \Gamma(t)
\end{aligned}$$

Thus, the interface is to be captured for all later time, by merely locating the set $\Gamma(t)$ for which φ vanishes. This deceptively trivial statement is of great significance for numerical computation, primarily because topological changes such as breaking and merging are well defined and performed "without emotional involvement".

The motion is analyzed by convecting the φ values (levels) with the velocity field $\vec{v}$. This elementary equation is

$$\frac{\partial \varphi}{\partial t} + \vec{v} \cdot \nabla \varphi = 0. \tag{1}$$

Here $\vec{v}$ is the desired velocity on the interface, and is arbitrary elsewhere.

51

Actually, only the normal component of v is needed: $v_n = \vec{v} \cdot \frac{\nabla\varphi}{|\nabla\varphi|}$, so (1) becomes

$$\frac{\partial\varphi}{\partial t} + v_n|\nabla\varphi| = 0. \tag{2}$$

In the next section we give simple and computationally fast prescriptions for (1) reinitializing the function φ to be the signed distance to Γ, at least near the boundary [SSO]; (2) smoothly extending the velocity field v_n off the front Γ [CMOS]; and (3) solving (2) only locally near the interface Γ, thus lowering the complexity of this calculation by an order of magnitude [PMZOK]. This makes the cost of level set methods competitive with boundary integral methods, in cases when the latter are applicable [HLOZ].

We emphasize that all this is easy to implement in the presence of boundary singularities, topological changes, and in 2 or 3 dimensions. Moreover, in the case in which v_n is a function of the direction of the unit normal, as in crystalline motion [OM], then equation (2) becomes the first order Hamilton-Jacobi equation

$$\frac{\partial\varphi}{\partial t} + |\nabla\varphi|\gamma\left(\frac{\nabla\varphi}{|\nabla\varphi|}\right) = 0 \tag{3}$$

for $\gamma = \gamma(\vec{n})$ a given function of the normal.

High order accurate, essentially non-oscillatory discretizations to general Hamilton-Jacobi equations including (3) were obtained in [OSh], see also [OSe,BO].

Theoretical justification of this method for geometric-based motion came through the theory of viscosity solutions for scalar time-dependent partial differential equations [CGG,ES]. The notion of a viscosity solution—which applies to a very wide class of these equations, including those derived from geometric-based motions—enables a user to have confidence that their computer simulations give accurate, unique solutions. A particularly interesting result is in [ESS] where motion by mean curvature as defined by Osher and Sethian in [OSe] is shown to be essentially the same motion as is obtained from the asymptotics in the phase field reaction diffusion equation. The motion in the level set method involves no superfluous stiffness as is required in phase field models. As was proven in [MBO2], this stiffness due to a singular perturbation involving a small parameter ϵ will lead to incorrect answers as in [Ko] without the use of adaptive grids [NPV]. This is unnecessary in order for the level set model to function.

An interesting variant of the level set method for geometry-based motion was introduced in [MBO1] as diffusion generated motion, and has now been generalized to forms known as convolution generated motion or threshold dynamics. This method splits the reaction diffusion model into two highly simplified steps. For an overview of this approach, see [RM].

2   The Level Set Dictionary and Technology

We list key terms and advances in technology and define them by their level set representation.

1. The interface boundary $\Gamma(t)$ is defined by: $\{x|\varphi(x,t) = 0\}$. The region $\Omega(t)$ bounded by $\Gamma(t)$: $\{x|\varphi(x,t) > 0\}$.

2. The unit normal $\vec{n}$ to $\Gamma(t)$ is given by

$$\hat{n} = -\frac{\nabla\varphi}{|\nabla\varphi|}.$$

3. The mean curvature k is defined by

$$k = -\nabla \cdot \left(\frac{\nabla\varphi}{|\nabla\varphi|}\right).$$

4. The Dirac delta function concentrated on an interface is

$$\delta(\varphi)|\nabla\varphi|,$$

where $\delta(x)$ is a one-dimensional delta function.

5. The characteristic function χ of a region $\Omega(t)$:

$$\chi = H(\varphi)$$

where

$$H(x) \equiv 1 \ \text{ if } x > 0$$
$$H(x) \equiv 0 \ \text{ if } x < 0.$$

6. The surface (or line) integral of a quantity $p(x,t)$ over Γ:

$$\int_{R^n} p(x,t)\delta(\varphi)|\nabla\varphi|dx.$$

7. The volume (or area) integral of $p(x,t)$ over Ω

$$\int_{R^n} p(x,t)H(\varphi)dx.$$

53

8. The distance reinitialization procedure:

Let $d(x,t)$, be the signed distance of x to the closest point on Γ. The quantity $d(x,t)$ satisfies $|\nabla d| = 1$, $d > 0$ in Ω, $d < 0$ in $(\bar{\Omega})^c$ and is the steady-state solution (as $\tau \to \infty$) to

$$\frac{\partial \psi}{\partial \tau} + \text{sgn}(\varphi)(|\nabla \psi| - 1) = 0 \qquad (4)$$

$$\psi(x,0) = \varphi(x,t)$$

where $\text{sgn}(x) = 2H(x) - 1$ is the signum function.

Note: in recent work [PMOZK] it was found that degeneracies in the initial data φ for (4)—such as extreme flatness/steepness (vanishing/infinite $\partial\varphi/\partial n$)—can be removed by preconditioning it via:

$$\varphi^{(1)}(x,0) = \frac{\varphi(x,t)}{|\nabla\varphi(\bar{x},t)|}$$

$$\varphi^{(2)}(x,0) = \frac{\varphi^{(1)}(x,0)}{|\nabla\varphi^{(1)}(x,0)|}$$

$$\psi(x,0) = \varphi^{(2)}(x,0)$$

Moreover, in order to define d in a band of width ϵ around Γ, we need only solve (4) for $\tau = 0(\epsilon)$. Thus the computational complexity of this construction is minimal.

9. Smooth extension of a quantity, e.g., v_n on Γ off Γ. Let the quantity be $p(x,t)$. Solve to steady state, $\tau \to \infty$

$$\frac{\partial q}{\partial \tau} + \text{sgn}(\varphi)\left(\frac{\nabla\varphi}{|\nabla\varphi|} \cdot \nabla q\right) = 0$$

$$q(x,0) = p(x,t).$$

Again, we need only solve this for $\tau = 0(\epsilon)$ in order to extend p to be constant in the direction normal to the interface in a tube of width ϵ. See, e.g., [PMOZK,CMOS].

10. Local level set method [PMOZK]. We may solve (2) in a neighborhood of Γ of width $m\Delta x$, where m is typically 5 or 6. Points outside of this neighborhood need not be updated by this motion. This algorithm works in "φ" space – so no intricate computer science is used. For details see [PMOZK]. Thus this local method works easily in the presence of topological changes and for the multi-phase problems described below.

Additionally, this method may be used to compute the distance to Γ, with any order of accuracy, with computational complexity which is of order N, the total number of points updated. In fact, the same is true for the solution of general geometric-based motion such as a curvature regularization of a first order Hamilton-Jacobi equation. In contrast, the fast marching algorithm introduced in [Se] applies to a much more restricted class of equations, those where the speed function is given à priori and does not change sign. It is also only first order accurate, with no simple extension to a higher order. High accuracy is important if we are computing a distance function which must have accurate gradients and second derivatives for the purpose of computing interface normals and curvature. Also, fast marching actually has greater formal complexity ($N \log N$) than the above PDE-based method, though in actual implementation it can be faster—especially for problems where the propagation speed varies by orders of magnitude over the domain.

11. Coupling to external physics in two-phase Navier-Stokes flow: [SSO,CHMO]

$$\vec{u}_t = -\vec{u} \cdot \nabla \vec{u} - \frac{\nabla p}{\rho(\varphi)} + \vec{g} + \frac{1}{R_e} \nabla \cdot \left(\frac{(2\mu D)}{\rho(\varphi)} \right) \tag{5}$$

$$- \left(\frac{1}{Bd} \right) \left(\nabla \left(\frac{\nabla \varphi}{|\nabla \varphi|} \right) \right) \nabla H(\varphi), \tag{6}$$

$$\nabla \cdot \vec{u} = 0$$

where $\vec{u} = (u, v)$ is the fluid velocity, $\rho = \rho(\varphi)$ and $\mu = \mu(\varphi)$ are the piecewise constant fluid densities and viscosities, D is the viscous stress tensor, $\vec{g}$ is the gravitational force, k is the curvature of the interface, H is the Heaviside function and $R_e, Bd, \frac{\rho_1}{\rho_2}$, and $\frac{\mu_1}{\mu_2}$ are the parameters defining a given flow.

This equation is coupled to the front motion through the level set evolution equation (1) with $\vec{v} = \vec{u}$. This involves defining the interface numerically as having the width of the approximation to the jump in the approximate Heaviside function, which is approximately $3\Delta x$ in [SSO].

12. Coupling to the external physics in Stefan problems [CMOS]. Solve:

$$\frac{\partial T}{\partial t} = \nabla^2 T, \quad x \not\in \partial \Omega = \Gamma(t) \tag{7}$$

$$v_n = [\frac{\partial T}{\partial n}], \quad \bar{x} \in \Gamma(t)$$

where [·] denotes the jump across the boundary, and

$$T = -\bar{\varepsilon}_c k(1 - A \cos(k_A \theta + \theta_0)) + \bar{\varepsilon}_v v_n(1 - A \cos(k_A \theta + \theta_0))$$

on $\Gamma(t)$, and where k is the curvature, $\theta = \cos^{-1} \frac{\varphi_x}{|\nabla \varphi|}$, and the constants A, k_A, $\varphi_0, \bar{\varepsilon}_c$, and $\bar{\varepsilon}_v$ depend upon the material being modeled.

55

We directly discretize the boundary conditions at Γ. To update T at grid nodes near the boundary, if the stencil for the heat equation would cross Γ (as indicated by nodal sign change in φ), we merely use dimension by dimension one-sided interpolation and the given boundary T value at a ghost node placed at $\varphi = 0$ (found by interpolation on φ) to compute T_{xx} and/or T_{yy} (never interpolating across the interface) rather than the usual three-point central stencils. The level set function φ is updated, after reinitialization to be distance, by (1), using the extension off the interface of v_n as defined in (7).

13. Multi-phase flow using exactly as many level set functions as there are phases, with applications to drops and bubbles [ZCMO,ZMOW]; a variational level set approach.

   Define an energy function involving the area (length) of each interface, the volume (area) of each phase (using $\delta(\varphi_i), H(\varphi_i)$). Apply gradient descent to this energy using time as the descent variable. Enforce a no-vacuum, no-overlap constraint. This leads to slightly coupled system of geometrically driven motion perturbed by the constraints [ZCMO]. Additional constraints such as volume preservation may also be enforced in order to compute falling drops and bubble motion [ZMOW]. Finally, inertial forces are added in [KMOS].

14. Topological regularization of ill-posed problems with applications to vortex motion in incompressible flow.

   In [HOS] we computed two- and three-dimensional unstable vortex motion without regularization other than that in the discrete approximation to $\delta(\varphi)$ – this is done over a few grid points. The key observation, first made in [HO], is that viewing a curve or surface as the level set of a function, and then evolving it with a perhaps unstable velocity field, prevents certain types of blow up from occurring. This is denoted "topological regularization". For example, a tracked curve can develop a figure 8 pattern, but a level set captured curve will pinch off and stabilize before this happens. For the set up (involving two functions), see [HO], where we perform calculations involving the Cauchy-Riemann equations. The motions agree until pinch off, when the topological stabilization develops.

   As an example, we consider the two-dimensional incompressible Euler equations, which may be written as

$$\begin{aligned} \omega_t + \vec{u} \cdot \nabla \omega &= 0 \\ \nabla \times \vec{u} &= \omega \\ \nabla \cdot \vec{u} &= 0 \end{aligned}$$

   We are interested in situations in which the vorticity is initially concentrated on a set characterized by the level set function φ as follows:

   Vortex patch: $\omega = H(\varphi)$

56

Vortex sheet: $\omega = \delta(\varphi)$ (strength of sheet is $\frac{1}{|\nabla\varphi|}$)

Vortex sheet dipole: $\omega = \frac{d}{d\varphi}\delta(\varphi) = \delta'(\varphi)$.

The key observation is that φ also satisfies (8) and ω can be recovered from (8). For example, for the vortex sheet case we solve

$$\varphi_t + \vec{u} \cdot \nabla\varphi = 0$$
$$\vec{u} = \begin{pmatrix} -\partial y \\ \partial x \end{pmatrix} \Delta^{-1}\delta(\varphi).$$

Off-the-shelf Laplace solvers may be used. See [HOS] for results involving two- and three-dimensional flows.

15. The Wulff shape as the asymptotic limit of a growing crystalline interface [OM].

    For an initial state consisting of any number of growing crystals in R^n, n arbitrary, moving outward with normal velocity $\gamma > 0$ which depends on the angle of the unit normal, the asymptotic growth shape is a Wulff crystal, appropriately scaled in time. This shape minimizes the surface energy, i.e., the surface integral of γ, for a given volume. The proof uses the level set idea and then analyzes the solution to (3) using the Hopf-Bellman formulas [BE]. This result was first conjectured by Gross in (1918) [Gr].

    Additionally, with the help of the Brunn-Minkowski inequality, we show that if we evolve a convex surface under the motion described in (3), the ratio to be minimized monotonically decreases to its minimum as time increases. Thus there is a new link between this hyperbolic surface evolution and this (generally nonconvex) energy minimization.

16. Other applications of the level set method include Hele-Shaw flow (slow flow through porous media) [HLOZ], generalized interpolation of curves and surfaces [ZOMK], the construction of (Wulff) minimal surfaces [CMO], generalized ray tracing [FEO], computer vision [CCCD], computer aided design [KB], and combustion [ABS].

3   The Island Dynamics Model for Epitaxial Growth

We have developed a new continuum model for the epitaxial growth of thin films [Ca, Me, CGMORVZ]. Since our model describes hundreds of moving and merging interfaces, the level set method is essential for practical calculation. Here we will briefly outline the model and the novel level set techniques for the computations. We also show representative numerical results.

57

3.1   Epitaxial Growth

Molecular Beam Epitaxy (MBE) is a method for growing extremely thin films of material. The essential aspects of this growth process are as follows: under vacuum conditions a flux of atoms is deposited on a substrate material, typically at a rate that grows one atomic monolayer every several seconds. When deposition flux atoms hit the surface, they bond weakly rather than bounce off. These surface "adatoms" are relatively free to hop from lattice site to site on a flat (atomic) planar surface. However, when they hop to a site at which there are neighbors at the same level, they form additional bonds which hold them in place. This bonding could occur at the "step edge" of a partially formed atomic monolayer, which contributes to the growth of that monolayer. Or, it could occur when two adatoms collide with each other. If the critical cluster size is one, the colliding adatoms nucleate a new partial monolayer "island" that will grow by trapping other adatoms at its step edges.

By these means, the deposited atoms become incorporated into the growing thin film. Each atomic layer is formed by the nucleation of many isolated monolayer islands, which then grow in area, merge with nearby islands, and ultimately fill in to complete the layer. Because the deposition flux is continually raining down on the entire surface, including the tops of the islands, a new monolayer can start growing before the previous layer is completely filled. Thus islands can form on top of islands in a "wedding cake" fashion, and the surface morphology during growth can become quite complicated.

The process has traditionally been modeled using Kinetic Monte Carlo methods and Molecular Dynamics methods. These follow the trajectories of adatoms hopping on the surface, with varying degrees of accuracy. Such direct simulation methods provide a wealth of insight into the growth process, but they are computationally expensive. Also, the focus on single atoms makes it difficult to capture the behavior at the longer length scales that are important for the performance of integrated circuit devices grown via MBE.

For example, Resonant Tunneling Diodes (RTDs) are a very fast switching component grown with MBE and used in satellite communications electronics. The active region in these devices is a thin film on the order of 50 monolayers thick, with a lateral dimension of tens of thousands of atoms. The performance characteristics of the RTD are dependent in a extremely sensitive way on the thickness, roughness, and general morphology of this film. For practical device growth, it is important to achieve a high degree of repeatability and controllability of these film properties. For device design, it is important to further develop an understanding of how they relate to the device characteristics.

3.2   The Island Dynamics Model

The Island Dynamics model is a continuum model designed to capture the longer length scale features that are likely to be important for the analysis and control of monolayer thin film growth. It is also intended to model the physics relevant to studying basic issues of surface morphology, such as the effects of noise on growth, the long time evolution of islands, and the scaling relationships between surface features (mean island area, step edge length, etc.) in various growth regimes (precoalescence, coalescence). Refer to the classic work of [BCF] for useful background on the modeling of the growth of material surfaces.

In the Island Dynamics model, we treat each of the islands present as having a unit height, but a continuous (step edge) boundary on the surface. This represents the idea that the films are atomic monolayers, so that height is discrete, but they cover relatively large regions on the substrate, so x-y are continuum dimensions. The adatoms are modeled by a continuous adatom density function on the surface. This represents the idea that they are very mobile, and so they effectively occupy a given site for some fraction of the time, with statistical continuity, rather than discretely.

Thus, the domain for the model is the x-y region originally defined by the substrate, and the fundamental dynamical variables for this model are

- The island boundary curves $\Gamma_i(t), i = 1, 2, \ldots, N$

- The adatom density on the surface $\rho(x, y, t)$

The adatom density ρ obeys a surface diffusive transport equation, with a source term for the deposition flux

$$\frac{\partial \rho}{\partial t} = \nabla \cdot (D \nabla \rho) + F,$$

where $F = F(x, y, t)$ is specified. During most phases of the growth, it is simply a constant. This equation may also include additional small loss terms reflecting adatoms lost to the nucleation of new islands, or lost to de-absorption off the surface. This equation must be supplemented with boundary conditions at the island boundaries. In the simplest model of Irreversible Aggregation, the binding of adatoms to step edges leaves the adatom population totally depleted near island boundaries, and the boundary condition is

$$\rho|_\Gamma = 0$$

More generally, the effects of adatom detachment from boundaries, as well as the energy barriers present at the boundary, lead to boundary conditions of the form

$$\left[A\rho + B \frac{\partial \rho}{\partial n} \right] = C$$

where C is given and $[\cdot]$ denotes the local jump across the boundary. In particular, note that ρ itself can have a jump across the boundary, even though it satisfies a diffusive transport equation. This simply reflects the fact that the adatoms on top of the island are much more likely to incorporate into the step edge than to hop across it and mix with the adatoms on the lower terrace, and vice versa.

The island boundaries Γ_i move with velocities $\vec{v} = v_n \vec{n}$, where the normal velocity v_n reflects the island growth. This is determined simply by local conservation of atoms: the total flux of atoms to the boundary from both sides times the effective area per atom, a^2, must equal the local rate of growth of the boundary, v_n:

$$v_n = -a^2 [\vec{q} \cdot \vec{n}]$$

(this assumes there is no particle transport along the boundary; more generally, there is a contribution from this as well) where $\vec{q}$ is the surface flux of adatoms to the island boundary and $\vec{n}$ is the local outward normal. In general, the net atom flux $\vec{q}$ can be expressed in terms of the diffusive transport, as well as attachment and detachment probabilities, all of which can be directly related to the parameters of Kinetic Monte Carlo models. In the special case of Irreversible Aggregation, $\vec{q}$ is simply the surface diffusive flux of adatoms

$$\vec{q} = -D\nabla\rho$$

To complete the model we include a mechanism for the nucleation of new islands. If islands nucleate by random binary collisions between adatoms (and if the critical cluster size is one), we expect the probability that an island is nucleated at a time t, at a site (x, y), scales like

$$P[dx, dy, dt] = \epsilon\rho(x, y, t)^2 dt \, dx \, dy.$$

This model describes nucleation as a site-by-site, time step-by-time step random process. A simplifying alternative is to assume the nucleation occurs at the continuous rate obtained by averaging together the probabilistic rates at each site. In this case, if we let $n(t)$ denote the total number of islands nucleated prior to time t, we have the deterministic rate equation

$$\frac{dn}{dt} = <\epsilon\rho^2>$$

where $< \cdot >$ denotes the spatial average. In this formulation, at each time when $n(t)$ reaches an new integer value, we nucleate a new island in space. This is carried out by placing it randomly on the surface with a probability weighted by ρ^2, so that the effect of random binary collisions is retained.

This basic model also has natural extensions to handle more complex thin film models. For example, additional continuum equations can be added to model the dynamics of the density of kink sites on the island boundaries, which is a microstructural property that significantly influences the local adatom attachment rates (see [CGMORVZ]). Also, we can couple this model to equations for the elastic stress that results from the "lattice mismatch" between the size of the atoms in the growing layers and the size of the atoms in the substrate.

Conversely, the above model has a particularly interesting extreme simplification. We can go to the limit where the adatoms are so mobile on the surface ($D \to \infty$) that the adatom density is spatially uniform, $\rho(x, y, t) = \rho(t)$. In this case, the loss of adatoms due to the absorbing boundaries is assumed to take on a limiting form proportional to the adatom density and the total length L of all the island boundaries, which can be written as a simple sink term

$$\frac{d\rho}{dt} = F - \lambda L \rho.$$

(This equation can be derived systematically from the conservation law for the total number of adatoms, $\int \rho$, that follows from the adatom diffusion equation. The above loss term is just a simplified model for the net loss of adatoms to the island boundaries.) Further, it is assumed the velocity takes on a given normal dependent limiting form, $v_n = v_n(\vec{n})$ (which implies that growing islands will rapidly assume the associated "Wulff shape" for this function $v_n(\vec{n})$ (as in [OM])). We have used this "Uniform Density" model to prototype the numerical methods, and to develop an understanding of how the island dynamics models are related to the continuum "rate equation" models that describe island size distribution evolution while using no information at all about the spatial interactions of the islands.

3.3   Level Set Methods for Island Dynamics

Much of the above model is formally a Stefan problem and many of the level set techniques required for this were developed in [CMOS] and can similarly be applied here; for example, the internal boundary condition discretization of the adatom diffusion equation, and the procedure for extending the interface velocity v_n to a velocity defined on all of space. Here we will only highlight the aspects of the level set method that were newly developed for the island dynamics model.

Representation   Islands can only merge if they are part of the same monolayer, and the islands on monolayer j must be on top of a larger island in monolayer $j - 1$. Since there is no overhang at step edges, it is also true that the boundaries of islands on different layers j, k will never cross. We can capture all this behavior conveniently in a level set representation by letting the $\phi = 0$ level represent the island boundaries of the first monolayer, the $\phi = 1$ level represent the island boundaries of the second monolayer, and in general, the $\phi = j - 1$ level represent the boundaries of islands in the jth monolayer, where ϕ is a smooth function. (The 2 level case of this was introduced in [CHMO] to handle immiscible fluids.) In this regard, ϕ is just a smooth version of the surface height function $h(x, y)$, which is integer valued and jumps at the boundaries Γ_i. Indeed, $h = (\phi)$, where (z) denotes the least integer greater than z. The advantage of computing with ϕ is that its smoothness allows us to solve the level set advection equation and compute normals, curvature, etc. of island boundaries much more accurately. However, note that there is no longer a simple canonical choice for such a smooth ϕ. In practice, we simply allow ϕ to evolve from the trivial initial $\phi = -0.5$, through the processes of nucleation and growth. But it is no longer convenient to reinitialize ϕ during the calculation.

61

Nucleation   Island nucleation is modeled simply by selecting a nucleation site on the grid, and increasing the ϕ values at this point (and a few neighboring points) by $+1$, which automatically introduces a new $\phi = j$ level consisting of the smallest loop that is representable on the grid. Such an island is born with a small, grid-dependent area, and in order to better conserve the total number of atoms, we include a loss term in the adatom transport equation proportional to the nucleation rate dn/dt and the area of this small newly seeded island. This is a small $O(dx^2)$ correction to the ideal equations, for the sake of better discrete conservation of atoms.

Connected Components   Gathering statistics in the island dynamics model requires counting the number of islands, and determining their individual areas, boundary lengths, etc. Individual islands are precisely the connected components of the $\phi = 0, 1, 2, \ldots$ levels. Thus we require an algorithm for identifying connected components of level sets. In contrast, many level set applications never require such distinctions.

A practical and fast algorithm can be based on the iterative propagation of an arbitrary component label on the grid, as follows: the goal is to label every connected component, and also label each grid point as being in a certain connected component. Starting from an arbitrary node, it is labeled as being in component 1. At each iteration, every labeled node passes its label to its neighbors that are (a) unlabeled and (b) are not separated from it by a $\phi = 0, 1, 2, \ldots$ boundary. Whenever there are no such neighbors found for any of the labeled nodes, we select any remaining unlabeled node and give it the label 2, and continue in this fashion. This process terminates when all nodes have been labeled. The connected component j simply consists of all nodes with the label j, and the total number of connected components N equals the highest value of the label used. This arbitrary labeling allows us to count and locate each of the $j = 1, 2, \ldots, N$ islands in the domain. Combined with the subgrid representation of the component boundaries implicit in ϕ, we can accurately compute all the individual island properties without ever having to apply any complicated decision procedures to locate components.

Penalty Formulation of Internal Boundary Conditions   In [CMOS], the internal boundary conditions for the Stefan problem were implemented in direct fashion, which was relatively simple in the level set formulation. However, it required the use of spatial difference stencils for the Laplacian that are one-sided, irregular, and include small cut cells (and hence require implicit timestepping to avoid CFL limitations) near the interface. It is desirable to have an even simpler level set-based method for Stefan problems that avoids all these complications due to irregular discretization.

To achieve this, we have implemented the $\rho|_\Gamma = 0$ condition of the Irreversible Aggregation model via a penalty formulation. In order to keep ρ near zero on Γ we simply add a strong spatial sink term, proportional to ρ and concentrated only on Γ; i.e., the sink is proportional to a delta function concentrated on Γ. Thus the diffusion equation with internal boundary conditions is replaced by the diffusion-sink equation

$$\frac{\partial \rho}{\partial t} = \nabla(D\nabla\rho) + F - K\rho\delta_\Gamma$$

with no added internal boundary conditions. Here δ_Γ is a delta function on the island boundaries, which is represented by the usual level set means ([CHMO], [SSO]). K is a large penalty constant, which should scale like D/ϵ, where ϵ is the width of the smoothed out delta function used in practice. This formulation is similar to the delta function source term formulations used to treat surface tension forces (which otherwise would be internal boundary jump conditions on the PDEs) in two-phase flow [CHMO], [SSO]. The advantage is that this form can be discretized using standard stencils, and solved using standard diffusion equation solvers (in particular, explicit timestepping can be used for the evolution, if we want the simplest possible implementation).

In the context of Stefan problems, this form can be derived more abstractly by viewing dynamics of ρ as a constrained steepest descent on the gradient energy of ρ, $E[\rho] = \int D|\nabla\rho|^2$, and then including the constraint $\rho_\Gamma = 0$ into the energy as a penalty term of the form $\int K\rho^2\delta_\Gamma$. Using this energy formulation, it is also possible to conveniently express more general Γ-boundary conditions as well, in terms of other penalties or as changes in D (diffusion barriers, mimicking the physics that yields such conditions) that are concentrated only on the island boundaries.

One nice side effect of this formulation is that if atoms attaching to the boundary are counted by integrating the local sink term—instead of the local gradient flux of ρ—we get this simple asymptotic expression for the boundary velocity

$$v_n \approx K\rho$$

on the island boundary. This is a pointwise evaluation rather than the less local $[\partial\rho/\partial n]$ from the internal boundary condition formulation. Note that ρ is nearly 0, but this is balanced by fact that K is very large. In the $K \to \infty$ limit, the correct and well-defined v_n is obtained.

3.4   Computational Results

Figure 1 shows the island boundary evolution for the simple Uniform Density model, in the case where the specified normal velocity is isotropic. The figures show the view one would have looking directly down at the surface, at coverages (ratio of number of atoms deposited on the surface to the number of lattice sites on the surface) of 10%, 50%, 100% and 130%. Islands on the first monolayer are shown with solid line boundaries, while those on the second monolayer have dashed line boundaries and those on the third (one such island is shown in the 130% coverage case) have dot-dashed line boundaries. These calculations were done on a 128×128 numerical grid.

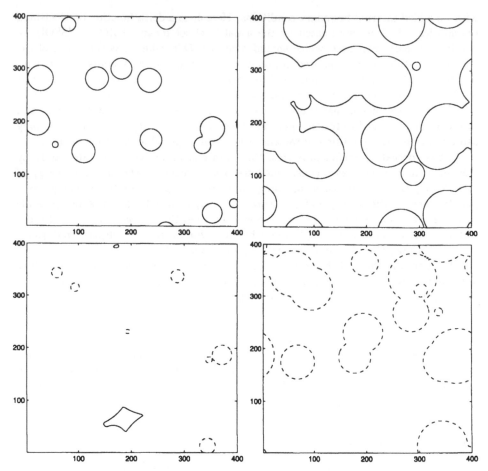

Figure 1: Island boundaries during growth in the simplified Uniform Density model, with isotropic normal velocity, at coverages of 10%, 50%, 100% and 130%. Islands on the first, second and third monolayers are shown with solid, dashed, and dot-dashed lines, respectively.

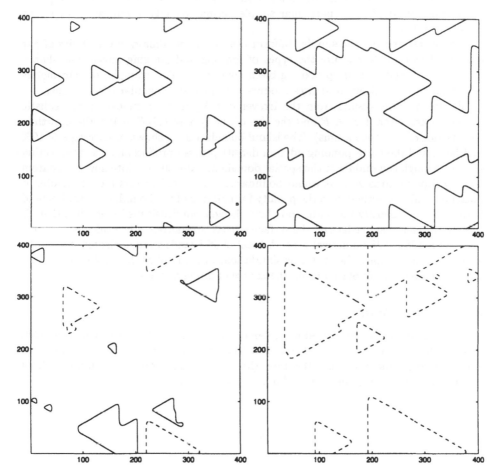

Figure 2: Island boundaries during growth in the simplified Uniform Density model, with anisotropic normal velocity, at coverages of 10%, 50%, 100% and 130%; line dashing indicates different monolayers. The islands are nucleated circular, but rapidly assume the triangular Wulff shape associated with their anisotropic velocity v_n.

65

Figure 2 shows a similar evolution (computed on a 256 × 256 grid), except the specified normal velocity now has a three-fold anisotropy, which causes the islands to naturally assume an asymptotically triangular ("Wulff") shape as they grow. We emphasize that this shape is not imposed on the growth—it is simply the asymptotic shape that results from a particular simple normal velocity specification $v_n(\vec{n})$, as proven in [OM].

These results from the simple Uniform Density model illustrate the ability of our numerical method to capture the types of growing and merging anisotropic shapes that are observed in real epitaxial growth conditions. For example, certain metals grown on certain silicon substrates produce very precise triangular island shapes.

Figure 3 shows results from the Irreversible Aggregation model, using realistic physical parameters for D, F and the size of the system ($D/F = 10^6$, the growth is on a lattice of 400 × 400 atoms). The island boundaries are shown at coverages of 10% and 50%, and the corresponding adatom density $\rho(x, y_0)$ profiles along a cross section $y = y_0$ through the middle of the spatial domain are also show. Note how the adatom density dips towards zero at island boundaries, as desired in this model. (Note: it does not vanish exactly due to the penalty formulation of the boundary condition used here.) The two density plots illustrate how the adatom density is lower when there is a large amount of step edges on the surface to soak up adatoms. The calculations were done on a 128 × 128 numerical grid. A total of 56 islands were nucleated during the growth of one complete layer in this simulation, which implies that a similar number of merger events took place during the course of filling in the layer.

3.5   Conclusion

Our recent results, as illustrated here, demonstrate that the Island Dynamics models include the desired physical features, and that the level set numerical methods can effectively solve these models. Together, they provide a promising new framework for doing analysis and control of thin film epitaxial growth.

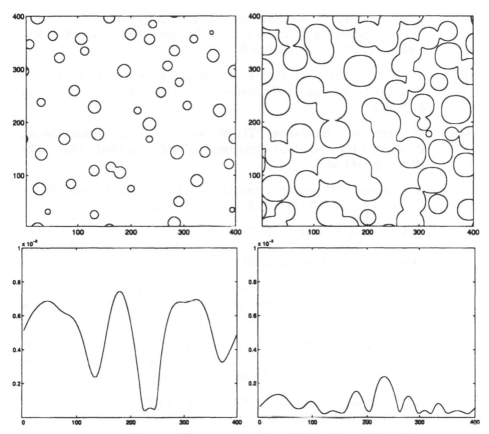

Figure 3: Island and adatom density evolution in the Irreversible Aggregation model. The surface is shown at 10% and 50% coverages; associated adatom densities are shown along the horizontal midline through the domain. Note the much lower mean density at 50% coverage, due to greater absorption at the many island boundaries.

References

[ABS] T. ASLAM, J. BDZIL AND D.S. STEWART, "Level Set Methods Applied to Modeling Detonation Shock Dynamics", *J. Comput. Phys.*, v. 126, (1996), pp. 390-409.

[BCF] W.K. BURTON, N. CABRERA AND F.C. FRANK, "The Growth of Crystals and the Equilibrium Structure of Their Surfaces", *Phil. Trans. Roy. Soc.*, London, Ser. A, (1951), pp. 243-299.

[BE] M. BARDI AND L.C. EVANS, "On Hopf's Formulas for Solutions of Hamilton-Jacobi Equations", *Nonlinear Analysis*, TMA, v. 8, (1984), pp. 1373-1381.

[BO] M. BARDI AND S. OSHER, "The Nonconvex Multidimensional Riemann Problem for Hamilton-Jacobi Equations", *SIAM J. on Math. Anal.*, v. 22, (1991), pp. 344-351.

[Ca] R. CAFLISCH, "The Island Dynamics Model for Epitaxial Growth", in VIP Kickoff Meeting DARPA/NSF, Stanford, CA 7/1/97.

[CCCD] V. CASELLES, F. CATTÉ, T. COLL AND F. DIBOS, "A Geometric Model for Active Contours in Image Processing", Report #9210, CEREMADE, Université Paris, Dauphine, (1992).

[CGG] Y.G. CHEN, Y. GIGA AND S. GOTO, "Uniqueness and Existence of Viscosity Solutions of Generalized Mean Curvature Flow Equations", *J. Differential Geom.*, v. 33, (1991), pp. 749-786.

[CGMORVZ] R. CAFLISCH, M. GYURE, B. MERRIMAN, S. OSHER, C. RATSCH, D. VVEDENSKY, AND J. ZINCK, "Island Dynamics and the Level Set Method for Epitaxial Growth", submitted to *Appl. Math. Letters* (1998).

[CHMO] Y.C. CHANG, T.Y. HOU, B. MERRIMAN AND S. OSHER, "A Level Set Formulation of Eulerian Interface Capturing Methods for Incompressible Fluid Flows", *J. Comput. Phys.*, v. 124, (1996), p. 449.

[CMO] L.T. CHENG, B. MERRIMAN AND S. OSHER, "A Variational Based Method for Constructing Minimal Wulff Surfaces", preprint, (1998).

[CMOS] S. CHEN, B. MERRIMAN, S. OSHER AND P. SMEREKA, "A Simple Level Set Method for Solving Stefan Problems", *J. Comput. Phys.*, v. 135, (1997), pp. 8-29.

[ES] L.C. EVANS AND J. SPRUCK, "Motion of Level Sets by Mean Curvature I", *J. Differential Geometry*, v. 33, (1991), pp. 635-681.

68

[ESS]   L.C. EVANS, H.M. SONER, AND P.E. SOUGANIDIS, "Phase Transitions and Generalized Motion by Mean Curvature", *Comm. Pure and Appl. Math.*, v. 65, (1992), pp. 1097-1123.

[FEO]   E. FATEMI, B. ENGQUIST AND S. OSHER, "Numerical Solution of the High Frequency Asymptotic Expansion for the Scalar Wave Equation", *J. Comput. Phys.*, v. 120, (1995), pp. 145-155.

[Gr]   R. GROSS, "Zur Theorie des Washstrums und Lösuns Forganges Kristalliner Materie", *Abhandl. Math.-Phys. Klasse Köngl. Sächs*, Wiss, v. 35, (1918), pp. 137-202.

[HLOZ]   T. HOU, Z. LI, S. OSHER AND H.-K. ZHAO, "A Hybrid Method for Moving Interface Problems with Applications to the Hele-Shaw Flow", *J. Comput. Phys.*, v. 134, (1997).

[HO]   E. HARABETIAN AND S. OSHER, "Regularization of Ill-Posed Problems via the Level Set Approach", *SIAM J. On Applied Math.*, to appear (1998), also UCLA CAM Report #95-41, (1995).

[HOSh]   E. HARABETIAN, S. OSHER AND C.-W. SHU, "An Eulerian Approach for Vortex Motion Using a Level Set Regularization Procedure", *J. Comput. Phys.*, v. 127, (1996), p. 15-26.

[KB]   R. KIMMEL AND A.M. BRUCKSTEIN, "Shape Offsets via Level Sets", CAD, v. 25, #3, (1993), pp. 154-162.

[KMOS]   M. KANG, B. MERRIMAN, S. OSHER AND P. SMEREKA, "A Level Set Approach for the Motion of Soap Bubbles with Curvature Dependent Velocity or Acceleration", UCLA CAM Report #96-19, (1996).

[Ko]   R. KOBAYASHI, "Modeling and Numerical Simulations of Dendritic Crystal Growth", *Physica D*, v. 63, (1993), p. 410.

[MBO1]   B. MERRIMAN, J. BENCE AND S. OSHER, "Diffusion Generated Motion by Mean Curvature", in AMS Selected Lectures in Math., The Comput. Crystal Grower's Workshop, edited by J. Taylor (Am. Math Soc., Providence, RI, 1993), p. 73.

[MBO2]   B. MERRIMAN, J. BENCE AND S. OSHER, "Motion of Multiple Junctions: A Level Set Approach", *J. Comput. Phys.*, v. 12, (1994), p. 334.

[Me]   B. MERRIMAN, "Level Set Methods for Island Dynamics", in VIP Kickoff Meeting, DARPA/NSF, Stanford, CA 7/1/97.

[NPV]   R.H. NOCHETTO, M. PAOLINI AND C. VERDI, "An Adaptive Finite Element Method for Two Phase Stefan Problems in Two Space Dimensions, Part II: Implementation and Numerical Experiments", *SIAM J. Sci and Stat. Comput.* v. 12, (1991), p. 1207.

[OM]   S. OSHER AND B. MERRIMAN, "The Wulff Shape as the Asymptotic Limit of a Growing Crystalline Interface", UCLA CAM Report #97-30 (1997), also *Asian J. Math.*, (1998).

[OSe]   S. OSHER AND J.A. SETHIAN, "Fronts Propagating with Curvature Dependent Speed: Algorithms Based on Hamilton-Jacobi Formulas", *J. Comput. Phys.*, v. 79, (1988), p. 12.

[OSh]   S. OSHER AND C.-W. SHU, "High Order Essentially Nonoscillatory Schemes for Hamilton-Jacobi Equations", SINUM, v. 28, (1991), pp. 907-922.

[PMOZK] D. PENG, B. MERRIMAN, H.-K. ZHAO, S. OSHER, AND M. KANG, "A PDE Based Fast Local Level Set Method", submitted to *J. Comput. Phys.*, (1998).

[RM]   S. RUUTH AND B. MERRIMAN, "Convolution Generated Motion and Generalized Huygen's Principles for Interface Motion", UCLA CAM Report, (1998).

[Se]   J.A. SETHIAN, "A Fast Marching Level Set Method for Monotonically Advancing Fronts", *Proc. Nat. Acad. Sci.*, v. 98, 4 (1996).

[SSO]   M. SUSSMAN, P. SMEREKA AND S. OSHER, "A Level Set Method for Computing Solutions to Incompressible Two-Phase Flow", *J. Comput. Phys.*, v. 119, (1994), pp. 146-159.

[ZCMO] H.-K. ZHAO, T. CHAN, B. MERRIMAN AND S. OSHER, "A Variational Level Set Approach to Multiphase Motion", *J. Comput. Phys.*, v. 127, (1996), pp. 179-195.

[ZMOW] H.-K. ZHAO, B. MERRIMAN, S. OSHER AND L. WANG, "Capturing the Behavior of Bubbles and Drops Using the Variational Level Set Approach", UCLA CAM Report #96-39 (1996), *J. Comput. Phys.*, (to appear).

[ZOMK] H.-K. ZHAO, S. OSHER, B. MERRIMAN AND M. KANG, "Dynamic Interpolation of Curves and Surfaces", UCLA CAM Report, (1998).

B. Merriman, R. Caflisch and S. Osher
Department of Mathematics
UCLA, Los Angeles, CA
USA

M. MIMURA

Free Boundary Problems Arising in Ecological Systems

1 Introduction

The study of interaction of many species arising in ecological systems has recently developed as a central problem in population ecology. In particular, problems of co-existence and exclusion of competing species have been theoretically investigated by using mathematical models. Especially, in order to understand spatial segregation of competing species, a variety of reaction-diffusion (RD) equation models have been proposed so far. Quite recently, it has been emphasized that the free boundary problem approach is very useful to know the dynamics of spatial segregation of competing species.

We are concerned in this paper with some new types of free boundary problems to track moving boundaries of segregating regions of competing species which move by diffusion.

We first introduce a RD system to describe the competitive interaction of n species $U_1, U_2, \cdots, U_n$, which is well used in mathematical ecology. Let $u_i(t, x)$ be the population density of the i-th species U_i $(i = 1, 2, \cdots, n)$ at time $t > 0$ and position $x \in \Omega$ in $\mathbf{R}^N$ $(N \geq 1)$. The system for $u_i(t, x)$ $(i = 1, 2, \cdots, n)$ is given by

$$u_{it} = d_i \Delta u_i + (r_i - a_i u_i - \sum_{j=1}^{n} b_{ij} u_j) u_i \quad (i = 1, 2, \cdots, n) \quad t > 0, \ x \in \Omega, \quad (1)$$

where d_i is the diffusion rate, r_i is the intrinsic growth rate, a_i and b_{ij} are, respectively, the intraspecific and the interspecific competiton rates $(i, j = 1, 2, ..., n)$. All of the rates are positive constants. Taking Ω to be bounded in $\mathbf{R}^N$, we impose the homogeneous Neumann boundary conditions on the boundary $\partial\Omega$

$$u_{i\nu} = 0 \quad (i = 1, 2, \cdots, n) \quad t > 0, \ x \in \partial\Omega, \quad (2)$$

where ν is the outerward normal unit vector on $\partial\Omega$. The initial conditions are given by

$$u_i(0, x) = u_{0i}(x) \geq 0 \quad (i = 1, 2, \cdots, n) \quad x \in \Omega. \quad (3)$$

The simplest system of (1.1) is the case when $n = 2$, that is,

$$\begin{cases} u_{1t} = d_1 \Delta u_1 + (r_1 - a_1 u_1 - b_1 u_2) u_1 \\ \\ u_{2t} = d_2 \Delta u_2 + (r_2 - b_2 u_1 - a_2 u_2) u_2 \end{cases} \quad t > 0, \ x \in \Omega \quad (4)$$

with the boundary conditions

$$u_{1\nu} = 0 = u_{2\nu} \quad t > 0, \quad x \in \Omega. \tag{5}$$

Ecologically suppose the situation where two species are strongly competing, that is, the interspecific competition rates are stronger than the intraspecific ones, which requires the following inequalities:

$$\frac{a_1}{b_2} < \frac{r_1}{r_2} < \frac{b_1}{a_2}, \tag{6}$$

where stable spatially constant equilibrium solutions of (1.4), (1.5) are $(u_1, u_2) = (r_1/a_1, 0)$ and $(0, r_2/a_2)$ only.

The qualitative properties of solutions of (1.4), (1.5) have been investigated in mathematical communities. The first contribution is that the stable attractor of (1.4), (1.5) consists of equilibrium solutions only (Hirsch[9], Matano and Mimura[13]). Therefore, we may focus our attention on the existence and stability of equilibrium solutions for the study of asymptotic behavior of solutions of (1.4),(1.5). Along this line, Kishimoto and Weinberger[12] showed that when Ω is convex, any spatially inhomogeneous equilibrium solutions are unstable even if they exist. This indicates that stable equilibrium solutions have to be only $(r_1/a_1, 0)$ and $(0, r_2/a_2)$, which ecologically implies that two strongly competing species can never coexist in convex domains. This is called *Gause's competitive exclusion* [7].

On the other hand, if the domain Ω is not convex, the structure of equilibrium solutions is not so simple but depends on the shape of Ω. In fact, if Ω takes suitable dumb-bell shape in 2-dimensions, there exist stable spatially inhomogeneous equilibrium solutions, which exhibit spatial segregation in a sense that the solution (u_1, u_2) takes nearly $(r_1/a_1, 0)$ in one subregion and $(0, r_2/a_2)$ in the other; that is, two competing species coexist with spatial segregation in their habitat.

Thus, the asymptotic behavior of solutions of (1.5) with (1.6) is made clear, depending on the shape of the domain Ω. However, it does not give any necessary information on transient dynamics of segregating regions of two competing species, no matter how each species results in either extinction or existence under competition.

The purpose of this paper is to derive several evolutional equations to describe the boundaries between segregating patterns of two or three competing species, by taking different scaling limits.

2 Spatial segregation limits for a two competing species system

2.1 Small diffusion case

In this section, we consider (1.4), (1.5) with (1.6) under the situation where the diffusion rates d_1 and d_2 are both sufficiently small (in other words, all of the other

r_i, a_i and b_i $(i = 1, 2)$ are sufficiently large). Then (1.4) is conveniently rewritten as

$$\begin{cases} u_{1t} = \varepsilon^2 \Delta u_1 + (r_1 - a_1 u_1 - b_1 u_2)u_1, \\ \\ u_{2t} = d\varepsilon^2 \Delta u_2 + (r_2 - b_2 u_1 - a_2 u_2)u_2 \end{cases} \quad t > 0, \quad x \in \Omega, \tag{7}$$

in which a sufficiently small parameter ε is included. By the bistability condition (1.7) and the smallness of ε, one can intuitively expect that the solution (u_1, u_2) of (2.1) tends to either $(r_1/a_1, 0)$ or $(0, r_2/a_2)$ for fixed $x \in \Omega$ so that the region Ω is generically separated into two subregions:

$$\Omega_1(t) = \{x \in \Omega | (u_1, u_2) \approx (r_1/a_1, 0) \ \ t > 0\}$$

and

$$\Omega_2(t) = \{x \in \Omega | (u_1, u_2) \approx (0, r_2/a_2) \ \ t > 0\},$$

where internal layer regions appear between $\Omega_1(t)$ and $\Omega_2(t)$. These subregions exhibit, respectively, segregating patterns of the two competing species. In order to study the dynamics of $\Omega_1(t)$ or $\Omega_2(t)$, we take the limit $\varepsilon \downarrow 0$ in (2.1) so that the internal layer regions become sharp interfaces, say $\Gamma(t)$, which is the boundaries between $\Omega_1(t) = \{x \in \Omega | (u_1, u_2) = (r_1/a_1, 0) \ \ t > 0\}$ and $\Omega_2(t) = \{x \in \Omega | (u_1, u_2) = (0, r_2/a_2) \ \ t > 0\}$. Using the singular limit analysis, Ei and Yanagida[5] derived the following evolutional equation to describe the motion of the interface $\Gamma(t)$:

$$V = -\varepsilon(\varepsilon L(d)(N - 1)\kappa + c) + o(\varepsilon^2), \tag{8}$$

where V is the normal velocity of the interface from $\Omega_1(t)$ and $\Omega_2(t)$, κ is the mean curvature of the interface, $L(d)$ is a positive constant depending on d where $L(1) = 1$. c is the velocity of the travelling front solution $(u_1, u_2)(x - ct)$ of the 1-dimensional system

$$\begin{cases} u_{1t} = \varepsilon^2 u_{1xx} + (r_1 - a_1 u_1 - b_1 u_2)u_1, \\ \\ u_{2t} = d\varepsilon^2 u_{2xx} + (r_2 - b_2 u_1 - a_2 u_2)u_2 \end{cases} \quad t > 0, \quad -\infty < x < \infty, \tag{9}$$

with the boundary conditions

$$(u_1, u_2)(t, -\infty) = (r_1/a_1, 0) \text{ and } (u_1, u_2)(t, \infty) = (0, r_2/a_2). \tag{10}$$

It is proved in Kan-on and Fang[11] that the velocity of the travelling wave solution of Problem (2.3),(2.4) is uniquely determined, depending on values of the parameters r_i, a_i and b_i $(i = 1, 2)$. In particular, if a_1 is a free parameter and the others are fixed to satisfy the inequalities (1.7), there exists a unique $a^* > 0$ such that $c = 0$ for $a_1 = a^*, c > 0$ for $a_1 > a^*$ and $c < 0$ for $a_1 < a^*$ [10]. This indicates that when ε is sufficiently small, the qualitative behavior of solutions of (2.1) is quite similar to one of the following bistable scalar reaction-diffusion equations with cubic nonlinearity

$$u_t = \varepsilon^2 L(d)\Delta u + u(1 - u)(u - a) \tag{11}$$

where a is some constant satisfying $0 < a < 1$.

When the special case $c = 0$, that is, u_1 and u_2 coexist in the 1-dimensional problem, (2.2) becomes the well-known equation of motion by mean curvature,

$$V = \varepsilon^2 L(d)(N - 1)\kappa, \tag{12}$$

which has been analytically and numerically investigated by many authors (see [6],[8], for instance). If $\Gamma(t)$ can be obtained by solving (2.2), it gives us the time evolution of segregating patterns of two competing species. One typical example of the dynamics of $\Gamma(t)$ of (2,6) in 2-dimensions follows. Suppose that $\Gamma(0)$ is given by an embedded closed curve in Ω such that $\Omega_1(0)$ is inside and $\Omega_2(0)$ is outside of $\Gamma(0)$. Then it is shown that u_1 is extinct in finite time and the whole domain is occupied by u_2, even if they can coexist in 1 dimension[8].

2.2   Large interspecific competition case

From a more ecological application, the situation can be considered where only the interspecific competition rates b_1 and b_2 are very large. In order to treat this situation, it is convenient to rewrite (1.4) as

$$\begin{cases} u_{1t} = \Delta u_1 + r_1(1 - u_1)u_1 - bu_1u_2, \\[2mm] u_{2t} = d\Delta u_2 + r_2(1 - u_2)u_2 - \alpha bu_1u_2 \end{cases} \quad t > 0, \quad x \in \Omega, \tag{13}$$

where b and α are positive constants. Here we assume that b is very large and all the other parameters are of order $O(1)$. The coefficient α means the competition ratio between two species U_1 and U_2; if $\alpha > 1$, U_1 has a competitive advantage, while if $\alpha < 1$, the situation is reversed.

In order to study how segregating patterns of two competing species depend on values of b, we demonstrate 2-dimensional numerical simulations of (2.6) with (1.6) in a rectangle domain in $\mathbf{R}^2$. We take b as a free parameter, leaving the others d_1, d_2, r_1, r_2 and α to be suitably fixed. Let b be not small but not so large, that is, competition between them is not so strong. As is shown in Fig. 1, one can observe weak spatial segregation of u_1 and u_2 where its overlapped zone is not so clear. When b becomes larger, the segregating pattern becomes clearer and the overlapped zone becomes quite narrow. We note here that no sharp interfaces appear and both of them are almost zero there, as in Fig. 2. One can thus expect that taking the limit $b \uparrow \infty$, u_1 and u_2 possess disjoint supports (habitats) with only a common curve, on which they are both zero. This curve, which is a free boundary, gives the boundary of segregating patterns of two competing species under very strong interspecific competition.

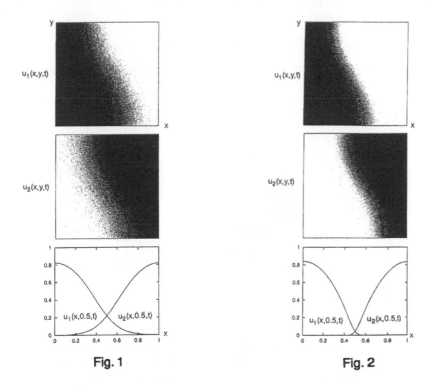

Fig. 1   Fig. 2

Motivated by the numerical simulations above, we derive the limiting system from (2.7) as $b \uparrow \infty$, which can be described as some kind of free boundary problem for the moving boundaries between segregating patterns of two competing species.

Let $\Gamma(t)$ be an interface which separates two subregions:

$$\Omega_1(t) = \{x \in \Omega, u_1 > 0 \text{ and } u_2 = 0\} \text{ and}$$

$$\Omega_2(t) = \{x \in \Omega, u_2 > 0 \text{ and } u_1 = 0\}$$

in Ω (see Fig. 3 in 2 dimensions, for instance). u_1 and u_2 satisfy, respectively,

$$\begin{cases} u_{1t} = d_1 \Delta u_1 + r_1(1 - u_1)u_1, & t > 0, x \in \Omega_1(t), \\ u_{2t} = d_2 \Delta u_2 + r_2(1 - u_2)u_2, & t > 0, x \in \Omega_2(t), \end{cases} \tag{14}$$

$$u_{1\nu} = 0 = u_{2\nu} \quad t > 0, x \in \partial\Omega. \tag{15}$$

75

On the interface,

$$u_1 = u_2 = 0, \quad t > 0, x \in \Gamma(t) \tag{16}$$

$$0 = -\alpha d_1 u_{1\sigma} - d_2 u_{2\sigma} \quad t > 0, x \in \Gamma(t) \tag{17}$$

where σ is the outward unit vector on the interface. The initial conditions are given by

$$u_i(0, x) = u_{i0}(x), \quad x \in \Omega_i(0) \quad (i = 1, 2) \tag{18}$$

$$\Gamma(0) = \Gamma_0. \tag{19}$$

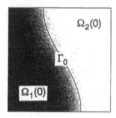

Fig. 3

The problem is to find $(u_1(t, x), u_2(t, x))$ and $\Gamma(t)$ which satisfy (2.8) - (2.13). One interesting remark is that the system (2.8) - (2.13) is quite similar to the classical two-phase Stefan problem except for the following two points: (i) The system (2.8) is described by the logistic growth equation which is well-known in theoretical ecology; (ii) If we compare (2.11) with the interface equation in the Stefan problem, it corresponds to the situation that the latent heat is zero, where the strength ratio α of the interspecific competition between u_1 and u_2 is contained. The proof of convergence is stated in [3]. For the limiting free boundary problem (2.8)-(2.13), we refer to the papers by Cannon and Hill[2] and Tonegawa[15].

The methods above can be extended to the similar problems for more numbers of competing species. Let us discuss here the RD system for three competing species.

3 Spatial segregation limits for a three competing species model

3.1 Slow diffusion case

In this section, we consider the following competing system for three species u_1, u_2 and u_3 in 2 dimensions. In a similar way to the 2-competing case discussed in Section

2.1, we treat

$$u_{it} = \varepsilon^2 d_i \Delta u_i + (r_i - a_i u_i - \sum_{j=1}^{3} b_{ij} u_j) u_i, \quad (i = 1, 2, 3) \quad t > 0, x \in \Omega, \qquad (20)$$

under the situation where ε is a sufficiently small parameter. We impose the homogeneous Neumann conditions (1.2) at the boundary. Assume that all of the interspecific competition b_{ij} $(i = 1, 2, 3)$ are suitably large such that stable equilibria of the diffusionless system of (3.1)

$$v_{it} = (r_i - a_i v_i - \sum_{j=1}^{3} b_{ij} v_j) v_i, \quad (i = 1, 2, 3) \quad t > 0 \qquad (21)$$

are $P_1 = (r_1/a_1, 0, 0)$, $P_2 = (0, r_2/a_2, 0)$ and $P_3 = (0, 0, r_3/a_3)$ only and other equilibria are unstable.

By this assumption with the help of numerical simulations, we find that any solution of (3.1) generically tends to one of the equilibria P_i $(i = 1, 2, 3)$, because ε is sufficiently small, so that there appear interfaces which divide the whole domain Ω into three subdomains Ω_i $(i = 1, 2, 3)$ where the solution (u_1, u_2, u_3) is close to one of the P_i $(i = 1, 2, 3)$. Here we should note that there may exist triple junctions where three interfacial curves intersect at one point.

Let us show some numerical simulations of (3.1). We fix $d_i = d = 1$ $(i = 1, 2, 3)$ in our computations. The first example is the completely symmetric case. Fig. 4 shows that the angles between any two interfacial curves are approximately equal and that the dynamics of interfaces is very slow. The second is the case which slightly loses symmetry. Fig. 5 shows the angles are not necessarily equal, but they continue to be constant in time. The dynamics is similar to the one in Fig. 4. The third is the cyclic and symmetric case. Because of the cyclic property, Fig. 6 demonstrates the occurrence of a rotating stationary spiral wave with three arms. The triple junction point is its center where the angles between any two interfacial curves are about equal. Finally, we consider a cyclic but non-symmetric case. Fig. 7 shows that stationary waves no longer occur but there are some clusters with small spirals. Here we should emphasize that our system is a non-gradient system so that it is possible to possess such cyclic property that generates spiral waves. This pattern is never generated by any gradient systems such as vector-valued Ginzburg-Landau equations with three well potentials([1], [14]).

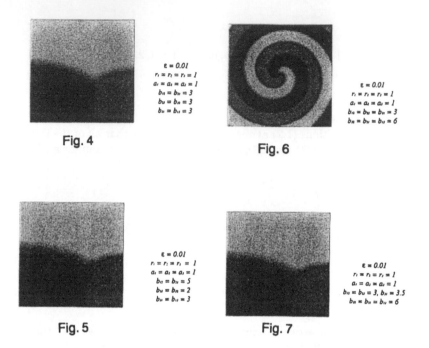

Fig. 4

$\varepsilon = 0.01$
$r_i = r_i = r_i = 1$
$a_i = a_i = a_i = 1$
$b_{ii} = b_{ii} = 3$
$b_{ii} = b_{ii} = 3$
$b_{ii} = b_{ii} = 3$

Fig. 6

$\varepsilon = 0.01$
$r_i = r_i = r_i = 1$
$a_i = a_i = a_i = 1$
$b_{ii} = b_{ii} = b_{ii} = 3$
$b_{ii} = b_{ii} = b_{ii} = 6$

Fig. 5

$\varepsilon = 0.01$
$r_i = r_i = r_i = 1$
$a_i = a_i = a_i = 1$
$b_{ii} = b_{ii} = 5$
$b_{ii} = b_{ii} = 2$
$b_{ii} = b_{ii} = 3$

Fig. 7

$\varepsilon = 0.01$
$r_i = r_i = r_i = 1$
$a_i = a_i = a_i = 1$
$b_{ii} = b_{ii} = 3, b_{ii} = 3.5$
$b_{ii} = b_{ii} = b_{ii} = 6$

In this section, we restrict our discussion to the first two examples above and derive the formula of the relation among three angles at the triple junction point. Let $\Gamma_1(t)$ be the interface between $\Omega_2(t)$ and $\Omega_3(t)$ and other $\Gamma_2(t)$ and $\Gamma_3(t)$ are similarly defined. Also, let θ_1 be the angle between the interfaces $\Gamma_2(t)$ and $\Gamma_3(t)$. θ_2 and θ_3 are similarly defined. Then the relation among θ_1, θ_2 and θ_3 is explicitly represented as

$$\frac{M_1}{\sin \theta_1} = \frac{M_2}{\sin \theta_2} = \frac{M_3}{\sin \theta_3} \text{ with } \theta_1 + \theta_2 + \theta_3 = 2\pi \tag{22}$$

for some constants M_i $(i = 1, 2, 3)$. We should remark that though our system does not fall in the framework of gradient systems, the formula (3.3) obtained in the limit $\varepsilon \downarrow 0$ is exactly same as the one in gradient systems[1]. Its derivation and more precise discussion to (3.1) are stated in [4].

3.2 Large interspecific competition case

We consider (1.1) with $n = 3$, which is a direct extension of the 2-competing species system (2.7). The system which we treat is

$$\begin{cases} u_{1t} = d_1 \Delta u_1 + r_1(1 - u_1)u_1 - bu_1 u_2 - \beta c u_1 u_3 \\ u_{2t} = d_2 \Delta u_2 + r_2(1 - u_2)u_2 - \alpha bu_1 u_2 - e u_2 u_3 \quad t > 0, \ x \in \Omega \\ u_{3t} = d_3 \Delta u_3 + r_3(1 - u_3)u_3 - c u_1 u_3 - \gamma e c u_2 u_3 \end{cases} \quad (23)$$

where d_i, r_i $(i = 1, 2, 3), b, c, e, \alpha, \beta$ and γ are positive constants. The boundary conditions are

$$u_{i\nu} = 0 \quad (i = 1, 2, 3) \quad t > 0, x \in \partial\Omega \quad (24)$$

where ν is the outward normal unit vector on $\partial\Omega$. Following the procedure used in Section 2.2, we assume that interspecific competition rates are very large. Because of three competing species, the situation is classified into the following three cases:

(i) Only b is sufficiently large and the other parameters c and e are of order $O(1)$;

(ii) Both b and c are sufficiently large and the other parameter e is of order $O(1)$;

(iii) All of b, c and e are sufficiently large.

Here we restrict our discussion to the case (i) where only u_1 and u_2 are very strongly competing, one can thus expect that u_1 and u_2 exhibit spatial segregation, while though u_3 competes with them, it is smoothly distributed in the whole domain. Let $\Gamma_{12}(t)$ be an interface which separates two subregions:

$$\Omega_{13}(t) = \{x \in \Omega, u_1, u_3 > 0 \text{ and } u_2 = 0\}$$

and

$$\Omega_{23}(t) = \{x \in \Omega, u_2, u_3 > 0 \text{ and } u_1 = 0\}.$$

Then (u_1, u_3) and (u_2, u_3), respectively, satisfy the following RD systems in $\Omega_{13}(t)$ and $\Omega_{23}(t)$:

$$\begin{cases} u_{1t} = d_1 \Delta u_1 + r_1(1 - u_1)u_1 - \beta c u_1 u_3 & t > 0, x \in \Omega_{13}(t) \\ u_{3t} = d_3 \Delta u_3 + r_3(1 - u_3)u_3 - c u_1 u_3 & t > 0, \quad x \in \Omega_{13}(t) \end{cases} \quad (25)$$

and

$$\begin{cases} u_{2t} = d_2 \Delta u_2 + r_2(1 - u_2)u_2 - e u_2 u_3 & t > 0, \quad x \in \Omega_{23}(t) \\ u_{3t} = d_3 \Delta u_3 + r_3(1 - u_3)u_3 - \gamma e c u_2 u_3 & t > 0, \quad x \in \Omega_{23}(t) \end{cases} \quad (26)$$

The interface equation is

$$u_1 = 0 = u_2, u_3 \in C^1(\Omega) \quad t > 0, x \in \Gamma_{12}(t) \quad (27)$$

$$0 = -\alpha d_1 u_{1\sigma} - d_2 u_{2\sigma} \quad t > 0, x \in \Gamma_{12}(t) \tag{28}$$

where σ is the outward unit vector on the interface. The initial conditions are given by

$$u_i(0, x) = u_{i0}(x) \ \ x \in \Omega_{i3}(0) \ \ (i = 1, 2) \text{ and } u_3(0, x) = u_{30}(x) \ \ x \in \Omega \tag{29}$$

$$\Gamma_{12}(0) = \Gamma_{012}. \tag{30}$$

One can thus obtain a new type of free boundary problem (3.6)-(3.11), which is different from the classical Stefan problem arising in solidification.

4 Concluding remark

We have introduced some free boundary problems which are derived from competition-diffusion systems by taking different scaling limits. We thus arrive at two familiar types of limiting problems: One is the equations of mean curvature motion and the other is the Stefan-like problem. Unfortunately, there have been many unsolved problems for these systems. These are future work for us.

References

[1] BRONSARD, L. AND REITICH, F. On three-phase boundary motion and the singular limit of a vector-valued Ginzburg-Laudau equation, *Archive of Rat. Mech. Math.* 124, 355-379 (1993)

[2] CANNON, J. R. AND HILL, C. D. On the movement of a chemical reaction interface, *Indiana U. Math. J.* 20, 429-454 (1970)

[3] DANCER, E. N., HILHORST, D., MIMURA, M. AND PELETIER, L. A. Spatial segregation limit to competition-diffusion systems, manuscript.

[4] EI, S.-I., IKOTA, R. AND MMURA, M. A three phase partition problem arising in a competition-diffusion system, in preparation.

[5] EI, S.-I. AND YANAGIDA, E. Dynamics of interfaces in competition-diffusion systems, *SIAM J. Appl. Math.* 54, 1355-1373 (1994).

[6] GAGE, M. AND HAMILTON, R. The heat equation shrinking convex plane curves, *J. Differential Geometry.* 23, 69-96 (1986).

[7] GAUSE, G. F. *The struggle for existence. A Classic of Mathematical Biology and Ecology*, The Williams & Wilkins Company, Baltimore, 1934.

[8] GRAYSON, M. The heat equation shrinks embedded plane curves to round points, *J. Differential Geometry* 26, 285-314 (1987).

[9] HIRSCH, M. W. Differential equations and convergence almost everywhere of strongly monotone semiflows, *PAM Technical Report*, University of California, Berkeley (1982).

[10] KAN-ON, Y. Parameter dependence of propagation speed of travelling waves for competition-diffusion equations, *SIAM J. Math. Anal.* 26, 340-363 (1995).

[11] KAN-ON, Y. AND FANG, Q. Stability of monotone travelling waves for competition-diffusion equations, *Japan J. Indust. Appl. Math.* 13, 343-349 (1996).

[12] KISHIMOTO, K. AND WEINBERGER, H. F. The spatial homogeneity of stable equilibria of some reaction-diffusion system on convex domains, *J. Differ. Equations* 58, 15-21 (1985).

[13] MATANO, H. AND MIMURA, M. Pattern formation in competition-diffusion systems in nonconvex domains, *Publ. RIMS, Kyoto Univ.*, 19, 1049-1079 (1983).

[14] STERNBERG, P. AND ZIEMER, W. P. Local minimizers of a three phase partition problem with triple junctions, preprint.

[15] TONEGAWA, Y. On the regularity of chemical reaction interface, to appear in *Comm. in PDE*.

Masayasu Mimura
Department of Mathematics
Hiroshima University
1-3-1 Kagamiyama, Higashi-Hiroshima, Japan

C. Pozrikidis

Overview of Dynamical Simulations of the Flow of Suspensions of Liquid Capsules and Drops

Abstract

We study the two-dimensional low-Reynolds number flow of suspensions of deformable particles including liquid drops and gas bubbles with constant surface tension, and capsules enclosed by elastic membranes. The investigations are based on large-scale numerical simulations conducted by boundary-integral methods. In the parametric studies, we vary the drop to suspended fluid viscosity ratio, the capillary number expressing the particle deformability, and the number of simulated particles per computational box. The results illustrate the macroscopic rheological properties of a suspension regarded as a continuum, the evolution of the suspended phase microstructure, and the apparent random motion of the individual particles.

1 Overview

Dynamical simulations of the flow of suspensions of deformable particles are relevant to, and provide us with insights into, a variety of engineering and biomedical processes involving two–phase and particulate flow. Applications include dispersion, homogenization, mixing and emulsification, as well as in blood flow in the circulation. The main objectives of the simulations are (a) to identify and analyze the physical mechanisms that govern the evolution of the microstructure and to elucidate the effects of particle constitution and interfacial composition; (b) to assess the significance of the flow boundaries with reference to the formation of particle-free lubrication zones; (c) to establish a relationship between the geometrical properties of the microstructure of the dispersed phase and the global properties of the suspension regarded as a continuum; (d) to describe the statistical properties of the dispersed phase in the context of statistical mechanics, and to analyze the apparent random motion of the individual particles in terms of hydrodynamic diffusion.

Until a few years ago, theoretical and computational studies of suspension rheology with deformable particles included investigations of dilute or highly concentrated systems (e.g., Kennedy et al. 1994). Numerical simulations in the range of intermediate concentrations with strong hydrodynamic interactions have become feasible only recently, thanks to the development of efficient numerical methods based on boundary integral formulations (e.g., Li et al. 1996, Loewenberg & Hinch 1996, Toose 1997, Charles & Pozrikidis 1998). It is worth noting that analogous numerical methods for suspensions of rigid spherical particles have been available for over fifteen years, and have been used to study a variety of systems. At the present time, extensive large-scale simulations of three-dimensional motions of suspensions of deformable particles are prohibited by excessive demands on computer memory and CPU time; one must necessarily compromise by studying two-dimensional flows.

In our research program we study the flow of suspensions of deformable particles at moderate and high volume fractions, under the premise of Stokes flow. The particle deformability and the interfacial mobility add a host of features that alter the behavior of the suspension in several fundamental ways (Li *et al.* 1996, Charles & Pozrikidis 1998). The simulations cover a broad range of conditions extending from the dilute limit where the particles translate, rotate, and deform as though they were in isolation, to the limit of high volume fractions where the suspension becomes concentrated and is either immobilized due to particle blockage, or resembles a foam; the motion of a foam is distinguished by the continuous formation and peeling off of thin liquid films joining at Plateau borders. Moreover, in the simulations we consider droplets with clean interfaces characterized by constant surface tension, droplets whose interfaces are populated by surfactants, and particles whose interfaces are viscoelastic membranes.

Our model two-dimensional suspensions contain random or regular distributions of particles dispersed in unbounded or bounded domains of flow. The motion can be generated by boundary motion, an imposed shear flow, or a pressure gradient. The dynamical simulations are conducted using the method of *interfacial dynamics*. The numerical method is an advanced implementation of the boundary integral formulation for Stokes flow incorporating several novel features, as will be discussed in Section 2.

In an early series of simulations, we studied the flow of suspensions of two-dimensional liquid drops in a channel that is bounded by two plane walls, as illustrated in Figure 1 (a) (Zhou & Pozrikidis 1993ab, 1994). The flow is driven either by the relative motion of the walls, or by an imposed pressure gradient. We considered the motion of single periodic files, multiple periodic files, and random arrangements of drops with up to twelve drops per periodic cell. The results helped us identify the physical mechanisms and consequences of drop-drop and drop-boundary interactions, and characterize the motion of the suspension in terms of an effective coefficient of shear viscosity and normal stress difference. In a parallel study, we performed a numerical investigation of the optical properties of a flowing emulsion using a ray tracing method (Pozrikidis & Sheth 1996).

In a second series of simulations, we considered homogeneous, doubly periodic simple shear flow in the absence of flow boundaries. These studies provided us with insights into a more general class of locally unidirectional shear flows; studies of flow in channels are pertinent to wall-bounded flows. On a more pragmatic level, the absence of boundaries simplifies the computation of the Green's function involved in the boundary-integral equation, and expedites the implementation of the numerical method by taking advantages of the Ewald summation method (Pozrikidis 1996). Last, the study of doubly periodic flows allows for comparisons with laboratory data, and models of concentrated emulsions and foams based on lubrication flow.

We carried out a series of simulations for highly concentrated doubly periodic emulsions (Li *et al.* 1994), and another set of simulations for periodically repeated random suspensions with 25 or 49 droplets per periodic box (Li *et al.* 1996), as illustrated in Figure 1 (b, c). In both cases, we assumed that the drop viscosity is equal to the ambient fluid viscosity. In the most recent series of simulations, we computed the motion of suspension of 25 drops per periodic box with drop fluid viscosity different than the suspending fluid viscosity (Charles & Pozrikidis 1998). Since high-viscosity drops behave like rigid particles, varying the drop fluid viscosity allows us to study the transition from emulsion-type to a suspension-type behavior with respect to rheology and dynamics of the microstructure.

In Section 2 we briefly describe the numerical method, and in Section 3 we indicate ongoing and future extensions of past work into new directions.

2   Numerical method

To illustrate the computational methodology, we consider the flow of a doubly periodic suspension of two-dimensional liquid capsules with viscosity $\lambda\mu$ suspended in an ambient fluid with viscosity μ, as depicted in Figure 1 (c). The motion is driven by an incident simple shear flow with shear rate G. Under the action of the incident flow, the periodic lattice deforms at a constant rate and in a recurrent manner with a period that is equal to $1/G$.

At vanishing Reynolds number, the motion of the fluid inside and outside the drops is governed by the linearized quasi-steady equation of motion and by the continuity equation, with appropriate values of physical constants for the suspended and suspending fluids (e.g., Pozrikidis 1997). The boundary conditions require that the velocity remain continuous across each interface, but the interfacial traction undergoes a discontinuity $\Delta f = (\sigma_1 - \sigma_2) \cdot n$; σ is the stress tensor, and the subscripts 1 or 2 label the two fluids. For liquid drops and gas bubbles with constant surface tension γ, $\Delta f = \gamma \kappa n$ where κ is interfacial curvature and n is the normal vector directed into the ambient fluid. When the interface has a more complex structure, Δf is given by a more involved constitutive equation that may involve the surface viscosity and moduli of elasticity. In the presence of surfactants, the surface tension is a function of the surfactant concentration whose evolution is governed by a convection-diffusion equation defined over the interface. The solution of the convection-diffusion equation can be obtained using a standard finite-difference or finite-volume method (Yon & Pozrikidis 1998).

Rendering all variables non-dimensional using as characteristic length the size of a periodic cell L, characteristic velocity GL, and characteristic stress μG, we find that the motion depends on the viscosity ratio λ and, in the case of liquid drops, on the capillary number $Ca = \mu GL/\gamma$. One or more elasticity numbers are introduced for capsules enclosed by elastic membranes. The problem is reduced to computing the

84

evolution of the suspension from a given initial configuration. This is classified as a free-boundary problem with evolving boundaries and discontinuous physical properties.

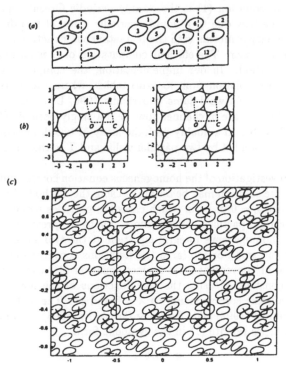

Figure 1. (a) Flow of a random suspension of drops in a chanel. (b) Motion of a doubly periodic concentrated emulsion of foam. (c) Flow of a doubly periodic concentrated suspension of liquid drops.

To compute the motion of the particles, we use the boundary integral method for Stokes flow. The main advantage of this method is that the dimension of the computational problem is reduced with respect to that of the physical problem by one unit, and the solution may be found using well-founded iterative procedures. The velocity u at a point x_0 that is located at an interface is computed by solving the following Fredholm integral equation of the second kind:

$$u_j(x_0) = \frac{2}{1 + \lambda} u_j^\infty(x_0) - \frac{1}{2\pi\mu} \frac{1}{1 + \lambda} \int_{Interfaces} \Delta f_i(x) G_{ij}(x, x_0)\, dl(x) \qquad (1)$$
$$+ \frac{\beta}{2\pi} \int_{interfaces}^{PV} u_i T_{ijk}(x, x_0) n_k(x)\, dl(x)$$

where u^∞ is the incident shear flow, $\beta = (1-\lambda)(1+\lambda)$, the line integrals are evaluated over the interfaces of all drops in a periodic cell, and PV denotes the principal value of the double-layer integral (Pozrikidis 1992).

The kernel G in equation (1) is the doubly periodic Green's function of Stokes flow representing the flow due to a doubly periodic array of two-dimensional point forces whose periodicity conforms with the instantaneous structure of the lattice; T is the associated stress tensor. The efficient computation of G is an important feature of the numerical method. In our implementation, the components of this tensor are computed by interpolation from look-up tables, where the entries of the table are produced in terms of Ewald sums (Pozrikidis 1996, Li et al. 1996, Charles & Pozrikidis 1998). In the case of channel flow, G is a singly periodic Green's function representing the flow due to a doubly periodic array of two-dimensional point forces in a channel that is bounded by two parallel walls, and T is the corresponding stress tensor.

A theoretical investigation of the homogeneous equation corresponding to equation (1) indicates that the solution may be obtained using the method of successive substitutions, except when $\lambda = 0$ or ∞ in which cases the particles reduce, respectively, to inviscid bubbles or rigid particles. Furthermore, due to the presence of eigenfunctions in the generalized homogeneous integral equation at the two aforementioned limits, when ∞ is either too small or too large the iterations converge but only slowly. To accelerate the rate of convergence, we replace equation (1) with a deflated integral equation that is constructed using the eigenfunctions of the homogeneous integral equation and its adjoint corresponding to $\lambda = 0$ and ∞. The deflation procedure is explained by Pozrikidis (1992) and Charles & Pozrikidis (1998). To illustrate a central computational task, we rewrite (1) in the symbolic form

$$q = Lq + F, \tag{2}$$

where q is the unknown function; in this case the velocity, L, is a compact linear integral operator, and F is a known forcing function. The numerical method involves tracing each interface with a set of marker points, solving the integral equation (1) for the velocity at the marker points using an iterative method, and advancing the position of the marker points to obtain the evolution of the interfaces. This procedure is concisely termed the *method of interfacial dynamics*. Specifically, with reference to equation (2), we guess the solution q, compute the right-hand side, and replace the guessed with the computed solution. The individual steps are as follows:

1. Represent each interface with a set of boundary elements that are defined in terms of a set of marker points.

2. Approximate the unknown function q over each boundary element using a truncated polynomial expansion in terms of properly defined surface variables.

3. Substitute the local expansions of the unknown function into the integral equation (1), and move the constant coefficients out of the interface integrals. In this manner, the integral equation is converted to an algebraic equation involving the coefficients.

4. Compute the unknown coefficients using a *collocation* method; this involves enforcing the discretized integral equation at collocation points over the interfaces, thereby producing a system of M linear algebraic equations.

5. Solve the derived system of linear algebraic equations for the coefficients of the local expansions by fixed-point iterations.

Additional features of the numerical implementation include (a) Adaptive marker-point redistribution around the interfaces to resolve regions of high curvature; (b) Repositioning of the centers of strongly interacting particles to avoid artificial coalescence and reduce the stiffness of the ODEs governing the maker point trajectories; (c) An optional feature is the use of the method of multi-pole expansions to expedite the computation of the hydrodynamic interactions of well-separated particles. We found, however, that the use of look-up tables outperforms this fast-summation method.

3 Future directions

Ongoing and future work aims at extending previous investigations with respect to flow conditions, geometrical arrangement, and number of suspended particles per periodic box, as well as exploring new systems. Present efforts point in three directions:

1. Study singly periodic systems with 25 or a higher number of particles randomly distributed within each periodic box, in pressure-driven channel flow. This is an extension of our earlier work with 12 particles per box and drop fluid viscosity equal to the ambient fluid viscosity. The higher number of drops allow us to investigate the properties of the suspension using methods of statistical mechanics. Preliminary results showed that the time for the suspension to reach statistical equilibrium is much longer in channel than in homogeneous shear flow.

2. Study the effect of interfacial constitution by carrying out simulations with suspensions of liquid capsules enclosed by elastic membranes; these are primitive models of red blood cells. Apart from revealing the significance of particle deformability, the simulations illustrate the properties of suspensions of rigid-like but flexible particles with non-isotropic shapes.

3. Study the effect of an insoluble surfactant. Marangoni stresses due to surface tension gradients attributed to surfactants can immobilize interfaces in depletion, or promote drop disintegration by means of tip streaming in accumulation. Thus they are expected to have a significant influence on the macroscopic behavior of suspensions.

This research is supported by the National Science Foundation, the American Chemical Society, and the SUN Microsystems corporation.

References

[1] CHARLES, R. & POZRIKIDIS, C. 1998 Effect of particle deformability on the flow of suspensions of liquid drops. *J. Fluid Mech.* **365**, 205-234

[2] KENNEDY, M., POZRIKIDIS, C. & SKALAK, R. 1994 Motion and deformation of liquid drops, and the rheology of dilute emulsions in shear flow. *Computers Fluids*, 23, 251-278.

[3] LI, X., ZHOU, H. & POZRIKIDIS, C. 1994 A numerical study of the shearing motion of emulsions and foams. *J. Fluid Mech.* 271, 1-26.

[4] LI, X., CHARLES, R. & POZRIKIDIS, C. 1996 Simple shear flow of suspensions of liquid drops. *J. Fluid Mech.* 320, 395-416.

[5] LOEWENBERG, M. & HINCH, E. J. 1996 Numerical simulation of a concentrated emulsion in shear flow. *J. Fluid Mech.* 321, 395-419.

[6] POZRIKIDIS, C. 1992 *Boundary Integral and Singularity Methods for Linearized Viscous Flow*. Cambridge University Press.

[7] POZRIKIDIS, C. 1996 Computation of periodic Green's functions of Stokes flow. *J. Eng. Math.* 30, 79-96.

[8] POZRIKIDIS, C. 1997 *Introduction to Theoretical and Computational Fluid Dynamics*. Oxford University Press.

[9] POZRIKIDIS, C. & SHETH, K. 1996 A note on light transmission through an evolving suspension of liquid drops. *Chem. Eng. Comm.* 148-150, 477-486.

[10] TOOSE, E. M. 1997 Simulation of the deformation of non-Newtonian drops in a viscous flow. *Doctoral Dissertation*, Universiteit Twente.

[11] YON, S. & POZRIKIDIS, C. 1997 A finite-volume / boundary-element method for flow past interfaces in the presence of surfactants, with application to shear flow past a viscous drop. *Computers & Fluids* **27**, 879-902.

[12] ZHOU, H. & POZRIKIDIS, C. 1993a The flow of suspensions in channels: single files of drops. *Phys. Fluids A*, 5(2), 311-324.

[13] ZHOU, H. & POZRIKIDIS, C. 1993b The flow of ordered and random suspensions of liquid drops in a channel. *J. Fluid Mech.* 255, 103-127.

[14] ZHOU, H. & POZRIKIDIS, C. 1994 Pressure-driven flow of suspensions of liquid drops. *Phys. Fluids*, 6, 80-94.

Costas Pozrikidis
Department of Applied Mechanics and Engineering Sciences
University of California, San Diego
La Jolla, Ca 92093-0411
USA

M. SAVAGE
Meniscus Roll Coating: Steady Flows and Instabilities

Free boundary problems are ubiquitous in liquid film coating. In particular, meniscus roll coating refers to a coating regime which arises when flow rates are small so that a "fluid bead" is located in the nip bounded by two moving rolls and two free surfaces.

In forward roll coating the rolls are contra-rotating and the flow consists of two large, closed recirculations around which a fluid transfer-jet or "snake" meanders in order to transport inlet flux to the upper roll. The pressure field is entirely subambient and dominated by capillary pressure at the upstream meniscus. Two mathematical models are used to describe this steady, two-dimensional flow. A small flux model is based on the lubrication approximation so that the fluid velocity is a combination of Couette and Poiseuille flows and pressure is given by Reynolds equation. A second model incorporates the full effects of curved menisci, nonlinear free surface conditions and a dynamic contact line. The finite element method is used to obtain numerical solutions for the velocity and pressure fields.

The stability of the two-dimensional base flow to small amplitude perturbations is then considered. It is shown that, for a given fluid and a given roll configuration, the upstream meniscus passes through the nip as the upper roll speed is gradually increased and an instability in the form of bead break soon follows.

1 Introduction

Figure 1 illustrates forward mode, roll coating in which two rolls move in the same direction through the nip with peripheral speeds U_1 and U_2. Fluid is picked up from the reservoir by the action of viscous lifting and a film of thickness H_I. Figure 2,

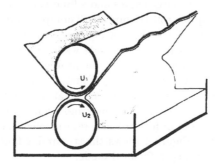

Figure 1. A two-roll coater with web operating in forward mode.

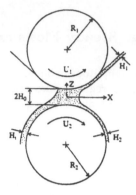

Figure 2. A cross section showing a fluid bead located in the nip.

enters the nip region where a "fluid bead" is located. Two uniform films of thickness H_2 and H_1 exit the bead – one attached to the lower roll and the other attached to the web which passes beneath the upper roll.

There are two coating regimes of practical significance, each identified according to the inlet feed condition. The "inlet flooded regime" arises when the thickness of the inlet film is greater than the minimum gap thickness, $H_i > 2H_0$ and gives rise to an "upstream bank", with the upstream meniscus located far from the nip. The "inlet starved regime" corresponds to the meniscus coating regime in which $H_i \ll 2H_O$ and the upstream meniscus is located close to the nip where it exerts a dominant influence on the flow dynamics within the bead. Meniscus roll coating is used in industry to produce extremely thin films of high quality (Gaskell and Savage (1997)). In an experimental investigation of forward roll meniscus coating (Malone (1992)) confirmed the presence of two large eddies and a fluid transfer or "snake" transferring fluid to the web on the upper roll (Figure 3). Thompson (1992) provided an initial theoretical investigation in which the coating bead was modelled as a rectangular, double lid-driven cavity with free surface side walls and negligible flow rate. This "zero-flux model" reduced to a boundary value problem for the stream function, the solution to which can be expressed as an eigenfunction series expansion. Subsequently a comprehensive treatment of the modelling and analysis of steady-state meniscus roll coating was given by Gaskell, Savage, Summers and Thompson (1995) and followed by an experimental investigation of forward and reverse meniscus roll coating (Gaskell, Innes and Savage (1998)). Figure 4 shows the streamline flow in the fluid bead of a meniscus, roll coater operating in forward mode.

Roll-coating flows are known to exhibit the ribbing instability which appears for some parameter values as a small amplitude, three-dimensional perturbation to a fluid-

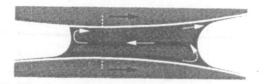

Figure 3. A dye trace showing, the path of the transfer-jet through the bead (first appeared in *J. Fluid. Mech.*).

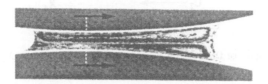

Figure 4. Laser illuminated particles revealing the flow in meniscus roll coating: forward mode with $R/H_0 = 100$, $S = 1.0$, $\lambda = 0.048$ and $Ca_2 = 2.3 \times 10^{-3}$ (first appeared in *J. Fluid. Mech.*).

air interface and is periodic in the third (axial) direction. Ribbing has been extensively researched in the case of inlet flooded coating. For meniscus coating, however, the main instability is bead break – a two-dimensional instability which arises when the speed of the upper roll is increased, and the upstream meniscus passes through the nip so that the entire coating bed is located downstream of the nip. At a critical speed ratio, S (where $S = U_1/U_2$), the upstream meniscus loses stability, and accelerates toward the downstream meniscus causing the bead to collapse (Gaskell, Kapur and Savage (1998)). The work presented here is to be found in greater depth and detail in the abovementioned papers.

2   Modelling/analysis for steady flows

Steady-state flow in the coating bead is governed by four nondimensional parameters: gap ratio G, capillary number Ca_2, speed ratio S and flow rate λ defined by

Gap Ratio $\qquad\qquad\qquad\qquad\qquad G = R/H_0 \qquad\qquad$ (1)

Capillary number $\qquad\qquad\qquad Ca_2 = \dfrac{\mu U_2}{\sigma} \qquad\qquad$ (2)

Speed ration $\qquad\qquad\qquad\qquad S = \dfrac{U_1}{U_2} \qquad\qquad$ (3)

Flow Rate $\qquad\qquad\qquad\qquad \lambda = \dfrac{H_i}{2H_0} \qquad\qquad$ (4)

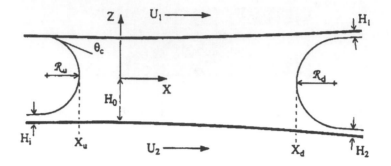

Figure 5. A schematic of the coating bead bounded by menisci at X_u and X_d.

where μ and σ represent the viscosity and surface tension of the fluid; U_1 and U_2 the upper and lower roll speeds; H_i, H_0 and R the inlet film thickness, the semiminimun gap thickness and an "average roll radius" ($\frac{2}{R} = \frac{1}{R_1} + \frac{1}{R_2}$ where R_1 and R_2 are radii of the upper and lower rolls). Figure 5 shows a schematic of the bead for $X_u \leq X \leq X_d$ for which the appropriate nondimensional lengths are

$$\bar{x} = \frac{X}{(2RH_0)^{\frac{1}{2}}} \quad \text{and} \quad \bar{z} = \frac{Z}{H_0} \tag{5}$$

Flow continuity requires

$$U_2 H_i = U_2 H_2 + U_1 H_1$$
$$H_i = H_2 + S H_1 \tag{6}$$

2.1 Small flux model; lubrification solution

With the coordinate system shown in Figure 5 the semi-gap width, $H(X)$, at station X is approximated by

$$H(X) = H_0 + \frac{X^2}{2R} \tag{7}$$

and the fluid beads are assumed to lie between X_u and X_d.

Locating X_d

For small capillary numbers ($Ca < 10^{-2}$ for meniscus coating) the position of the downstream meniscus, X_d, can be accurately predicted via the work of Landau and

92

Levich (1942) who related the uniform film thickness H_1, H_2 to the radius of curvature, R_d, of the downstream meniscus;

$$\frac{H_1}{R_d} = 1.34 \left(\frac{\mu U_1}{\sigma}\right)^{2/3} \tag{8}$$

$$\frac{H_2}{R_d} = 1.34 \left(\frac{\mu U_2}{\sigma}\right)^{2/3} \tag{9}$$

Since $H_1 \ll H(X_d)$, $H_2 \ll H(X_d)$ then $H(X_d)$ is a suitable approximation for R_d;

$$R_d \approx H_d = H_0 + \frac{X_d^2}{2R} \tag{10}$$

$$R_d = H_0 \left(1 + \bar{d}^2\right) \tag{11}$$

With H_1 and H_2 given by (8) and (9) and R_d by (11) then substituting into (6) yields an equation for $\bar{d}$:

$$1 + \bar{d}^2 = \frac{2\lambda}{1.34(Ca_2)^{2/3}} \left(1 + S^{5/3}\right)^{-1} \tag{12}$$

Hence for a given geometry and specified roll speeds, λ, Ca_2 and S are all known and $\bar{d}$ is given by (12). In addition, equations (8) and (9) give rise to a two-thirds power law for film thickness ratio,

$$\frac{H_1}{H_2} = S^{2/3} \tag{13}$$

as predicted by Ruschak (1985).

Locating X_u

If U and W are velocity components in the X and Z directions and $P(X, Z)$ is pressure, the flow is described by Stokes equation assuming that the effects of gravity and fluid inertia are negligible

$$0 = \frac{\partial P}{\partial X} - \mu \nabla^2 U \tag{14}$$

$$0 = \frac{\partial P}{\partial Z} - \mu \nabla W \tag{15}$$

$$0 = \frac{\partial U}{\partial X} + \frac{\partial W}{\partial Z} \tag{16}$$

Assuming lubrication theory to be valid (the flows are effective one-dimensional, $W/U \ll 1$ and $\frac{\partial}{\partial X} \ll \frac{\partial}{\partial Z}$) equations (14)-(16) reduce to

$$\frac{\partial P}{\partial x} = \mu \frac{\partial^2 U}{\partial Z^2} \tag{17}$$

$$\frac{\partial P}{\partial Z} = 0 \tag{18}$$

The solution for $U(X, Z)$ satisfying $U = U_1$ on $Z = H(X)$ and $U = U_2$ on $Z = -H(X)$ is

$$U = \frac{1}{2\mu}\frac{dP}{dX}(Z^2 - H^2) + \frac{(U_1 - U_2)}{2}\frac{Z}{H} + \frac{(U_1 + U_2)}{2} \tag{19}$$

where $P = P(X)$ is independent of Z as result of (18).

The flux, Q, per unit axial length is given by

$$Q = \int_{-H}^{H} U\, dZ = -\frac{2}{3\mu}\frac{dP}{dX}H^3 + (U_1 + U_2)H \tag{20}$$

which, in turn, gives rise to Reynolds equation for $P(X)$:

$$\frac{dP}{dX} = \frac{3\mu}{2}\left[\frac{(U_1 + U_2)}{H^2} - \frac{U_2 H_i}{H^3}\right] \tag{21}$$

Lubrication assumptions break down where the W components of velocity become insignificant – which is in a thin layer adjacent to each free surface. We shall ignore the thickness of such layers and assume, therefore, that the solution of Reynolds equation is over the domain $X_u \leq X \leq X_d$ and that fluid pressure at each end point is given by capillary pressure at the corresponding mniscus.

Hence

$$P(X = X_u)) = -\frac{\sigma}{R_u}, \qquad P(X = X_d) = -\frac{\sigma}{R_d} \tag{22}$$

where R_u and R_d are approximated by the full gap width $2H(X_u)$ and semi-gap width $H(X_d)$, respectively,

$$R_u = 2H_0(1 + \bar{B}^2), \qquad R_d = H_0(1 + \bar{d}^2) \tag{23}$$

The aim now is to solve (21) subject to boundary conditions (23).

Writing

$$\bar{x} = \tan\alpha \text{ so that } \bar{b} = \tan\alpha_b, \quad \bar{d} = \tan\alpha_d \tag{24}$$

the following expression for α_b is obtained:

$$\frac{\sqrt{2}}{3Ca_2(R/H_0)^{\frac{1}{2}}}[\cos^2\alpha_b - 2\cos^2\alpha_d] = \left(\frac{1 + S}{2}\right)[\sin 2\alpha_d - \sin 2\alpha_b + 2\alpha_d - 2\alpha_b]$$
$$- \lambda\left[\frac{\sin 4\alpha_d - \sin 4\alpha_b}{8} + \frac{3(\alpha_d - \alpha_b)}{2} + \sin 2\alpha_d - \sin 2\alpha_b\right] \tag{25}$$

In the meniscus coating regime, flow rates are small ($\lambda \ll 1$) as is also the modified capillary number $Ca_2(R/H_0)^{\frac{1}{2}} \ll 1$ and so the left-hand side of (25) is comparable to the first term on the right-hand side – thus indicating that capillary pressure at the menisci is comparable to hydrodynamics pressures within the fluid bead.

With λ, S and $Ca_2(R/H_0)^{\frac{1}{2}}$ specified, $\alpha_d(\bar{x}_d)$ is given by (12) and (α_b) by (25). The results are discussed in Section 2.3.

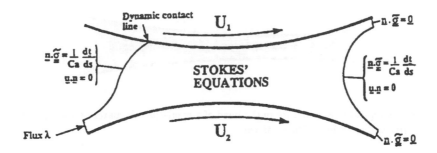

Figure 6. The nonlinear boundary value problem for Stokes flow in the bead; $\underline{n}$ and $\underline{t}$ are unit normal and tangent vectors and s is the free surface arc length.

2.2 Refined, small flux model / C.F.D. solution

The mathematical model 2.1 is refined by including curved menisci, nonlinear free-surface boundary condition and a dynamic contact line with an imposed slip length and an apparent contact angle suggested by experiment. The full nonlinear boundary problem is illustrated in Figure 6 where the boundary conditions are shown and the equation of motion can be written in divergence form:

$$\nabla \cdot T, \quad \nabla \cdot U = 0$$

where

$$T = -PI + \mu[\nabla U + (\nabla U)^\top] \tag{26}$$

and I are the stress and unit tensors.

Nondimensional variables $\tilde{x}$, $\tilde{z}$, $\tilde{u}, \tilde{w}$, $\tilde{p}$ and $\tilde{\sigma}$ are defined by

$$\tilde{x} = \frac{X}{H_0}; \quad \tilde{z} = \frac{Z}{H_0}; \quad \tilde{u} = \frac{U}{U_2}; \quad \tilde{w} = \frac{W}{U_2};$$
$$\tilde{p} = \frac{pH_0}{\mu U_2}; \quad \tilde{\sigma} = \frac{TH_0}{\mu U_2}; \tag{27}$$

and a numerical (finite elements solution is obtained using the Galerkin, weighted residual formulation for equations (26). This finite element approach is ideally suited, due to its topological flexibility, to the solution of a two - free surface problem with curved free surfaces. Since the shape of the menisci are unknown à priori, an appropriate parameter representation is via the spine method of Kistler and Scriven (1983) together with a combination of multiple origins and spine interdependency.

Additional equations for the unknown "spine heights" are obtained from the weighted residuals of the kinematic equation. The flow domain is tessellated using triangular V6/P3 elements which satisfy the LBB stability condition (Babuska

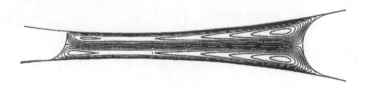

Figure 7. A typical flow structure in the bead of a meniscus roll coater operating in forward mode (first appeared in *J. Fluid. Mech*).

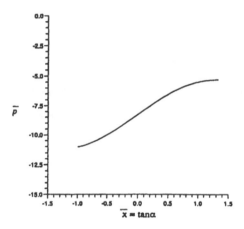

Figure 8. A typical pressure distribution, along the centerline $Z = 0$ of a meniscus roll coater operating in forward mode.

and Aziz 1972). The rate of convergence is optimal and no "locking" (Hughes 1987), occurs. The global stiffness matrix is solved using the frontal method (Hood 1976) and Newton iteration gives second order convergence within six iterations.

2.3 Results

A two-roll coater with $R/H_0 = 100$, flow rate $\lambda = 0.075$ and an apparent contact angle of 95° was selected. Finite element solutions are shown in Figure 7 for speed ratio $S = 1$, and capillary number $Ca_2 = 2.25 \times 10^{-3}$. Figure 7 shows the double eddy structure of the coating bead with each eddy having a separatrix, subeddies and a, saddle point at the nip. A typical pressure profile across the fluid bead is shown in Figure 8 for S= 1 in which a key feature is that $\frac{dP}{dX}$ is everywhere positive. For a flow rate of $\lambda = 0.075$, Figure 9 shows the variation of H_1/H_2 with speed ratio S and includes a comparison with equation (13) derived from the work of Landau and Levich. Agreement between finite element predictions and equation (13) is very good

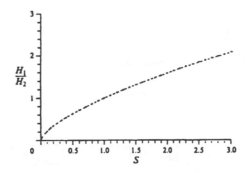

Figure 9. H_1/H_2 against S for $R/H_0 = 100$ and $\lambda = 0.075$; — Landau-Levich equation (13); • finite element solution (first appeared in *J. Fluid. Mech*).

over the whole range of S.

A comparison between finite element solutions and analytical results from the small flux (lubrication) model for the position and extent of the fluid bead are shown in Figure 10. For a gap ratio, $R/H_0 = 100$ and a flow rate $\lambda = 0.2$ agreement is again extremely good. As the speed ratio (and hence Ca) increases, the extent of the fluid bead is seen to contract as both menisci approach the nip. At a certain capillary number the upstream meniscus reaches and subsequently passes through the nip. There is then a small range of capillary numbers over which the 'fluid bead' is located downstream of the nip and yet remains stable. A critical capillary number is then reached beyond which it is no longer possible to obtain steady-state solutions. The fluid bead has now collapsed and this instability is the subject of Section 3.

3 Instability

Figure 11(a)-(c) illustrates the sequence of events just prior to instability and bead break. With the gap ratio, flow rate and lower roll speed held constant the upper roll speed (and hence S) is gradually increased. At $S = 1.0$ figure 11(b) the bead lies entirely to the right of the nip and as S is further increased, the bead continues to contract until a critical speed ratio, $S^c = 1.05$, is reached. The upstream meniscus then becomes unstable; the meniscus rapidly approaches the downstream meniscus causing the bead to collapse and break up. Experiments by Kapur (1 997) suggest that bead break is a two-dimensional instability which can be analysed at two levels: via a stability hypothesis and a linearised stability analysis, respectively.

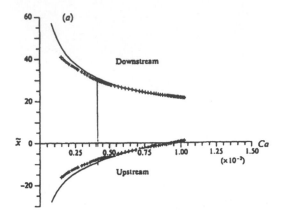

Figure 10. A comparison between analytical solutions (—) and finite elements (+) for the locations of the menisci as functions of capillary number with $R/H_0 = 100$ and $\lambda = 0.2$ (first appeared in *J. Fluid. Mech*).

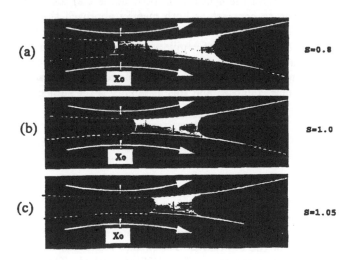

Figure 11. The location and extent of the fluid bead for (a) $S = 0.8$ (b) $S = 1.0$ and (c) $S = 1.05$ with $R/100 = 100$ and $\lambda = 0.15$.

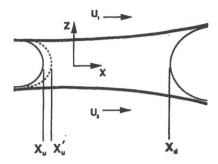

Figure 12. A cross section showing the displacement of the upstream meniscus from X_u to X_u^1.

3.1 Stability hypothesis

Figure 12 shows a two-dimensional cross section of the bead with the upstream and downstream menisci at X_u and X_d, respectively. The stability hypothesis is simply a local analysis which considers the stability of a meniscus by its response to a small perturbation. The only equation required is the balance of fluid and capillary pressure

$$P + \frac{\sigma}{R_u} = 0 \quad \text{at} \quad X = X_u \tag{28}$$

The meniscus is then perturbated to a new position X_u^1 given by

$$X_u^1 = X_u + \epsilon \tag{29}$$

and hence the force acting on the interface (in the negative X direction) is

$$\left(P + \frac{\sigma}{R_u}\right)_{X_u^1} = \left(P + \frac{\sigma}{R_u}\right)_{X_u} + \epsilon \frac{d}{DX}\left(P + \frac{\sigma}{R_u}\right)_{X_u} + 0(\epsilon^2) \tag{30}$$

Stability of the meniscus at X_u requires

$$\frac{d}{dX}\left(P + \frac{\sigma}{R_u}\right) < 0 \tag{31}$$

Approximating the radius of curvature by the gap thickness as in Section 2.1, $R_u \approx H(X_u)$ and therefore, the criterion for stability is

$$\frac{dP}{dX} - \frac{\sigma}{H^2}\frac{dH}{dX} > 0 \quad \text{at} \quad X_u \tag{32}$$

99

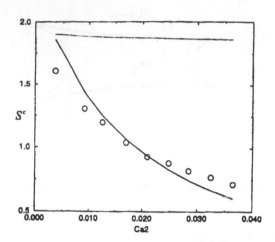

Figure 13. Critical speed ratio S^c as a function of Ca_2 for $R/100 = 100$ and $\lambda = 0.15$; — theoretical curves ∘ experimental data points.

Observations

- In forward mode, meniscus roll coating with $\lambda \ll 1$ and $S > 1$

$$\frac{dP}{dX} > 0 \qquad \text{as illustrated in Figure 8.}$$

- $\dfrac{dH}{dX} < 0$ for $X_u < 0$

- $\dfrac{dH}{dX} > 0$ for $X_u > 0$

Conclusions

1. If $X_u < 0$, the upstream meniscus remains stable.

2. $X_u > 0$ is a *necessary condition* for instability. The bead must lie entirely downstream of the nip.

3. Onset of instability arises when

$$\frac{dP}{dX} = \frac{\sigma}{H^2} \frac{dH}{dX}$$

100

which gives an equation for the critical speed ratio S^c

$$Ca_2 \sqrt{\frac{R}{H_0}}(1 + S^c) = \frac{\sqrt{2}}{3}\frac{X_u}{1 - \lambda} \tag{33}$$

For a gap ratio $R/H_0 = 100$ and flow rate $\lambda = 0.1$, Figure 13 compares prediction for S^c against capillary number Ca_2 with some recent experimental data. Agreement is surprisingly good over this parameter range and indicates that the stability hypothesis is extremely useful for predicting the stability of a fluid-air interface to two-dimensional perturbations.

The analytical details of a full linearised perturbation analysis are not given here but will appear in a journal paper (*Physics of Fluids* 1998). This analysis confirms (32) as the criteria for stability by generating a growth rate for perturbations which is proportional to $(\frac{\sigma}{H^2}\frac{dH}{dX} - \frac{dP}{dX})$.

References

[1] BABUSKA, I. & AZIZ, A.K. 1972. Lectures on the mathematical foundations of the finite element method. In Mathematical Foundations of the Finite Element Method with Applications to Partial Differential Equations (ed. A.K. Aziz). Academic.

[2] GASKELL, P.H., SAVAGE, M.D., SUMMERS, J.L. & THOMPSON, H.M. 1995. Modelling and Analysis of Meniscus Roll Coating. J. Fluid Mech. 298, 113-137.

[3] GASKELL, P.H. & SAVAGE, M.D. 1997. Meniscus Roll Coating in Liquid Film Coating (eds. S.F. Kistler & P.M. Schweitzer). Chapman and Hall.

[4] GASKELL, P.H, INNES, G. & SAVAGE, M.D. 1998. An Experimental Investigation of Meniscus Roll Coating J. Fluid Mech., Vol. 355, 17-44.

[5] GASKELL, P.H, KAPUR, N. & SAVAGE, M.D. 1998. Bead Break Instability in Meniscus Roll Coating. Submitted to Phys. of Fluids.

[6] HOOD P. 1976. Frontal Solution Program for Unsymmetric Matrices. Intl. J. Numer. Meth. Engng. 10, 379-399.

[7] HUGHES, T.R.J. 1987. The Finite Element Method; Linear Static and Dynamic Finite Element Analysis. Prentice-Hall.

[8] KAPLIR, N. 1997. (Private Communication)

[9] KISTLER, S.F & SCRIVEN, L.E. 1983. Coating Flows. In Computational Analysis of Polymer Processing (eds. J.R.A. Pearson & S.M. Richardson), 243-299. Appl. Sci. Publishers.

[10] LANDAU, L. & LEVICH, B. 1942. Dragging of a Liquid by a Moving Plate. Acta Physicochimica URSS XVII (1 -2), 42.

[11] MALONE, B. 1992. An Experimental Investigation of Roll Coating Phenomena. PhD. Thesis. University of Leeds.

[12] RUSCHAK, K.J. 1985. Coating Flows. Ann. Rev. Fluid Mech. 17, 65-89.

[13] THOMPSON, H.M. 1992. A Theoretical Investigation of Roll Coating Phenomena. PhD Thesis. University of Leeds.

Mike Savage
Dept of Physics and Astronomy
University of Leeds
Leeds LS2 9JT, U.K.

Part 2.

**Mathematical Developments of
Free Boundary Problems**

C.-M. Brauner, J. Hulshof, A. Lunardi and C. Schmidt-Lainé

Instabilities, Bifurcation and Saddle-Points in Some FBPs in Combustion

1 Introduction

In the modelling of combustion and flames, the stability question is of major importance. Detonations are, by nature, unstable. Exchange of stability often arises in premixed flames, as shown by the classical thermo-diffusive model for the Lewis number close to 1. In this framework, the famous Kuramoto-Sivashinsky equation exhibits cellular instabilities. As far as diffusion flames are concerned, chambered flames present S-shaped response curves for arbitrary Lewis numbers. Away from laminar combustion, turbulence may stabilize the flame and lead to extinction/reignition phenomena. Solid combustion should not be forgotten: clearly, instabilities in the combustion of solid propellants can endanger the rocket propulsion.

Although there is a large literature on instabilities in combustion, most of the references are restricted to a rather formal approach, namely, a linear stability analysis based on the derivation of a so-called "dispersion relation". Such an analysis neglects all the curvature effects stemming from the corrugation of the flame front and is irrelevant, mathematically speaking. Among the few mathematical papers involving a deep nonlinear stability analysis, let us mention the bifurcation theorem in [10]. The authors give a full rigorous treatment of the thermo-diffusive model, when the activation energy is finite.

At the limit of high activation energy, the flame zone becomes a thin interface which can be viewed as a free boundary. At this front, several types of jump conditions may occur: usually the unknown (temperature, mass fraction, etc.) is continuous, in certain circumstances it is given, while the jump in the gradient is either given or determined by some nonlinear relation. This discontinuity concentrates the reaction term at the free boundary. As a basic example we mention the Near Equidiffusional Flame (NEF) theory, see [7]; the equidiffusional case gives rise to a one-phase paradigm model, see the survey [14]:

$$\begin{cases} u_t = \Delta u, & x < \xi(t,y), \ y \in \omega, \\ u = u_*, \ \frac{\partial u}{\partial n} = 1, & x = \xi(t,y), \ y \in \omega, \\ \frac{\partial u}{\partial n} = 0, & x < \xi(t,y), \ y \in \partial\omega, \\ u(-\infty, y, t) = 0, & y \in \omega. \end{cases} \tag{1.1}$$

Here u_* is a positive constant often normalized at 0. We denote by ω a bounded domain in $\mathbf{R}^{N-1}$ with smooth boundary $\partial\omega$.

Problem (1.1) admits as equilibria a family of $1D$-travelling waves (TW) solutions $u(x,y,t) = U(x + ct)$, which are (orbitally) stable, see [2]. In the context of spherical flames, the model problem is formulated in a bounded domain Ω_t:

$$\begin{cases} u_t = \Delta u, & \text{in } \Omega_t, \\ u = u_*, \ \frac{\partial u}{\partial n} = 1 & \text{on } \partial \Omega_t. \end{cases} \quad (1.2)$$

However, problem (1.2) does not admit trivial equilibria. It is called a focusing problem because the domain vanishes in finite time, see [14, 9, 12].

Our purpose is to present a general approach to the stability question for FBPs such as the TW problem and the problem in a bounded domain above, generalizing methods developed for special cases by the authors (see the reference list). Although we are motivated by combustion problems, we emphasize that our method can be applied to other FBPs in addition to those with "combustion type" free boundary conditions $u = 0$ and $\frac{\partial u}{\partial n} = g$ at the interface.

The principle of the method will be given in the next section, for simplicity in a bounded domain only, see [4] for details. It is based on a reduction to a fixed domain Ω and the elimination of the front, thanks to the splitting

$$u = U + \Phi \cdot \nabla U + w, \quad (1.3)$$

where Φ measures the local $C^{2+\alpha}$-difference between the free boundary $\partial \Omega_t$ and $\partial \Omega$. The price to pay is nonlocal terms which make the equation for w fully nonlinear. However, the reformulation in terms of w enables us to apply abstract results from [13] and to establish the principle of linearized stability, bifurcation and existence of stable and unstable manifolds.

As examples we exhibit a saddle-point property for focusing flames in Section 3 and in Section 4 a two-phase version of the TW problem with a Burgers term, namely, a Deflagration to Detonation Transition (DDT) model proposed by Stewart and Ludford [8]. In dimension $N = 2$, the planar TW becomes unstable at a critical value u_*^c of u_*, at which an infinite number of bifurcated branches of nonplanar TWs accumulate (see [1] for details).

2 A general procedure for nonlinear stability analysis

We consider free boundary problems of the form

$$u_t = \mathcal{L}u + f, \quad x = (x_1, \ldots, x_N) \in \Omega_t, \quad t > 0, \quad (2.1)$$

with free boundary conditions

$$u = 0 \quad \text{and} \quad \frac{\partial u}{\partial n} = g \quad \text{on} \quad \partial \Omega_t. \quad (2.2)$$

Here Ω_t is a bounded domain in $\mathbf{R}^N$ with (free) boundary $\partial \Omega_t$, $\mathcal{L}$ is a uniformly elliptic operator on $\mathbf{R}^N$ with smooth coefficients, and f and g are given smooth functions defined on the whole of $\mathbf{R}^N$. We are interested in the stability properties of equilibria of (2.1)(2.2). We assume that the pair (Ω, U) is a smooth equilibrium, namely,

$$\mathcal{L}u + f = 0 \text{ in } \Omega, \quad u = 0 \text{ and } \frac{\partial u}{\partial n} = g \text{ on } \partial\Omega, \tag{2.3}$$

with Ω bounded and $\partial\Omega$ in $C^{2+\alpha}$ and $U \in C^{2+\alpha}(\overline{\Omega})$. We consider (2.1)(2.2) with initial data Ω_0 and u_0 close (in the $C^{2+\alpha}$ sense) to Ω and U. The main structural condition is that g does not vanish on $\partial\Omega$.

Since $\partial\Omega$ is smooth, there exists a smooth outward pointed normal

$$\nu : \partial\Omega \to S^{N-1}, \tag{2.4}$$

where $S^{N-1} = \{x \in \mathbf{R}^N : |x| = 1\}$. For $\delta > 0$ we define a map

$$X : \partial\Omega \times [-\delta, \delta] \to \mathbf{R}^N, \quad X(\phi, r) = \phi + r\nu(\phi). \tag{2.5}$$

If δ is sufficiently small, then (2.5) defines a bijection to a compact neighbourhood $N(\delta\Omega)$ of $\delta\Omega$.

Next we assume that on some time interval I, $\partial\Omega_t$ is close to $\partial\Omega$ in the sense that in terms of the coordinates r and ϕ defined by (2.5), $\partial\Omega_t$ is given by

$$\partial\Omega_t = \{x = \phi + r\nu(\phi) \in N(\delta\Omega) : r = s(\phi, t)\}, \tag{2.6}$$

where $s : \partial\Omega \times I \to [-\delta_1, \delta_1]$ with $0 < \delta_1 < \delta$ is a smooth function. Thus $r < s(\phi, t)$ in $\Omega_t \cap N(\partial\Omega)$.

For the coordinate transformation from Ω_t back to Ω we denote by ξ a point in Ω and by x a point in Ω_t. The time coordinate of the problem on Ω is denoted by τ and the local coordinates near $\partial\Omega$ of ξ are denoted by z and θ, i.e., $\xi = \theta + z\nu(\theta)$. We transform the free boundary problem by setting

$$x = \xi + \alpha(z)s(\theta, \tau)\nu(\theta) \text{ (whence } r = z + \alpha(z)s(\theta, \tau) \text{ and } \phi = \theta \text{) and } t = \tau. \tag{2.7}$$

Here $\alpha : [-\delta, \delta] \to [0, 1]$ is a smooth mollifier. For the transformation to be a diffeomorphism s has to be restricted to a small interval, say $(-\epsilon, \epsilon)$, so as to ensure that the map $z \in [-\delta, \delta] \to z + \alpha(z)s \in [-\delta, \delta]$ is a diffeomorphism.

In what follows it will be convenient to write

$$\Phi(\xi, \tau) = \alpha(z)s(\theta, \tau)\nu(\theta). \tag{2.8}$$

This function Φ contains all the information about the $C^{2+\alpha}$-difference between the fixed boundary Ω and the free boundary Ω_t and remarkably will not appear in the final formulation. We see that x, t, ∇_x and $\frac{\partial}{\partial t}$ transform as

$$x = \xi + \Phi(\xi, \tau), \quad t = \tau, \quad \nabla_x = (I + \nabla_\xi\Phi)^{-1}\nabla_\xi, \quad \frac{\partial}{\partial t} = D_\tau = \frac{\partial}{\partial \tau} - \frac{\partial\Phi}{\partial \tau}\cdot(I + \nabla_\xi\Phi)^{-1}\nabla_\xi. \tag{2.9}$$

107

Here we denote by $\nabla\Phi$ a matrix which has the gradients of the components of Φ as columns, i.e., $\nabla\Phi$ is the transpose of the Jacobian matrix representing the derivative $D\Phi$ of Φ. If we write $u(x,t) = \hat{u}(\xi,\tau)$, we can rewrite the action of $\mathcal{L}$ on u as

$$(\mathcal{L}u)(x,t) = (\hat{\mathcal{L}}\hat{u})(\xi,\tau) = (\hat{\mathcal{L}}_0\hat{u})(\xi,\tau) + (\hat{\mathcal{L}}_1\hat{u})(\xi,\tau) + (\hat{\mathcal{L}}_2^r\hat{u})(\xi,\tau), \qquad (2.10)$$

where $\hat{\mathcal{L}}_0$ is the original operator $\mathcal{L}$ with x replaced by ξ and $\hat{\mathcal{L}}_1$ is an operator with coefficients linear in Φ and its first and second order ξ-derivatives. The remainder term $\hat{\mathcal{L}}_2^r$ has coefficients bounded by a multiple of $|\Phi|^2 + |\nabla_\xi\Phi|^2 + |\nabla_\xi\nabla_\xi\Phi|^2$. This is easily seen from the expansion

$$(I + \nabla_\xi\Phi)^{-1} = I - \nabla_\xi\Phi + (\nabla_\xi\Phi)^2(I + \nabla_\xi\Phi)^{-1}. \qquad (2.11)$$

Likewise we get an expansion for the time derivative which reads

$$\frac{\partial u(x,t)}{\partial t} = D_\tau u(\xi,\tau) = \frac{\partial\hat{u}}{\partial\tau} - \frac{\partial\Phi}{\partial\tau}\cdot\nabla_\xi\hat{u} + \frac{\partial\Phi}{\partial\tau}\cdot(\nabla_\xi\Phi)(I + \nabla_\xi\Phi)^{-1}\nabla_\xi\hat{u}. \qquad (2.12)$$

The transformation of Ω_t to Ω also acts on the equilibrium U itself and this has to be taken into account when we expand $\hat{u}$ near U. Observing that

$$U(\xi + \Phi(\xi,\tau)) = U(\xi) + (\nabla_\xi U(\xi))\cdot\Phi(\xi,\tau) + R_2(\xi,\Phi(\xi,\tau)), \qquad (2.13)$$

the expansion (2.13) leads us to look for a solution $u(x,t) = \hat{u}(\xi,\tau)$ of the form

$$u(x,t) = \hat{u}(\xi,\tau) = U(\xi) + (\nabla_\xi U(\xi))\cdot\Phi(\xi,\tau) + w(\xi,\tau). \qquad (2.14)$$

Comparing the zero order terms simply gives

$$\frac{\partial U(\xi)}{\partial\tau} - (\hat{\mathcal{L}}_0 U)(\xi) = f(\xi),$$

while the first order terms

$$\left(\frac{\partial}{\partial\tau} - \hat{\mathcal{L}}_0\right)((\nabla_\xi U(\xi))\cdot\Phi(\xi,\tau)) - \left(\frac{\partial\Phi}{\partial t}\cdot\nabla_\xi + \hat{\mathcal{L}}_1\right)U(\xi) = (\nabla_\xi f(\xi))\cdot\Phi(\xi,\tau), \quad (2.15)$$

reduce to

$$-(\hat{\mathcal{L}}_0(\nabla_\xi U(\xi))\cdot\Phi(\xi,\tau)) - \hat{\mathcal{L}}_1 U(\xi) = (\nabla_\xi f(\xi))\cdot\Phi(\xi,\tau). \qquad (2.16)$$

All terms of order zero and order one in Φ involving $U(\xi)$ and its derivatives cancel. The resulting expansion reads

$$\frac{\partial w}{\partial\tau} - \hat{\mathcal{L}}_0 w = \left(\frac{\partial\Phi}{\partial\tau}\cdot\nabla_\xi + \hat{\mathcal{L}}_1\right)w + \left(-\frac{\partial\Phi}{\partial\tau}\cdot(\nabla_\xi\Phi)(I + \nabla_\xi\Phi)^{-1}\nabla_\xi + \hat{\mathcal{L}}_2^r\right)w \qquad (2.17)$$

$$+\left(\frac{\partial\Phi}{\partial\tau}\cdot\nabla_\xi + \hat{\mathcal{L}}_1\right)(\Phi\cdot U) + \left(-\frac{\partial\Phi}{\partial\tau}\cdot(\nabla_\xi\Phi)(I + \nabla_\xi\Phi)^{-1}\nabla_\xi + \hat{\mathcal{L}}_2^r\right)(U + \Phi\cdot\nabla_\xi U) + R_3(\xi, \Phi(\xi,\tau)).$$

108

The linear part (the left-hand side) is now in the form we want, but the higher order terms contain derivatives of both Φ and w, including $\frac{\partial \Phi}{\partial \tau}$. To eliminate Φ from (2.17) we have to transform the free boundary conditions.

The first free boundary condition $u = 0$ on $\partial \Omega_t$ implies that

$$w(\xi, \tau) = -(\nabla_\xi U(\xi)) \cdot \Phi(\xi, \tau) = -g(\xi) s(\xi, \tau) \quad \text{for } \xi \in \partial \Omega. \qquad (2.18)$$

For the normal derivative we have that

$$\frac{\partial}{\partial n} = n \cdot \nabla_x = \nu \cdot \nabla_\xi - \nabla_\xi s \cdot \nabla_\xi + (\nabla_\xi s \cdot \nabla_\xi \Phi + \rho \cdot)(I + \nabla_\xi \Phi)^{-1} \nabla_\xi. \qquad (2.19)$$

Since g is expanded as

$$g(\xi + \Phi(\xi, \tau)) = g(\xi) + (\nabla_\xi g(\xi)) \cdot \Phi(\xi, \tau) + R_4(\xi, \Phi(\xi, \tau)), \qquad (2.20)$$

with again R_4 smooth and bounded by a multiple of $\Phi(\xi, \tau)^2$, we can compute the condition following from $\frac{\partial u}{\partial n} - g = 0$. The new condition becomes

$$\frac{\partial w}{\partial \nu} + s\left(\frac{\partial^2 U}{\partial \nu^2} - \frac{\partial g}{\partial \nu}\right)w = R_5(\xi, s, \nabla_\xi s) + (\nabla_\xi s \cdot \nabla_\xi)w + R_6(\xi, s, \nabla_\xi s)\nabla_\xi w. \qquad (2.21)$$

Now we can eliminate the free boundary terms s, Φ and their derivatives from the problem. We begin with equation (2.17) and evaluate it at $\partial \Omega$ using (2.18). In the resulting equation we isolate all the terms containing $\frac{\partial s}{\partial \tau}$ to obtain an equation of the form

$$-g(\xi)\frac{\partial s}{\partial \tau} = \nu(\xi)\frac{\partial s}{\partial \tau}R_7(\xi, w, \nabla_\xi w, \nabla_\xi \nabla_\xi w) - R_8(\xi, w, \nabla_\xi w, \nabla_\xi \nabla_\xi w), \qquad (2.22)$$

where R_7 and R_8 are smooth with $R_7(\xi, 0, 0, 0) = R_8(\xi, 0, 0, 0) = 0$. Since we have assumed that g is nonzero, it follows that $\frac{\partial s}{\partial \tau}$ can be expressed as

$$\frac{\partial s}{\partial \tau} = \frac{R_8(\xi, w, \nabla_\xi w, \nabla_\xi \nabla_\xi w)}{R_7(\xi, w, \nabla_\xi w, \nabla_\xi \nabla_\xi w)\nu(\xi) + g(\xi)}, \quad \xi = \theta \in \partial \Omega.$$

Next we substitute this back into (2.17) to convert all the other terms with Φ and its ξ-derivatives into the traces of w and its (tangential) ξ-derivatives. This finally yields our final formulation: problem (2.1),(2.2) is rewritten for w as the fully nonlinear problem

$$w_t = \mathcal{L}w + \mathcal{F}(w, Dw, D^2 w) \quad \text{on } \Omega, \quad t > 0, \qquad (2.23)$$

with boundary conditions also fully nonlinear,

$$\mathcal{B}w = \mathcal{G}(w, Dw) \quad \text{on } \partial \Omega. \qquad (2.24)$$

We derive that the local behaviour of solutions near an equilibrium (Ω, U) is determined by the *linearized* operator L defined by

$$Lv = \mathcal{L}v \quad \text{for } v \in D(L) = \{v \in \cap_{p>1} W^{2,p}(\Omega), \mathcal{L}v \in C(\overline{\Omega}), \mathcal{B}v = 0 \text{ on } \partial\Omega\}, \qquad (2.25)$$

where

$$Bv = \frac{\partial v}{\partial n} + \frac{1}{g}(\frac{\partial g}{\partial n} - \frac{\partial^2 U}{\partial n^2})v \tag{2.26}$$

is the linearized boundary operator. The reformulation of (2.1)(2.2) as the fully nonlinear problem (2.23)(2.24) enables us to perform a *nonlinear* stability analysis of the free boundary problem.

Our first result concerns the local existence.

Theorem 2.1 *For every $T > 0$ there are r, $\rho > 0$ such that if w_0 satisfies the compatibility condition $Bw_0 = \mathcal{G}(w_0, Dw_0)$ and $\|w_0\|_{C^{2+\alpha}(\overline{\Omega})} \leq \rho$, problem (2.23)-(2.24) with initial data $w = w_0$ has a solution $w \in C^{1+\alpha/2,2+\alpha}([0,T] \times \overline{\Omega})$. Moreover w is the unique solution in $B(0,r) \subset C^{1+\alpha/2,2+\alpha}([0,T] \times \overline{\Omega})$.*

For the stability result, i.e., for the description of the stable and unstable manifolds, we have to exclude the existence of eigenvalues with zero imaginary part:

$$\sigma(L) \cap i\mathbf{R} = \emptyset. \tag{2.27}$$

Let P be the spectral projection associated to the subset of $\sigma(L)$ with positive real part.

Theorem 2.2 *Assume that the spectrum of L contains elements with positive real part and that (2.27) holds.*

(i) There exists a unique local unstable manifold of the form

$$\varphi : B(0, r_0) \subset P(C(\overline{\Omega})) \mapsto (I - P)(C^{2+\alpha}(\overline{\Omega})),$$

φ Lipschitz continuous, differentiable at 0 with $\varphi'(0) = 0$.

(ii) There exists a unique local stable manifold of the form

$$\psi : B(0, r_1) \subset \{w_0 \in (I - P)(C^{2+\alpha}(\overline{\Omega})) : Bw_0 = \mathcal{G}(w_0, Dw_0)\} \mapsto P(C(\overline{\Omega})),$$

ψ Lipschitz continuous, differentiable at 0 with $\psi'(0) = 0$.

Clearly, if the spectrum is contained in $\{z : Rez < -\delta\}$ for some $\delta > 0$, then $P = 0$ and the stability theorem follows from (ii).

3 The focusing flame problem

The behaviour of solutions of the focusing problem from the introduction (with $u_* = 0$) is exhibited by self-similar solutions of the form

$$u(x,t) = \sqrt{T - t}\, f(\eta), \quad \eta = \frac{|x|}{\sqrt{T - t}}, \quad \Omega_t = \{|x| < b\sqrt{T - t}\} \tag{3.1}$$

with $f(\eta)$ satisfying

$$f''(\eta) + \frac{N-1}{\eta}f'(\eta) + \frac{1}{2}f = \frac{1}{2}\eta f' \quad \text{for } 0 \le \eta \le b; f'(0) = f(b) = 0, \ f'(b) = 1. \quad (3.2)$$

The stability analysis of the self-similar profile fits in the framework presented above. Transforming the problem to self-similar variables

$$\tilde{x} = \frac{x}{(T-t)^{\frac{1}{2}}}, \ \tilde{t} = -\log(T-t), \ \tilde{u}(\tilde{x},\tilde{t}) = \frac{u(x,t)}{(T-t)^{\frac{1}{2}}}, \ \tilde{\Omega}_{\tilde{t}} = \{\tilde{x} : x \in \Omega_t\}, \quad (3.3)$$

omitting the tildes, we arrive at

$$u_t = \mathcal{L}u = \Delta u - \frac{1}{2}x \cdot \nabla u + \frac{1}{2}u, \quad x \in \Omega_t \ \text{ with } \ u = 0, \quad \frac{\partial u}{\partial n} = 1 \ \text{ on } \partial\Omega_t. \quad (3.4)$$

The self-similar solution (3.1) is transformed by (3.3) into an equilibrium

$$U(x) = f(|x|), \ \Omega = B_b = \{x \in \mathbf{R}^N : |x| < b\}, \quad (3.5)$$

with $f(\eta)$ defined by (3.2). We cannot expect (3.5) to be stable because the original problem is invariant under translations in x and t. Thus if we apply a small shift to (3.5), we obtain another self-similar solution which is transformed by (3.3) into a solution which starts close to (3.5) but moves away from it. As a consequence, we have that L defined by (2.25) with

$$\mathcal{L}v = \Delta v - \frac{1}{2}x \cdot \nabla v + \frac{1}{2}v \ \text{ on } B_b, \quad (3.6)$$

and (2.26) with

$$Bv = \frac{\partial v}{\partial n} + (\frac{N-1}{b} - \frac{b}{2})v = 0 \ \text{ on } \partial B_b \quad (3.7)$$

must have unstable eigenvalues: one with a 1-dimensional eigenspace spanned by a radial eigenfunction corresponding to a shift in t, and another with an N-dimensional eigenspace corresponding to N shifts in x. Thus the equilibrium (3.5) is unstable.

Using power series it can be shown that these two eigenvalues are the only positive eigenvalues and that the remaining part of the spectrum consists of negative eigenvalues. This may be interpreted by saying that although the equilibrium (3.5) is unstable, its profile does have a stability property: the unstable manifold consists exactly of the images under (3.3) of shifts in space and time of (3.5) and is therefore given by the parametrization

$$(\epsilon_1, \epsilon_2, t) \mapsto \sqrt{1 + \epsilon_2 e^t} \, U(\frac{x - \epsilon_1 e^{\frac{1}{2}t}}{\sqrt{\epsilon_2 e^t + 1}}). \quad (3.8)$$

The (local) image of (3.8) under the transformation of (2.1),(2.2) to (2.23),(2.24) by means of the transformation in Section 2 is the unstable manifold given by Theorem 2.2 for this particular case. This is very much like the travelling wave case where the center manifold of the equilibrium is given by the translates of the equilibrium.

111

4 Travelling waves for a two-phase problem

In papers [5], [6] we have studied the stability of the planar travelling wave solutions to a free boundary problem with jump conditions at the interface defined by $\zeta = \xi(t, y)$,

$$
\begin{cases}
u_t(t, \zeta, y) = \Delta u(t, \zeta, y) + u(t, \zeta, y)u_\zeta(t, \zeta, y), \ t \geq 0, \ \zeta \neq \xi(t, y), \ y \in \omega, \\
u(t, \xi(t, y), y) = u_*, \ [\partial u/\partial n](t, \xi(t, y), y) = -1, \ t \geq 0, \ y \in \omega, \\
\partial \xi/\partial n(t, y) = 0, \ t \geq 0, \ y \in \partial\omega, \\
\partial u/\partial n(t, \zeta, y) = 0, \ t \geq 0, \ \zeta \neq \xi(t, y), \ y \in \partial\omega, \\
u(t, -\infty, y) = 0, \ u(t, \infty, y) = u_\infty, \ t \geq 0, \ y \in \omega,
\end{cases}
\tag{4.1}
$$

where $\omega \subset \mathbf{R}^{N-1}$ is a bounded open set with regular boundary $\partial\omega$, and u_*, u_∞ are given positive numbers. By $[f](\xi)$ we mean $f(\xi^+) - f(\xi^-)$ at a point of possible discontinuity ξ.

N-dimensional TW solutions are special solutions of (4.1), such that $\xi(t, y) = -ct + s(y)$, $u(t, \zeta, y) = U(\zeta + ct, y)$. Therefore, setting $z = \zeta + ct$, the triplet (c, s, U) satisfies

$$
\begin{cases}
cU_z(z, y) = \Delta U(z, y) + U(z, y)U_z(z, y), \ z \neq s(y), \ y \in \omega, \\
U(s(y), y) = u_*, \ [\partial U/\partial n](s(y), y) = -1, \ y \in \omega, \\
\partial U/\partial n(x, y) = 0, \ x \neq s(y), \ y \in \partial\omega, \\
\partial s/\partial n(y) = 0, \ y \in \partial\omega, \\
U(-\infty, y) = 0, \ U(+\infty, y) = u_\infty, \ y \in \omega.
\end{cases}
\tag{4.2}
$$

For certain values of the parameters (precisely, for $u_\infty > \sqrt{2}$, $2/u_\infty < u_* < 2/u_\infty + u_\infty$) problem (4.2) admits a (unique up to translations) $1D$ or "planar" TW solution $c = c_0$, $s \equiv 0$, $U(z, y) \equiv U_0(y)$. It is important to remark that the speed c_0 is independent of u_* and it is given by

$$
c_0 = \frac{u_\infty}{2} + \frac{1}{u_\infty},
\tag{4.3}
$$

while U_0 satisfies

$$
\begin{cases}
c_0 U_0'(z) = U_0''(z) + U_0(z)U_0'(z), \ z \neq 0, \\
U_0(0) = u_*, \ [U_0'](0) = -1, \\
U_0(-\infty) = 0, \ U_0(+\infty) = u_\infty.
\end{cases}
\tag{4.4}
$$

This planar TW is orbitally stable under small (suitably smooth) perturbations iff

$$
2/u_\infty < u_* \leq u_*^c = u_\infty/2 + 1/u_\infty + \sqrt{(u_\infty/2 + 1/u_\infty)^2 - 1}.
$$

The instability for $u_* > u_*^c$ is due to a positive eigenvalue $\tilde{\lambda}$ of the linear operator L associated to (4.1), such that $\lim_{u_* \downarrow u_*^c} \tilde{\lambda} = +\infty$. Note that $\tilde{\lambda}$ does not cross the imaginary axis as u_* crosses u_*^c.

We prove in [1] that in dimension $N = 2$ there exists a sequence of bifurcation points $u_*^k \downarrow u_*^c$ giving rise to branches of nonplanar travelling waves (c, s, U) bifurcating

112

from the "trivial branch" $(c_0, 0, U_0)$. Their speeds c depend on the corrugation of their fronts: if $\omega = (-1, 1)$, then

$$c = c_0 + \frac{1}{2u_\infty} \int_{-1}^{1} \left(\sqrt{(1 + (s'(y))^2)} - 1 \right) dy > c_0. \tag{4.5}$$

The sequence u_*^k is determined by the relation $\tilde{\lambda}(u_*^k) = \lambda_k$, where $-\lambda_k = -k^2\pi^2/4$ is the ordered sequence of the negative eigenvalues of the second order derivative in $(-1, 1)$ associated with the Neumann boundary condition.

The fact that a sequence of bifurcation points accumulates at u_*^c contributes to the understanding of the sharp instability phenomenon occurring at $u_* = u_*^c$.

The proof is based on two main tricks: first, as in Section 2 about the time depending case, we are able to eliminate the front s by an appropriate splitting of the unknown U; then another difficulty arises, namely, at $u_* = u_*^k$, the kernel of the linearized operator around U_0 is $2D$ and its range has codimension 2. We take advantage of the translation invariance to reduce the dimension of the kernel, and simultaneously we eliminate c reducing the codimension of the range. So we turn our problem of bifurcation of fronts to a classical bifurcation problem from simple eigenvalues, which eventually gives infinitely many branches of nonplanar TW solutions.

References

[1] C.-M. BRAUNER & A. LUNARDI: *Bifurcation of nonplanar travelling waves in a free boundary problem*, Preprint Math. Appl. Bordeaux, 97022 (1997).

[2] C.-M. BRAUNER, A. LUNARDI & CL. SCHMIDT-LAINÉ: *Multidimensional stability analysis of planar travelling waves*, Appl. Math. Letters **7** (1994), 1-4.

[3] C.M. BRAUNER, J. HULSHOF & CL. SCHMIDT-LAINÉ: *The saddle point property for focusing selfsimilar solutions*, Proc. A.M.S. 127, 473-479 (1999).

[4] C.M. BRAUNER, J. HULSHOF & A. LUNARDI: *A general approach for stability in free boundary problems*, Preprint Mathematical Institute Leiden W98-29 (1998).

[5] C.M. BRAUNER, A. LUNARDI & CL. SCHMIDT-LAINÉ: *Stability of travelling waves with interface conditions*, Nonl. Anal. **19**, (1992) 455-474.

[6] C.-M. BRAUNER, A. LUNARDI & CL. SCHMIDT-LAINÉ: *Stability of travelling waves in a multidimensional free boundary problem*, Nonlinear Analysis, to appear.

[7] J.D. BUCKMASTER & G.S.S. LUDFORD: *Theory of laminar flames*, Cambridge University Press, Cambridge, 1982.

[8] D. S. STEWART & G. S. S. LUDFORD, *The acceleration of fast deflagration waves*, Z.A.M.M., **63** (1983), 291-302.

[9] V.A. GALAKTIONOV, J. HULSHOF & J.L. VAZQUEZ: *Extinction and focusing behaviour of spherical and annular flames described by a free boundary problem*, Journal Math. Pures et Appl. **76**, (1997) 563-608.

[10] L. GLANGETAS & J.M. ROQUEJOFFRE: *Bifurcations of travelling waves in the thermo-diffusive model for flame propagation*, Arch. Rat. Mech. Anal., **134** (1996), pp. 341-402.

[11] D. HILHORST & J. HULSHOF: *An elliptic-parabolic problem in combustion theory: convergence to travelling waves*, Nonl. Anal. 17, pp. 519-546, 1991

[12] D. HILHORST & J. HULSHOF: *A free boundary focusing problem*, Proc. AMS **121**, (1994) 1193-1202.

[13] A. LUNARDI: Analytic Semigroups and Optimal Regularity in Parabolic Problems, Birkhäuser, Basel, 1995.

[14] J.L. VAZQUEZ: *The Free Boundary Problem for the Heat Equation with Fixed Gradient Condition*, Proc. Int. Conf. "Free Boundary Problems and Applications", Zakopane, Pitman Res. Notes Math. 363, Longman, 1996.

Claude-Michel Brauner
Mathématiques Appliquées de Bordeaux, Université Bordeaux I
33405 Talence Cedex, France

Josephus Hulshof
Mathematical Department of the Leiden University
Niels Bohrweg 1, 2333 CA Leiden, The Netherlands

Alessandra Lunardi
Dipartimento di Matematica, Università di Parma
Via D'Azeglio 85/A, 43100 Parma, Italy

Claudine Schmidt-Lainé
CNRS UMR 128, Ecole Normale Supérieure de Lyon
69364 Lyon Cedex 07, France

V. A. GALAKTIONOV, S. I. SHMAREV AND J. L. VAZQUEZ*

Regularity of Solutions and Interfaces to Degenerate Parabolic Equations. The Intersection Comparison Method

Abstract

We present a method of analysis of interface dynamics and regularity suitable for application to nonlinear parabolic equations with finite speed of propagation. Typical examples are the equations

$$u_t = (u^m)_{xx}, \qquad u_t = (u^m)_{xx} \pm u^p,$$

with parameters $m > 1$, $p > 0$. Such questions have been successfully studied for the first equation, the well-known porous medium equation, and also for the the second one (a reaction-diffusion model) in the range of parameters $m > 1$, $p \geq 1$. These cases are characterized physically by the advancing character of the interface movement (so-called heating waves), and mathematically by good a priori estimates. Our method applies to the study of the strong absorption case of the second equation, with minus sign and $0 < p < 1$, for which previous methods only gave little information and where the interfaces may move forward and backward (heating and cooling waves). It allows establishment of the interface equation and proof of the Lipschitz continuity of the interfaces. The method is based on intersection comparison with travelling waves. We give the detailed outline of the analysis of the simpler subcase $m+p \geq 2$. The method can be applied to a large class of nonlinear parabolic equations and conservation laws with free boundaries, but it is restricted to problems posed in one space dimension or multi-dimensional problems with radial symmetry.

1 Introduction

A number of degenerate parabolic equations of the general form

$$u_t = A(x, t, u, u_x, u_{xx}) \tag{1.1}$$

have in principle the curious property of *finite speed of propagation*, absent from standard parabolic equations. Let us assume to be specific that we consider the Cauchy problem for such an equation posed in the domain $Q = \{(x, t) : -\infty < x < \infty, t > 0\}$ with initial data

$$u(x, 0) = u_0(x), \quad x \in \mathbf{R}, \tag{1.2}$$

*Authors partially funded by DGICYT Project PB94-0153.

and also that we restrict ourselves to nonnegative solutions $u = u(x,t)$. Then we want to consider equations for whenever the initial function u_0 is compactly supported and the solution $u(\cdot, t)$ is also compactly supported as a function of x for all $t > 0$. More precisely, the positivity set $\Omega(t) = \{x \in \mathbf{R} : u(x,t) > 0\}$ is bounded for all $t > 0$, limited by two *free boundaries*, also called *interfaces*, defined as

$$\zeta(t) = \inf\{x \in \mathbf{R} : u(x,t) > 0\}, \qquad \eta(t) = \sup\{x \in \mathbf{R} : u(x,t) > 0\}. \tag{1.3}$$

Precisely speaking, these are the *outer* interfaces, since there could be other inner interfaces corresponding to holes in the domain of positivity.

Two basic questions in the qualitative theory of those equations are the laws governing the motion and the smoothness of the interfaces and the solutions near them. This subject arose with the study of the simplest models of such equations, the Stefan problem and the Porous Medium Equation, and a large literature has been collected about it. For the former equation we refer to [KS], where the C^∞ regularity of the interfaces is proved. For the latter a method has been proposed beginning with the work of Aronson [Ar1], and a successful answer has been found after more than a decade of work, cf. [AB, CF, AV, An2]: the interfaces are analytic after a possible initial resting period (the waiting time $t_w \geq 0$) and the dynamics is described by Darcy's law, which for the right interface reads: for $t > t_w$

$$\eta'(t) = - \lim_{x \to \eta(t)-0} \pi_x(x,t), \qquad \pi = \frac{m}{m-1} u^{m-1}. \tag{1.4}$$

It is an elegant method which can be generalized; it requires, however, some good conditions on the equation, since it is strongly based on a priori estimates of second (and higher) derivatives.

We present below a technique that is effective in answering such questions in situations where previous methods failed. The method is presented in the framework of the study of the equations of the form

$$u_t = (u^m)_{xx} - u^p, \tag{1.5}$$

with positive exponents m and p. Assuming that initial data are given by a continuous, nonnegative and bounded function u_0, the Cauchy problem (1.5),(1.2) has a unique continuous nonnegative weak solution $u(x,t)$, [Ka]. In order to have finite propagation, we impose the slow-diffusion restriction

$$m > 1. \tag{1.6}$$

Our concern here is the description of the solutions and free boundaries of the Cauchy problem in the range

$$0 < p < 1,$$

which is called the *strong absorption* range since the problems of concern for us happen for values of $u \approx 0$ and the absorption coefficient $u^p/u \to \infty$. The solutions of such

116

equations possess a number of specific properties, absent for this equation for $p \geq 1$ or for the porous medium equation:

(i) For any bounded initial function the solution vanishes identically after a finite time, T_e.

(ii) Depending on the shape of the compactly supported initial function $u_0(x)$, the support of the solution may expand or shrink with time, or even change alternatively in both ways. See an explicit example in [Ke].

(iii) A new feature of strong absorption equations, called *pulse splitting*, was explained by Kamin and Rosenau in [RK]. It means that a solution which at some time t_1 has a connected support (it is a simple pulse) may exhibit a support with two or more connected components at a later time t_2.

A preliminary study of the support evolution in the strong aborption situation is performed by Chen et al. in [CMM]. A formal analysis of the interface behaviour is done in [RK].

2 Method and main results

The problem now is to establish the smoothness of the interfaces of the solutions to such problems, the smoothness of the solution near the interfaces, and the relationship between both. We propose a new approach for the study of such questions. The method is based on the idea that the local analysis of the solutions of nonlinear parabolic equations like (1.5) near an interface can be done by means of *intersection comparison* with the family of *travelling wave solutions* of the equation. To this effect, we need a complete classification of travelling wave solutions which for equation (1.5) has been performed in [HV] in the whole range of exponents. Gilding and Kersner [GK] have recently established that for a wide class of nonlinear parabolic equations the property of finite speed of propagation can be characterized in terms of the family of travelling wave solutions. The comparison technique, usually called Intersection Comparison or Lap Number Theory, roughly says that the number of sign changes between two solutions of equations like (1.5) is nonincreasing in time, see Theorem 4.1 below. We are able to show in this way that the behaviour of the solutions near the interface, and in particular its local velocity, are determined in first approximation by a travelling wave profile.

The methods of investigation we propose here are explained via consideration of equation (1.5) in the range of parameters

$$m + p - 2 \geq 0. \tag{2.1}$$

Let us summarize our main results for a continuous, nonnegative, finitely supported and bell-shaped initial function u_0.

• We prove that the interface function $\eta(t)$ is left and right differentiable at every instant $(0, T_e)$ and that the following limit relations hold: for every $t_0 \in (0, T_e)$

$$\lim_{t \nearrow t_0} D^+\eta(t) = \lim_{t \nearrow t_0} D^-\eta(t) = D^-\eta(t_0), \qquad \lim_{t \searrow t_0} D^+\eta(t) = \lim_{t \searrow t_0} D^-\eta(t) = D^+\eta(t_0).$$

Therefore, the velocity of the interfaces is well-defined, at least laterally. In principle it can be infinite.

• We obtain the formulae which express the interface velocity in terms of the solution profile, so-called *interface equations*, for both expanding and receding waves. (See Theorems 4.3 and 4.4.) The equations for cooling waves are never given by Darcy's law, while the equation for heating waves follows Darcy only if $m + p > 2$.

• The next step is to control the *basic interface regularity*. We prove that the solutions of problem (1.5)-(1.2) have *Lipschitz continuous interfaces* in any compact time interval $0 < t_1 \le t \le t_2 < T_e$ (Theorem 6.1).

Let us remark that an explicit example given in [Ke] has a flat interface at the extinction time, i.e., $\eta'(T_e-) = -\infty$, $\zeta'(T_e-) = \infty$. This is a general phenomenon for extinction points: the study of the behaviour near extinction is a separate subject that for $m + p = 2$ has been investigated in [GV1], where the precise parabolic shape for the interface is derived. It is remarked in that paper that the flat interface profile occurs also for $m + p > 2$, $p < 1$. On the other hand, the regularity at $t = 0$ depends on the initial data, as is usual in free boundary problems for parabolic equations. The claim that the initial profile is bell-shaped guarantees that so is the solution profile at each instant $t \in (0, T_e)$.

The techniques used in the proofs represent an extension or alternative to the method of study of interface regularity proposed for the porous medium equation that has been commented above. We have been unable to derive the properties of the shrinking interfaces appearing in the strong absorption case using only the standard techniques of such classical analysis, hence the need for the introduction of new ideas. For a complete description of the present method with full proofs we refer the reader to the paper [GSV].

Let us finally mention that the idea of investigating the qualitative properties of nonlinear diffusion equations by means of intersection comparison with sufficiently rich families of special solutions, not necessarily the travelling waves, was performed by two of the authors in [GV2] in order to study the properties of convexity and concavity.

3   Travelling wave solutions

Our analysis of interface behaviour and regularity is based on comparison with a family of travelling wave solutions (TWs for short), which are solutions of (1.5) of the form

$$u(x,t) = f(\zeta), \qquad \zeta = x - \lambda t, \tag{3.1}$$

where $\lambda \in \mathbf{R}$ is the velocity of the wave. These solutions have been studied and classified in [HV] for different parameters m, p and λ. Let us recall the main facts we will need from such a study. Equation (1.5) admits TW solutions with different behaviours in the different ranges of the exponents m and p. The profile function $f(\zeta)$ solves the problem

$$-\lambda f' - (f^m)'' + f^p = 0. \tag{3.2}$$

In order to study interfaces we need the travelling wave to land on the level $f = 0$ on one end of the real line. We can normalize such waves by asking that $f(\zeta) > 0$ for $\zeta < 0$ and $f(\zeta) = 0$ for $\zeta = 0$. New TWs of the same kind can be obtained then by shifting in ζ, $\hat{f}(\zeta) = f(\zeta - \zeta_0)$, or by reflection, $\hat{f}(\zeta) = f(-\zeta)$. This last operation inverts the direction of the movement.

According to [HV], for each value of $\lambda \in \mathbf{R}$ there exists exactly one normalized travelling wave $f_\lambda(\zeta)$, and the family $\{f_\lambda\}$ depends continuously on λ (and p and m). There are three ranges of values of the parameter p in which the travelling wave family behaves in a similar fashion, namely, $p \geq 1$, $2 - m \leq p < 1$ and $p < 2 - m$. In the range of interest for the present study, $p + m \geq 2$, the family of travelling waves can be divided into three types:

1) TWs with positive velocity $\lambda > 0$; the interface is given by $\eta(t) = \lambda t$, so that the support expands with constant speed $\eta'(t) = \lambda > 0$; at the interface the profile of the solutions of this type satisfies

$$(f^{m-1})'(0-) = \lim_{\zeta \nearrow 0}(f^{m-1})'(\zeta) < 0. \tag{3.3}$$

2) TW with zero velocity, $\lambda = 0$. Then $\zeta = x$ and the corresponding stationary solution $f(\zeta)$ is given by the explicit formula

$$f(\zeta) = c(m,p)\,\zeta^{2/(m-p)}. \tag{3.4}$$

3) TWs with negative velocity $\lambda < 0$; in this event the support of $f(\zeta)$ shrinks with constant velocity $\eta'(t) = \lambda < 0$; the profile $f(\zeta)$ satisfies

$$(f^{1-p})'(0-) < 0. \tag{3.5}$$

All these profiles are monotone decreasing functions of ζ for $\zeta < 0$ with $f(\zeta) \to \infty$ as $\zeta \to -\infty$. Let us discuss next the interface equations; in the limit case $m + p = 2$ the following equation is satisfied at $\zeta = 0$:

$$\lambda = -\frac{m}{m-1}(f^{m-1})'(0-) + \frac{(m-1)}{(f^{m-1})'(0-)}. \tag{3.6}$$

119

Thus, a unique formula gathers all three types of TWs and there is a one-to-one correspondence between velocities $\lambda \in \mathbf{R}$ and slopes $z = -f^{m-1}(0-) > 0$.

On the contrary, for $m + p > 2$ (with $m > 1$, $p < 1$) the behaviour at the interface is different for heating and cooling fronts. Thus, for $\lambda > 0$ we have

$$\lambda = -\frac{m}{m-1}(f^{m-1})'(0-), \tag{3.7}$$

for $\lambda < 0$ we have

$$\lambda = (1-p)\frac{1}{(f^{1-p})'(0-)}, \tag{3.8}$$

and for $\lambda = 0$ both right-hand side limits coincide and are also zero. The one-to-one correspondence between velocities and slopes does not hold with the previous definitions, but it can be restored if we define the slope for negative λ and we change it into $(1 - p)/(f^{1-p})'(0-)$, while keeping it as $(f^{m-1})'(0-)$ for $\lambda \geq 0$. All these formulas come from equation (3.2) after using L'Hospital's rule, cf. [HV, Section 7].

Let us remember for comparison the situation in the other exponent ranges: for $p \geq 1$ there are no TWs of type 3 with negative velocity and shrinking support (cooling type), while for $p + m < 2$ there exist TWs for all velocities but the behaviour of their interfaces is in all cases of the type 2 described above, cf. [HV]. More precisely, in the case $m + p < 2$ the profile function $f(\zeta)$ is given by the formula

$$f^{(m-p)/2}(\zeta) = c_0\zeta\left(1 - c_1\lambda\zeta^\alpha + \ldots\right),$$

where $\alpha = (2 - m - p)/(m - p) > 0$ and c_0 and c_1 are constants depending only on m and p. Hence, in this case the slope of the TW solution is defined not by the first term of the asymptotic expansion, as in the regular case $m + p \geq 2$, but by the second term. That is the principal feature of the problem in this range of the exponents which require a special study.

4   The interface equation

We come now to the central ideas of the method. We assume from now on that u_0 is continuous, nonnegative and compactly supported. Under such conditions it is proved in the standard references that there exists a unique weak solution of the Cauchy problem for equation (1.5) which is continuous, nonnegative, C^∞ smooth in its domain of positivity and has compact support as a function of x for every $t > 0$ until the extinction time T_e. The standard maximum principle applies. However, we shall use in an essential way the special comparison result called Intersection Comparison that we state next (see [Sa, Ma, SGKM] for proofs. The idea goes back to Sturm, 1836).

120

Theorem 4.1 *Let u and v be two solutions of the Cauchy problem (1.5), (1.2) and let J(t) be the number of sign changes of the difference u − v at time t. If this number is finite for some t_0, it is equal or less than $J(t_0)$ for any $0 < t_0 < t < T_e$.*

This comparison technique has proved to be useful in the study of blow-up and extinction problems, cf. also [GV1, GV3]. We will assume for convenience of presentation that u_0 has only one maximum (it is bell-shaped) since then there are only two interfaces (the outer interfaces), $\Omega(t) = (\zeta(t), \eta(t))$ and both interfaces meet at $t = T_e$. We will comment in Section 7 on the modifications that occur when the solution has a finite number of humps.

The analysis of the travelling waves suggests that the behaviour of the profiles near an interface in the range $m + p \geq 2$ is best expressed in terms of derivatives of the following functions:

$$\pi(x,t) = \frac{m}{m-1} u^{m-1}(x,t), \qquad w(x,t) = \frac{1}{1-p} u^{1-p}(x,t).$$

The reader will immediately observe that both functions are proportional for $m+p = 2$ while for $m + p > 2$, the exponents are ordered $m - 1 > 1 - p > 0$. This will have as a consequence a difference in the formulation of the results of both cases, following what happened for TWs.

For a general solution the interface behaviour is also related to the lateral space derivatives of π and w at $x = \eta(t)$. We need a word about notation. We will be dealing with first-order lateral derivatives. Thus, the interface equation gives a formula for the right derivative of the interface at a time $t_0 > 0$:

$$D_t^+ \eta(t_0) = \lim_{t \downarrow t_0} \frac{\eta(t) - \eta(t_0)}{t - t_0}.$$

The limit as $t \uparrow t_0$ gives the left derivative, $D_t^- \eta(t_0)$. The subscript t is not really necessary in this context. A subscript is needed when we perform a partial differentiation like

$$D_x^- w(x_0, t_0) = \lim_{x \uparrow x_0} \frac{w(x, t_0) - w(x_0, t_0)}{x - x_0}.$$

Similarly, we define the right derivative $D_x^+ w$.

Theorem 4.2 *Let u be a solution of problem (1.5), (1.2). Then for every $0 < t < T$*

1) *there exists a left derivative of π at $x = \eta(t)$, $-\infty \leq D_x^- \pi(\eta(t), t) \leq 0$,*

2) *there exists a left derivative of w at $x = \eta(t)$, $-\infty \leq D_x^- w(\eta(t), t) \leq 0$.*

Proof. We introduce as a basic tool the idea of intersection comparsion with TWs. It is done in the same way for π and w. Let us examine in detail the limit $D_x^- w$.

We fix an arbitrary instant $t_0 \in (0, T)$ and assume for contradiction that the limit $D_x^- w(\eta(t_0), t_0)$ does not exist. Then there must exist sequences $\{x_n\}$, $\{y_n\}$ tending to $\eta(t_0-)$, such that

$$\lim_{n\to\infty} \frac{\eta(t_0) - x_n}{w(x_n, t_0)} = \sigma, \qquad \lim_{n\to\infty} \frac{\eta(t_0) - y_n}{w(y_n, t_0)} = \delta$$

with $\sigma, \delta \geq 0$, $\sigma < \delta$ (the strict inequality). Take some $\gamma \in (\sigma, \delta)$ and consider the travelling wave solution of the third kind $f_\gamma(\zeta)$ with interface passing through the point $(\eta(t_0), t_0)$ and propagating with the speed $\gamma \in (\sigma, \delta)$. According to our assumption the profile of the solution $u(x, t)$ oscillates at time t_0 about the profile of the travelling wave solution f_γ so that the difference of these profiles has an infinite number of sign changes near $x = \eta(t_0)$.

The next step is to consider what happens at some instant $t_* < t_0$. By slightly shifting γ we can always assume that the interfaces of $u(x, t)$ and $f_\gamma(\zeta)$ do not coincide at t_*. We now recall the following general assertion [An1, Theorem D]: *a solution to a linear uniformly parabolic equation can only have a finite number of zeros at any instant $t > 0$.* We apply the result to the difference $w = u(x, t) - f_\gamma(\zeta)$, which satisfies the linear parabolic equation following from (1.5),

$$w_t - m\left(\left[\int_0^1 (\theta u + (1 - \theta)f_\gamma)^{m-1}\, d\theta\right] w\right)_{xx} + p\left[\int_0^1 (\theta u + (1 - \theta)f_\gamma)^{p-1} d\theta\right] w = 0.$$

Since at $t = t_*$ the interfaces of both solutions are separated and they have quite different behaviour as $x \to -\infty$, the sign changes of w must be concentrated in the domain of strict parabolicity of the corresponding equation for $t \approx t_*$, hence they are finite at $t = t_*$. On the other hand, w has infinitely many sign changes at the later instant t_0. This contradicts the basic result of Intersection Comparison theory, which says that the number of space intersections of the profiles of different solutions to the same parabolic equation cannot increase with time, cf. Theorem 4.1. We thus conclude that there must exist a limit, $D_x^- w(\eta(t_0), t_0)$, finite or infinite. The existence of a limit of $(\eta(t_0) - x)/\pi(x, t_0)$ as $x \to \eta(t_0-)$ can be proved by the same arguments via comparison with the travelling solutions of the first kind (see Section 3). $\square$

Lemma 4.1 *If $m+p > 2$, then at least one of the quantities $D_x^- \pi(\eta(t), t)$, $1/D_x^- w(\eta(t), t)$ is equal to zero.*

Proof. Assume that $D_x^- \pi(\eta(t), t) = \beta < 0$. Then $u(x, t) \sim (\eta(t) - x)^{1/(m-1)}$ and $w(x, t) \sim (\eta(t) - x)^{(1-p)/(m-1)}$ near the interface, whence $1/D_x^- w = 0$ at $x = \eta(t) - 0$. The other case is proved in the same way. $\square$

Theorem 4.3 (Interface Equation I) *Let $m + p = 2$. For every $0 < t < T_e$ there exists limit $D_x^- \pi = D_x^- \pi(\eta(t), t) \le 0$ and the interface equation has the following form:*

$$D_t^+ \eta(t) = -D_x^- \pi + \frac{m}{D_x^- \pi}. \tag{4.1}$$

Proof. There is no essential difference with the standard proof for the PME, cf [Ar1]. Thanks to the previous result the limit of $D_x^- \pi$ exists, finite or infinite, as $x \to \eta(t) - 0$. See whole details in [GSV]. □

Theorem 4.4 (Interface Equation II) *Let $m + p > 2$. For every $0 < t < T$ there exists $D_t^+ \eta(t) \in [-\infty, \infty)$, and the interface equation has the following forms. If $D_t^+ \eta(t) \ge 0$, then*

$$D_t^+ \eta(t) = -D_x^- \pi(\eta(t), t), \tag{4.2}$$

while for $D_t^+ \eta(t) \le 0$, then

$$D_t^+ \eta(t) = \frac{1}{D_x^- w(\eta(t), t)}. \tag{4.3}$$

Proof. Fix an arbitrary point $t_0 \in (0, T)$. Because of Lemma 4.1, at every instant $t > 0$ only one of the two terms on the right-hand part of (4.1) can be nonzero. Therefore, we consider separately three independent situations: 1) $\pi_x(\eta(t_0), t_0) < 0$; 2) $1/w_x(\eta(t_0), t_0) < 0$; 3) $\pi_x(\eta(t_0), t_0) = 1/w_x(\eta(t_0), t_0) = 0$ (the derivatives π_x, w_x at $x = \eta(t_0)$ are understood as the left derivatives). In each of these cases the proof is rather standard and only uses the maximum principle. □

Using Lemma 4.1 for $m + p > 2$ and noting that for $p + m = 2$ $D_x^- w = (1/m) D_x^- \pi$, we can write the three interface equations (4.1)-(4.3) in a common form.

Corollary 4.1 (Common interface equation) *For all $m > 1$, $0 < p < 1$ with $m + p \ge 2$ we have*

$$D_t^+ \eta(t) = -\left(\pi_x - \frac{1}{w_x} \right) (\eta(t), t), \tag{4.4}$$

where the terms on the right-hand side are understood as side derivatives, $D_x^- \pi$, $D_x^- w$.

We use the conventions $1/0 = \infty$, $1/\infty = 0$.

5 Lateral derivatives of $\eta(t)$

The equation of the interface has given us important information about the movement of the interface by relating the right-hand derivative $D^+ \eta(t)$ to the profile at every time. In order to study the interface regularity we need to also control the left-hand derivative. The methods are very similar to those of Theorem 4.2.

Theorem 5.1 *For every $0 < t < T_e$, there exists $D^-\eta(t)$.*

Proof. We compare the solution $u(x,t)$ with the travelling wave solutions $f_\lambda(\zeta)$, $\zeta = x - \lambda(t_0 - t)$, $\lambda \neq 0$, to equation (1.5). The interfaces of these travelling wave solutions pass through the point $(t_0, \eta(t_0))$. Next, let us introduce the function

$$\xi(t) = \frac{\eta(t_0) - \eta(t)}{t_0 - t}.$$

The function $\xi(t)$ has a limit (finite or infinite) as $t \to t_0$, with $t < t_0$. Otherwise, there must exist two sequences $\{t_k\}$ and $\{\theta_k\}$ such that t_k, $\theta_k \to t_0 - 0$ as $k \to \infty$ and

$$\xi(t_k) \geq \alpha, \qquad \xi(\theta_k) \leq \gamma \tag{5.1}$$

with some α and γ, two different constants that we may also assume different from zero and infinity. This means that the interface $x = \eta(t)$ of the solution $u(x,t)$ has infinitely many points of intersection with the interface of the travelling wave solution $f_\lambda(\zeta)$, $\lambda = (\alpha + \gamma)/2$. We may now make recourse to the results of [GV1, Section 8]. Repeating their arguments, it is easy to see that *each oscillation of the function $\eta(t)$ about the line $x = \eta(t_0) + \lambda(t_0 - t)$ is accompanied by the disappearance of a point of the space intersection (i.e., sign change of the difference) of the profiles of $u(x,t)$ and $f_\lambda(\zeta)$.* Assumption (5.1) thus results in the following assertion: the profiles of $u(x,t)$ and $f_\lambda(\zeta)$ have infinitely many intersections at any instant $t_1 \in (0, t_0)$. The latter is impossible; however, if we consider the equation for the difference $\omega = u - f_\lambda$, we see that as a consequence of the fact that the supports of f_λ and $u(x,t)$ do not coincide, the function ω, a solution to a parabolic equation, must have infinitely many local maxima concentrated in the domain of strict parabolicity of that equation. Since the number of space intersections can only be finite (see [An1, Theorem D]), we arrive at the contradiction to (5.1). This proves that $D^-\eta(t_0) = -\lim_{t \to t_0 - 0} \xi(t)$ must exist. $\square$

Our next step is to establish lateral continuity for the side derivatives. First, we establish the right-continuity of $D^-\eta(t)$.

Theorem 5.2 *For every $t_0 \in (0, T_e)$*

$$\lim_{t \uparrow t_0} D^+\eta(t) = \lim_{t \uparrow t_0} D^-\eta(t) = D^-\eta(t_0). \tag{5.2}$$

The proof of this theorem is rather lengthy and is therefore omitted here. The idea of the proof consists of performing the Intersection Comparison Principle not with a single TW solution but with *two families* of TW solutions of which the solutions of either family propagate with the same speed but land at different points.

In a similar way we prove

124

Theorem 5.3 *For every $t_0 \in (0, T_e)$*

$$\lim_{t \downarrow t_0} D^+ \eta(t) = \lim_{t \downarrow t_0} D^- \eta(t) = D^+ \eta(t_0). \tag{5.3}$$

6 Bounded velocity for heating and cooling waves

Theorem 6.1 *Under the assumptions made on u_0, the interface function $\eta(t)$ of the solution to problem (1.5)-(1.2) is a Lipschitz continuous function for $0 < \tau \le t \le T < T_e$.*

The proof relies upon the interface equation and suitable estimates of the derivatives of the space profile. To estimate the term π_x at the interface (which is responsible for the velocity of the heating waves), we use the technique long known for the solutions of the PME and which happens to apply perfectly in the case under study. To analyze the term which controls the velocity of the cooling waves we consider a sequence of solutions of regularized non-degenerate problems and estimate the derivative of $w(x, t)$ by analyzing the equation it satisfies. It is worth noting that this analysis is performed via consideration of the level curves of the solutions. The analysis assumes that the solution has an x-profile with only one hump.

7 Multi-hump solutions

Let us now devote our attention to the situation where the solution has many humps, i.e., for $t > 0$ the function $u(\cdot, t)$ has more than one (local) maximum. According to [CMM] the number of local maxima (and local minima) is finite for $t > 0$ and, moreover, the number of intervals of monotonicity is a nonincreasing function of time, the discontinuity occurs when either a maximum meets a neighbouring minimum in an isolated way or when a number of some such collisions happen simultaneously. It is clear that the number of connected components of $\Omega(t)$, $n(t)$, is equal or less than the number of maxima of $u(\cdot, t)$, $n(t) \le p(t)$. It is also known that $n(t)$ is not necessarily decreasing because of the phenomenon of pulse splitting. When such phenomena occur, the definition of right (or left) interface implies jumping at any time where one of the humps disappears to the next hump which still continues to exist. We call such times *local extinction times*.

The results of previous sections can be revised in order to extend them to the present geometrical situation. We get the following result.

Theorem 7.1 *Under the assumptions that u_0 is continuous, nonnegative, bounded and compactly supported, the interface $x = \eta(t)$ of the solution to problem (1.5),(1.2) is a Lipschitz continuous curve in the interval $0 < t < T_e$ minus the points of local extinction of a connected component of Ω, where it can have a jump discontinuity.*

The jump discontinuity can be avoided if per chance the connected component of $\Omega(t)$ situated next to the left was moving toward the local extinction point fast enough so as to arrive at the point precisely at the local extinction time. This is a non-generic phenomenon, so discontinuity is the common rule. If it does not happen, then discontinuity of the derivative is the expected case.

The same applies to the left interface and the inner interface. It happens also at every starting point of a new inner interface.

8 Extensions

The method we presented in the previous sections can be extended to a large class of nonlinear evolution equations of the form

$$u_t = \phi(u)_{xx} + \psi(u, u_x) \tag{8.1}$$

under suitable assumptions on the structure functions ϕ and ψ. In particular, ϕ can be a continuous nondecreasing function with $\phi(0) = 0$ and $\phi'(u) > 0$ for every $u > 0$. Certain conditions on ϕ and ψ ensure the existence of interfaces, [GK]. This type includes equation (1.5). Instead of concentrating on the maximal generality for which the method applies, because there are still some open gaps in our understanding, we will comment on some important particular cases. Some of these cases have been studied already and our method gives just an alternative and unified approach to such problems.

To begin with, the Porous Medium Equation, $u_t = (u^m)_{xx}$ falls under the scope of our method. The set of travelling waves is given explicitly by the formulae $\pi = c(ct - x)_+$. The waves with initial support in the half-line $x < 0$ exist only for $c > 0$, i.e., all waves are heating waves. For general solutions interfaces are only expanding or (temporarily) stationary. The results of Sections 4 to 7 apply easily to give the Lipschitz regularity of the interfaces and Darcy's interface equation, once the mention of cooling waves is eliminated. The original proofs are due to Caffarelli and Friedman, [CF]. Indeed, the analysis can be continued to prove that moving interfaces are analytic but this is out of our scope. The basic results are easily extended to the generalized porous medium (or filtration) equation

$$u_t = \phi(u)_{xx} \tag{8.2}$$

under suitable assumptions on ϕ like ϕ is a smooth and increasing function for $u > 0$ with $\phi(0) = 0$ and behaves in a suitable sublinear way as $u \to 0$, cf. [CF, DV1, GV2].

The results on the Lipschitz regularity and interface equation can be extended to equation (1.5) for the weak absorption case $p \geq 1$. The original proofs are performed in [HV] for $p \geq m$ and in [SV] for $1 < p < m$. More generally, they apply for a diffusion-reaction equation of the form

$$u_t = (u^m)_{xx} + f(u), \tag{8.3}$$

where $m > 1$ and f is a locally Lipschitz continuous function, not necessarily mono-tone, as studied in [SGKM]. The case $f(u) = +u^p$ with $p < 1$ is studied in [DV1, DV2]. Then there are interfaces only for $m + p \geq 2$ and uniqueness is not guaranteed, so additional concepts like minimal and maximal solutions have to be introduced. We can use the same approach for the diffusion-convection model

$$u_t = \phi(u)_{xx} + \psi(u)_x, \tag{8.4}$$

with suitable assumptions of ϕ and ψ. We refer to [MPS, Chapter 3]. This includes, in particular, conservation laws of the form

$$u_t = \psi(u)_x, \tag{8.5}$$

for which the propagation properties are well-known. In the power cases $\phi(u) = u^m$ and $\psi(u) = u^p$ our methods apply literally when $m > 1$ and $p \geq 1$. But for $p < 1$ a different analysis is needed, which bears a resemblance with the diffusion-absorption case $m + p < 2$ which we have left out of our present analysis. For the analysis of monotone interfaces of this convective problem we refer to [MPS, Chapter 3].

Another way of generalizing consists of replacing the nonlinear diffusion term $(u^m)_{xx}$ with more general terms like the p-laplacian and considers, for instance,

$$u_t = (|u_x|^{p-2}u_x)_x \tag{8.6}$$

with $p > 2$ so that interfaces arise, cf. [Ka, EV]. This extension has a complete formal resemblance but also presents regularity problems with which the analysis must be carefully deal.

The principle of intersection comparison is formulated for one-dimensional prob-lems. The extension of this technique to treat multi-dimensional problems is an important open problem.

9   Higher regularity

Once Lipschitz continuity is established, a natural question is to obtain higher reg-ularity, and in particular to determine whether there is a maximal regularity or the solutions and interfaces are C^∞ smooth, or even analytic, with respect to their vari-ables. The answer is well-known for the porous medium equation, cf. [An2] and its references. We will not discuss this problem here since it involves different techniques. Let us announce that for equation (1.5) in the critical case $m + p = 2$ and under the above assumptions on u_0 the interfaces are C^∞ smooth and the pressure is a C^∞ function in the positivity set and up to the free boundaries for $0 < t < T_e$, cf. [GSV2].

References

[An1] S. ANGENENT, The zero set of a solution of a parabolic equation, *J. reine angew. Math.* **390** (1988), 79–96.

[An2] S. ANGENENT, Analyticity of the interface of the porous media equation after the waiting time, Proc. Amer. Math. Soc. **102** (1988), 329-336.

[Ar1] D.G. ARONSON, Regularity properties of flows through porous media: The interface, *Arch. Rat. Anal. Mech.* **37** (1970), 1–10.

[Ar2] D.G. ARONSON, The porous medium equation, in "Nonlinear Diffusion Problems", Lecture Notes in Math. **1224**, A. Fasano and M. Primicerio eds., Springer Verlag, New York, 1986. Pp. 12–46.

[AB] D.G. ARONSON, PH. BÉNILAN, Régularité des solutions de l'équation des milieux poreux dans $\mathbf{R}^n$, *C. R. Acad. Sci. Paris. Ser. I* **288** (1979), 103–105.

[ACV] D.G. ARONSON, L.A. CAFFARELLI, J.L. VAZQUEZ, Interfaces with a corner-point in one-dimensional porous medium flow, *Comm. Pure Applied Math.* **38** (1985), 375–404.

[AV] D.G. ARONSON, J.L. VAZQUEZ, Eventual C^∞-regularity and concavity of flows in one-dimensional porous media, *Arch. Rat. Mech. Anal.* **99** (1987), 329–348.

[CF] L.A. CAFFARELLI, A. FRIEDMAN, Regularity of the free boundary of the one-dimensional flow of gas in a porous medium, *Amer. J. Math.* **101** (1979), 1193–1218.

[CMM] X.-Y. CHEN, H. MATANO, M. MIMURA, Finite-point extinction and continuity of interfaces in a nonlinear diffusion equation with strong absorption, *J. reine angew. Math.* **459** (1995), 1–36.

[DV1] A. DE PABLO, J.L. VAZQUEZ, Regularity of solutions and interfaces of a generalized porous medium equation, *Ann. Mat. Pura Applic. (IV)* **158** (1991), 51-74.

[DV2] A. DE PABLO, J.L. VAZQUEZ, Travelling waves and finite propagation for a reaction-diffusion equation, *J. Diff. Eqns.* **93** (1991), 19-61.

[EV] J.R. ESTEBAN, J.L. VAZQUEZ, Homogeneous diffusion in $\mathbf{R}$ with power-like diffusivity, *Arch. Rat. Mech. Anal.* **103** (1988), 39-80.

[F] A. FRIEDMAN, On the regularity of the solutions of nonlinear elliptic and parabolic systems of partial differential equations, *J. Math. Mech.* **7** (1985), 43–59.

[GSV] V.A. GALAKTIONOV, S. SHMAREV, J.L. VAZQUEZ, Regularity of interfaces in diffusion processes under the influence of strong absorption, *Preprint Univ. Autónoma Madrid*, 1997.

[GSV2] V.A. GALAKTIONOV, S. SHMAREV, J.L. VAZQUEZ, Higher regularity of interfaces in nonlinear diffusion equations, *work in preparation*.

[GV1] V.A. GALAKTIONOV, J.L. VAZQUEZ, Extinction for a quasilinear heat equation with absorption I. Technique of intersection comparison, *Comm. Partial. Differ. Equat.* **19** (1994), 1075–1106.

[GV2] V.A. GALAKTIONOV, J.L. VAZQUEZ, Geometrical properties of the solutions of one-dimensional nonlinear parabolic equations, *Math. Annalen* **303** (1995), 741–769.

[GV3] V.A. GALAKTIONOV, J.L. VAZQUEZ, Necessary and sufficient conditions for complete blowup and extinction for one-dimensional quasilinear heat equations, *Arch. Rat. Mech. Anal.* **129** (1995), 225–244.

[GK] B. GILDING, R. KERSNER, The characterization of reaction-convection-diffusion processes by travelling waves, *J. Diff. Eqns.* **124**, n.1 (1996), 27–79.

[HV] M.A. HERRERO, J.L. VAZQUEZ, The one-dimensional nonlinear heat equation with absorption: regularity of solutions and interfaces, *SIAM J. Math. Anal.* **18** (1987), 149–167.

[Ka] A.S. KALASHNIKOV, Some problems of the qualitative theory of non-linear, degenerate second-order parabolic equations, *Russian Math. Surveys* **42** (1987), 169–222.

[Ke] R. KERSNER, On the behaviour of temperature fronts in media with non-linear heat conductivity under absorption. *Moscow Univ. Math. Bull.* (translation of *Vestnik Moskov. Univ.*, Ser. I, *Mat. Mekh.*) **33** (1978), 35–41.

[KS] D. KINDERLEHRER, G. STAMPACCHIA, *An Introduction to Variational Inequalities and their Applications*, Academic Press, New York, 1980.

[Ma] H. MATANO, Nonincrease of the lap number for a one-dimensional semilinear parabolic equation, *J. Fac. Sci. Univ. Tokyo*, Sect. IA **29** (1982), 401-440.

[MPS] A.M. MEIRMANOV, V.V. PUKHNACHOV, S.I. SHMAREV, *Evolution Equations and Lagrangian Coordinates*, Walter de Gruyter, Berlin, 1997.

[RK] P. ROSENAU, S. KAMIN, Thermal waves in an absorbing and convecting medium. *Physica* **8D** (1983), 273–283.

[SGKM] A.A. SAMARSKII, V.A. GALAKTIONOV, S.P. KURDYUMOV, A.P. MIKHAILOV, *Blow-up in Quasilinear Parabolic Equations*, Walter de Gruyter, Berlin, 1995 (Russian ed.: Nauka, Moscow, 1987).

[Sa] D.H. SATTINGER, On the total variation of solutions of parabolic equations, *Math. Ann.* **183** (1969), 78-92.

[SV] S.I. SHMAREV, J.L. VAZQUEZ, The regularity of solutions of reaction-diffusion equations via Lagrangian coordinates, *Nonlin. Diff. Eqns and Appl.* **3** (1996), 465-497.

Victor A. Galaktionov
Dept. of Mathematical Sciences
University of Bath,
Bath BA2 7AY, UK

Sergei I. Shmarev
Lavrentiev Institute of Hydrodynamics
630090 Novosibirsk, RUSSIA and
Departamento de Matemáticas
Universidad de Oviedo
33007 Oviedo, SPAIN

Juan L. Vazquez
Departamento de Matemáticas,
Universidad Autónoma de Madrid
28049 Madrid, SPAIN

M. Guedda, D. Hilhorst and M. A. Peletier
Blow-Up of Interfaces for an Inhomogeneous Aquifer

1. Introduction

The evolution of the interface between salt and fresh water in underground aquifers can be modelled by the equation

$$\rho(x)u_t = \operatorname{div}(u(1-u)\nabla u) \text{ for } (x,t) \in \mathbf{R}^N \times \mathbf{R}^+, \tag{1.1}$$

in one and two space dimensions. The unknown u represents the height of the interface, scaled to vary between 0 and 1. The density function $\rho : \mathbf{R}^N \to \mathbf{R}^+$ is a combination of several coefficients: the porosity and permeability of the aquifer, the viscosity of the water and the density difference between the salt and fresh water. In Section 2 we give a derivation of the model that leads to this equation.

In [14] Van Duijn considers the equation

$$u_t - \big(u(1-u)\,u_x\big)_x = 0. \tag{1.2}$$

Because the diffusion coefficient $u(1-u)$ vanishes whenever $u=0$ or $u=1$, equation (1.2) exhibits behaviour very much similar to that of the classical porous medium equation $u_t = (u^m)_{xx}$; compactly supported disturbances of the equilibrium solutions $u \equiv 0$ and $u \equiv 1$ propagate with finite speed, giving rise to interfaces. One of the properties of these interfaces is their bounded speed; they move outward, but remain finite for all time. We shall see later that when a density function $\rho(x)$ is added as in (1.1) that declines to zero at a certain rate, this property is lost and interfaces become unbounded in finite time. The more general equation

$$u_t - \big(D(u)\phi(u_x)\big)_x = 0, \tag{1.3}$$

where the functions D and ϕ have properties somewhat closer to their physical equivalents in (2.8), was the subject of investigations by Bertsch, Van Duijn, Esteban, Hilhorst, and Zhang; they proved existence and uniqueness within a certain class of solutions for several types of problems, as well as the large-time asymptotic behaviour [15], and regularity properties of the solution and the interface [16], [3].

We are interested in the influence of the variation of ρ on the qualitative behaviour of solutions of (1.1). More specifically, when ρ tends to zero for large values of $|x|$, the large-time behaviour of the solutions and the interfaces can change fundamentally. We mention two cases of interest:

- If $\rho \in L^1(\mathbf{R}^N)$, then the solutions with finite mass do not decrease to zero as t tends to infinity, but converge to a constant value;

• For appropriate functions ρ, the interfaces between the sets $\{u = 0\}$, $\{0 < u < 1\}$ and $\{u = 1\}$ disappear in finite time.

In Section 3 we investigate these phenomena and give characterisations for the cases in which they occur. In the sequel we shall only sketch the proofs; the interested reader can find the details in [7].

2. Background of the problem

Consider a horizontal aquifer of even thickness h, confined between two impermeable layers, which will be represented by the domain $\Omega = \{(x, y, z) \in \mathbb{R}^3 : 0 < z < h\}$. The aquifer is fully saturated with water with a varying salt content, giving rise to spatial variations in the mass density δ of the water. The water is supposed to flow according to Darcy's law,

$$\frac{\mu}{\kappa} q + \nabla p + \delta g e_z = 0, \tag{2.1}$$

subject to the incompressibility condition

$$\text{div } q = 0. \tag{2.2}$$

Here q denotes the specific discharge and p the pressure; g is the acceleration of gravity, the parameter μ denotes the dynamic viscosity, and both are assumed constant; κ is the permeability, and is assumed dependent on the spatial variables x and y, but not on z and t (time). The unit vector e_z is supposed to point upward. Typically we are interested in the situation where the porosity ε varies in space, and especially where ε approaches zero for large $|x|$ and $|y|$. In such circumstances it is reasonable to expect κ to vary in space, too.

Let Ω contain a layer of salt water with constant density δ^s, underneath a layer of fresh water with density δ^f, with a sharp interface separating the two. We introduce the function $u : \mathbb{R}^2 \to [0, h]$, denoting the height of the interface, such that $\Omega \cap \{z < u(x, y)\}$ is filled with salt water and $\Omega \cap \{z > u(x, y)\}$ with fresh water. At every time t equations (2.1) and (2.2) determine the discharge field q, and from the evolution of the discharge one can deduce the evolution of the interface. Our goal is to derive an equation for u that describes the evolution of the interface but which does not explicitly refer to the discharge q. We set for brevity $\gamma = (\delta^s - \delta^f)g$.

A local analysis of the situation near the interface, as in [5], [1], or [8], shows that conservation of mass and continuity of the pressure p require certain conditions on q and the inclination angle of the interface. If we denote the discharge at the interface on the salt side and and on the fresh side by q^s and q^f, these conditions read

$$(q^s - q^f) \cdot \nu = 0, \quad (q^s - q^f) \cdot \tau = -\frac{\kappa\gamma}{\mu} \tau \cdot e_z, \quad (q^s - q^f) \cdot \sigma = 0, \tag{2.3}$$

where $\boldsymbol{\nu}$ is any vector normal to the interface, $\boldsymbol{\tau}$ is the unit vector tangential to the interface in the direction of steepest ascent, and $\boldsymbol{\sigma} = \boldsymbol{\nu} \times \boldsymbol{\tau}$. If u is differentiable, and if we denote the components of $\operatorname{grad} u$ by u_x and u_y, then $\boldsymbol{\nu}, \boldsymbol{\tau}, \boldsymbol{\sigma}$ are parallel to $(-u_x, -u_y, 1)^T, (u_x, u_y, u_x^2 + u_y^2)^T, (-u_y, u_x, 0)^T$, respectively.

We introduce a two-dimensional discharge $\boldsymbol{Q} = (Q_x, Q_y)^T$ with

$$Q_x(x,y) = \int_0^h \boldsymbol{q}(x,y,z) \cdot \boldsymbol{e}_x\, dz \quad \text{and} \quad Q_y(x,y) = \int_0^h \boldsymbol{q}(x,y,z) \cdot \boldsymbol{e}_y\, dz.$$

The incompressibility condition $\operatorname{div} \boldsymbol{Q} = 0$ (in which the divergence is taken in $\mathbf{R}^2$) is automatically satisfied, and by integrating Darcy's law over the height of the aquifer it follows that

$$\frac{\mu}{\kappa} \boldsymbol{Q} + \nabla P = 0 \quad \text{in} \quad \mathbf{R}^2,$$

where $P(x,y) = \int_0^h p(x,y,z)\, dz$ and where the gradient is taken in $\mathbf{R}^2$. We combine these two to find

$$\operatorname{div}\left(\kappa(x,y)\nabla P(x,y)\right) = 0 \quad \text{in} \quad \mathbf{R}^2.$$

With appropriate conditions on the behaviour of P at infinity, this implies $P \equiv 0$ and therefore, $\boldsymbol{Q} \equiv 0$ in $\mathbf{R}^2$.

It is common in the literature on hydrology to further simplify the problem by adopting the *Dupuit assumption*: within each flow region, fresh and salt, the x- and y-components of $\boldsymbol{q}$ are assumed to be independent of z, i.e., constant on vertical lines. If we denote these constant components by q_x^s and q_y^s for the salty region, and q_x^f and q_y^f for the fresh water region we can write

$$u q_x^s + (1-u) q_x^f = Q_x = 0, \tag{2.4}$$
$$u q_y^s + (1-u) q_y^f = Q_y = 0. \tag{2.5}$$

Solving (2.3), (2.4), and (2.5) for q_x^s and q_y^s we find

$$\begin{pmatrix} q_x^s \\ q_y^s \end{pmatrix} = \frac{\kappa \gamma}{\mu}(1-u)\frac{\nabla u}{1 + |\nabla u|^2}. \tag{2.6}$$

The motion of the interface is governed by the equation ([1], eq. (9.5.60))

$$\varepsilon \frac{\partial u}{\partial t} + \operatorname{div} \boldsymbol{Q}^s = 0, \tag{2.7}$$

where ε denotes the (space-dependent) porosity and $\boldsymbol{Q}^s = (Q_x^s, Q_y^s)^T$ is the total discharge of salt water:

$$Q_x^s(x,y) = \int_0^{u(x,y)} \boldsymbol{q}(x,y,z) \cdot \boldsymbol{e}_x\, dz \quad \text{and} \quad Q_y^s(x,y) = \int_0^{u(x,y)} \boldsymbol{q}(x,y,z) \cdot \boldsymbol{e}_y\, dz.$$

133

With the Dupuit assumption $Q_x^s = uq_x^s$ and $Q_y^s = uq_y^s$, and with equation (2.6) we can write equation (2.7) in the form

$$\varepsilon(x,y)\frac{\mu}{\gamma}\frac{\partial u}{\partial t} - \text{div}\left(\kappa(x,y)\,u(1-u)\frac{\nabla u}{1+|\nabla u|^2}\right) = 0. \qquad (2.8)$$

Since we shall mainly be interested in solutions u with relatively small gradients, we replace the quotient $\nabla u/(1+|\nabla u|^2)$ by ∇u. Furthermore, we shall only consider either one-dimensional or two-dimensional axially symmetric solutions, and therefore suppose that all data are either one-dimensional or axially symmetric. In the two-dimensional case, equation (2.8) reduces to

$$\varepsilon(r)\frac{\mu}{\gamma}u_t - \frac{1}{r}\left(r\kappa(r)\,u(1-u)u_r\right)_r = 0, \qquad (2.9)$$

where $r^2 = x^2 + y^2$ and subscripts denote differentiation. If we introduce a new space variable $\tilde{r}$, defined by

$$\tilde{r} \stackrel{\text{def}}{=} \int_0^r \frac{ds}{\kappa(s)},$$

then (2.9) takes the form

$$\rho(\tilde{r})u_t + \frac{1}{\tilde{r}}(\tilde{r}u(1-u)u_{\tilde{r}})_{\tilde{r}} = 0, \qquad (2.10)$$

in which $\rho(\tilde{r}) = (\mu/\gamma)\,\varepsilon(r)\kappa(r)$. We shall henceforth drop the tilde from $\tilde{r}$. In one space dimension, the equation becomes

$$\rho(x)u_t - \left(u(1-u)u_x\right)_x = 0. \qquad (2.11)$$

The derivation presented above is an extension of the one given in [8] to the case of constant coefficients ε and κ.

3. Statement of results

We will be concerned with the problem

$$\text{(P)}\left\{\begin{array}{ll} \rho(x)u_t = \text{div}(u(1-u)\nabla u) & \text{in } \mathbb{R}^N \times \mathbb{R}^+, \\ u(x,0) = u_0(x) & \text{for } x \in \mathbb{R}^N. \end{array}\right. \qquad (3.1)$$

The definition of a weak solution to (3.1) is similar to the one in [2] and can be found in [7]. For constant functions ρ this problem is well studied. When $\rho > 0$ is allowed to approach zero for large values of $|x|$, however, the qualitative behaviour of solutions can change fundamentally. The first result that we prove in this direction is the following:

THEOREM A (LARGE-TIME BEHAVIOUR). *Let N be equal to either one or two, $\rho \in C(\mathbf{R}^N) \cap L^\infty(\mathbf{R}^N), \rho > 0$ on $\mathbf{R}^N$, and let $u_0 \in C(\mathbf{R}^N), 0 \leq u_0 \leq 1$ and $\rho u_0 \in L^1(\mathbf{R}^N)$. Let u be the solution of Problem* (P) *with initial data u_0. Then*

$$u(t) \to \bar{u} \stackrel{def}{=} \frac{\displaystyle\int_{\mathbf{R}^N} \rho(x) u_0(x)\, dx}{\displaystyle\int_{\mathbf{R}^N} \rho(x)\, dx} \quad as \quad t \to \infty,$$

uniformly on compact subsets of $\mathbf{R}^N$.

This result clearly generalized the case of constant ρ, in which a solution with finite initial mass decays to zero as t tends to infinity.

A result which now is classical states that when ρ is constant, solutions of Problem (P) exhibit so-called interfaces, curve in the x,t-plane separating the region $\{u = 0\}, \{0 < u < 1\}$, and $\{u = 1\}$, provided that the initial condition has an appropriate form. When ρ is supposed positive but no longer constant, the same holds for small time; however, it is possible that the interfaces disappear in finite time. this was first pointed out for a different equation by Kamin and Kersner [9]. We wish to investigate the behaviour of solutions of Problem (P) when ρ tends to zero at infinity.

For reasons of clarity we restrict ourselves to the interfaces between the regions $\{u = 0\}$ and $\{u > 0\}$. It follows from the inversion $\tilde{u} = 1 - u$ that all arguments can be transposed in the interface between the regions $\{u = 1\}$ and $\{u < 1\}$. A solution u of (P) for $N = 1$ is said to exhibit finite time blow-up if its support is bounded from above initially and there exists a time T such that supp $u(t)$ is unbounded from above for all time $t > T$. For the formulation of Theorem B we shall need an auxilliary density function σ defined by

$$\sigma(x) = \min\{\rho(\xi) : 0 \leq \xi \leq x\},$$

the reason being that the function σ is monotonic while ρ need not be.

THEOREM B (BLOW-UP IN ONE DIMENSION). *Let $\rho \in C^1(\mathbf{R}) \cap L^\infty(\mathbf{R}), \rho > 0$ on $\mathbf{R}$, and let $u_0 \in L^\infty(\mathbf{R}), 0 \leq u_0 \leq 1$. Let u be a solution of* (3.1) *for $N = 1$, with initial data u_0, such that the support of u is bounded from above at time $t = 0$. Then the following implications hold:*

1. $\displaystyle\int_0^\infty x\rho(x)\, dx < \infty \implies$ *finite time blow-up;*

2. $\displaystyle\int_0^\infty x\sigma(x)\, dx = \infty \implies$ *no finite time blow-up.*

It follows immediatly that

COROLLARY A. *If $\rho(x)$ is non-increasing function of x for $0 < x < \infty$,*

then $\int_0^\infty x\rho(x)\,dx < \infty \iff$ *finite time blow-up.*

Note that the behaviour of ρ and u_0 toward $-\infty$ has no influence on the (qualitative) behaviour of the upper boundary of the support. We can apply this theorem once for $\{x > 0\}$ and once for $\{x < 0\}$ with independent results.

Using the Comparison Principle we can extend this result to a statement on a strip $\Omega = \mathbb{R} \times (-1, 1)$ with Neumann boundary conditions, with a density function ρ that does not depend on the vertical coordinate $\rho(x, y) = \rho(x)$ on Ω.

Consider the problem

$$\begin{cases} \rho u_t = \operatorname{div}(u(1-u)\nabla u) & \text{in } Q_T = \Omega \times [0, T], \\ \frac{\partial u}{\partial \nu} = 0 & \text{on } \partial\Omega \times [0, T], \\ u = u_0 & \text{at } t = 0. \end{cases} \tag{3.2}$$

We prove the following result.

THEOREM C (BLOW-UP IN A 2D STRIP). *Let the initial condition u_0 be such that $u_0(x, y) = 1$ for small x and $u_0(x, y) = 0$ for large x. Let $\gamma_0(t)$ denote the interface between $\{u > 0\}$ and $\{u = 0\}$ at time t:*

$$\gamma_0(t) = \operatorname{supp} u(t) \cap \{(x, y) \in \Omega : u(x, y, t) = 0\}.$$

Then the following statements hold:

1. *If $\int_0^\infty x\rho(x)\,dx < \infty$, then the interface γ_0 will run off to infinity in finite time;*

2. *If $\int_0^\infty x\sigma(x)\,dx = \infty$, then the interface γ_0 will remain bounded for all finite time.*

A similar statement holds for the interface γ_1 between the sets $\{u = 1\}$ and $\{u < 1\}$.

A different way of extrapolating the one-dimensional results is by considering the two-dimensional radially symmetric problem and transforming the ensuing (one-dimensional) equation to an equation of the form (1.1). In this case the auxiliary density function σ is different:

$$\sigma(r) = \min_{0 \le \xi \le r} \xi^2 \rho(\xi).$$

We establish the following result:

THEOREM D (BLOW-UP IN 2D, RADIALLY SYMMETRIC). *Let u be a solution of (1.1) with initial condition u_0. Suppose that both ρ and u_0 are radially symmetric, and that supp u_0 is compact.*

1. *If $\displaystyle\int_1^\infty \rho(r)r\log r\, dr < \infty$ and $0 \in Int(supp\, u_0)$, then the support of u ceases to be compact in finite time;*

2. *If $\displaystyle\int_1^\infty \sigma(r)\frac{\log r}{r}\, dr = \infty$, then the support of u will be compact for all time.*

References

[1] J. BEAR. *Dynamics of Fluids in Porous Media.* Dover Publications, Inc, New York, 1972.

[2] M. BERTSCH, R. KERSNER, AND L. A. PELETIER. Positivity versus localization in degenerate diffusion equations. *Nonlinear Analysis, Theory, Methods & Applications*, 9(9):987–1008, 1985.

[3] M. BERTSCH, C. J. VAN DUIJN, J. R. ESTEBAN, AND HONGFEI ZHANG. Regularity of the free boundary in a doubly degenerate parabolic equation. *Comm. in Partial Diff. Eqns.*, 14(3):391–412, 1989.

[4] E. DIBENEDETTO AND V. VESPRI. On the singular equation $\beta(u)_t = \Delta u$. Technical Report 1233, IMA, 1994.

[5] J. H. EDELMAN. *Over de berekening van grondwaterstromingen.* PhD thesis, Technische Hogeschool Delft, 1947.

[6] D. EIDUS. The Cauchy problem for the non-linear filtration equation in an inhomogeneous medium. *J. Diff. Eqns.*, 84:309–318, 1990.

[7] M. GUEDDA, D. HILHORST AND L.A. PELETIER. Disappearing Interfaces in Nonlinear Diffusion. to appear in *Advances in Mathematical Sciences and Applications*.

[8] G. DE JOSSELIN DE JONG. The simultaneous flow of fresh and salt water in aquifers of large horizontal extension determined by shear flow and vortex theory. In A. Verruijt and F. B. J. Barends, editors, *Proceedings of Euromech 143*, pages 75–82, Rotterdam, 1981.

[9] S. KAMIN AND R. KERSNER. Disappearance of interfaces in finite time. *Meccanica*, 28:117–120, 1993.

[10] S. KAMIN AND P. ROSENAU. Propagation of thermal waves in an inhomogeneous medium. *Comm. Pure Appl. Math.*, 34:831–852, 1981.

[11] S. KAMIN AND P. ROSENAU. Non-linear diffusion in a finite mass medium. *Comm. Pure Appl. Math.*, 35:113–127, 1982.

[12] O. A. LADYŽENSKAJA, V. A. SOLONNIKOV, AND N. N. URAL'CEVA. *Linear and Quasi-linear Equations of Parabolic Type*, volume 23 of *Translations of Mathematical Monographs*. American Mathematical Society, 1968.

[13] M. A. PELETIER. A supersolution for the porous media equation with nonuniform density. *Appl. Math. Lett.*, 7(3):29–32, 1994.

[14] C. J. VAN DUIJN. On the diffusion of inmiscible fluids in porous media. *SIAM J. Math. Anal.*, 10(3):486–497, 1979.

[15] C. J. VAN DUIJN AND D. HILHORST. On a doubly nonlinear diffusion equation in hydrology. *Nonlinear Analysis, Theory, Methods & Applications*, 11(3):305–333, 1987.

[16] C. J. VAN DUIJN AND HONGFEI ZHANG. Regularity properties of a doubly degenerate equation in hydrology. *Comm. in Partial Diff. Eqns.*, 13(3):261–319, 1988.

M. Guedda
Faculté de Mathématiques et d'informatique
Univesité de Picardie Jules Verne
33, rue Saint-Leu 80039 Amiens, France

D. Hilhorst
Université de Paris Sud, Laboratoire d'Analyse Numérique d'Orsay
Bâtiment 425, 91405, Orsay France

M. A. Peletier
Delft University of Technology, P.O. Box 5031
2600 GA, Delft, The Netherlands

B. GUSTAFSSON

An Exponential Transform and Regularity of Free Boundaries in Two Dimensions

1   Introduction

In this paper, which is entirely based on joint work [G-P] with Mihai Putinar, we present a new method for proving regularity of free boundaries in obstacle type problems in two dimensions, in the case of analytic data. Complete details on all matters discussed here are given in [G-P].

Let us state the problem first for any number $N \geq 2$ dimensions. Let $\Omega \subset \mathbf{R}^N$ be a bounded open set. The gravitational, or Newtonian, field produced by Ω considered as a body with density one is

$$\widehat{\Omega}(x) = c_N \int\limits_{\Omega} \frac{x - y}{|x - y|^N} \, dy$$

for an appropriate constant c_N, so that $-\text{div}\,\widehat{\Omega} = \chi_\Omega$, $\text{curl}\,\widehat{\Omega} = 0$. In particular, $\widehat{\Omega}$ is a harmonic vector field (i.e., is both divergence and curl free) outside $\overline{\Omega}$.

It is a simple consequence of the Cauchy-Kovalevskaya theorem that if part of $\partial\Omega$ is a real analytic hypersurface, then the exterior field $\widehat{\Omega}\big|_{\overline{\Omega}^c}$ has a continuation as a harmonic vector field across that part of $\partial\Omega$ into Ω. Our main concern is the problem of proving a converse statement:

Regularity Problem. Let $x_0 \in \partial\Omega$, $r > 0$ and assume there exists a harmonic vector field f in the ball $B = B_r(x_0)$ such that

$$\widehat{\Omega} = f \qquad \text{on} \qquad B \setminus \Omega.$$

Prove that $\partial\Omega \cap B$ is contained in a real analytic variety and describe all possible types of singularities of it.

This problem is closely related to the corresponding regularity problem for the obstacle problem. Indeed, in the latter problem the set Ω (then called the "noncoincidence set") is of the form $\Omega = \{u > 0\}$ for some function $u \geq 0$ satisfying $\Delta u = \chi_\Omega$ in a neighbourhood of $\partial\Omega$. In particular, $u \in C^{1,\alpha}$ for all $0 < \alpha < 1$ and both u and ∇u vanish on Ω^c (∇u because u attains its minimum on Ω^c).

It follows that $\widehat{\Omega} + \nabla u = \widehat{\Omega}$ outside Ω, $\text{div}\left(\widehat{\Omega} + \nabla u\right) = -\chi_\Omega + \nabla u = 0$ and $\text{curl}\,\widehat{\Omega} = 0$ everywhere. Hence

$$f = \widehat{\Omega} + \nabla u \tag{1.1}$$

provides the continuation of $\widehat{\Omega}$ as a harmonic vector field.

Thus, if Ω is the noncoincidence set of an obstacle problem, then Ω admits continuation of the gravitational field. The converse is not always true: if $\widehat{\Omega}$ has a continuation f, then trying to solve (1.1) for u it may come out that u is not zero (or any other fixed constant) on Ω^c, and even if it is, it may become negative in parts of Ω.

To our knowledge, the regularity problem is not completely solved when $N \geq 3$, even if Ω happens to be the noncoincidence set for an obstacle problem. The best known results, due to D. Kinderlehrer, L. Nirenberg, L. Caffarelli and others (see [F, Ro, Ca1, Ca2] for overviews and more references), state for the obstacle problem, that $\partial\Omega$ is a real analytic hypersurface at x_0 provided Ω^c satisfies a certain thickness condition there (which is fulfilled, e.g., if Ω^c contains a cone at x_0). On the other hand, even in two dimension it is known that (analytic) singularities of $\partial\Omega$ may occur at points where the thickness condition is not satisfied [Sch, Sa1, Sa2].

In two dimensions the regularity problem, even in a formulation weaker than stated above, has been completely solved by M. Sakai [Sa1]: $\partial\Omega$ is a real analytic set, smooth except for a few explicit types of possible analytic singularities, namely, inward cusps, double points, isolated points and certain degenerate cases.

Sakai's proof is long and technical. The difficulty lies in excluding the possibility that infinitely many components of Ω^c cluster at x_0, except in the trivial way that they all lie on an analytic arc passing through x_0.

Here we shall introduce a new tool in regularity theory, an exponential transform originally created by Carey and Pincus [C-Pi, M-P] in an operator theoretic context. The exponential transform provides an elegant and conceptually simple way of solving the regularity problem in two dimensions. Indeed, it directly provides a good defining function for the free boundary.

The connection between the exponential transform and topics close to free boundary problems (e.g., quadrature domains) was discovered by M. Putinar in a series of papers [P1, P2, P3].

2   Definition and elementary properties of the exponential transform

From now on we stick to two dimensions and identify $\mathbf{R}^2$ with the complex plane $\mathbf{C}$. Moreover, we modify the definition of $\widehat{\Omega}$ by a complex conjugation so that it coincides with the Cauchy transform of χ_Ω. Thus we set, for any bounded open set $\Omega \subset \mathbf{C}$,

$$\widehat{\Omega}(z) = -\frac{1}{\pi} \int_\Omega \frac{dA(\zeta)}{\zeta - z},$$

$dA = dx\,dy$ denoting area measure. Then

$$\overline{\partial}\widehat{\Omega} = \chi_\Omega \quad \text{in} \quad \mathbf{C} \tag{2.1}$$

140

in the sense of distributions $\left(\bar{\partial} = \bar{\partial}_z = \partial/\partial\bar{z} = \frac{1}{2}\left(\frac{\partial}{\partial x} + i\frac{\partial}{\partial y}\right)\right)$.

Definition 2.1 *The* exponential transform *of Ω is the function $E : \mathbb{C} \times \mathbb{C} \to \mathbb{C}$ defined by*

$$E(z, w) = \exp\left[-\frac{1}{\pi}\int_\Omega \frac{dA(\zeta)}{(\zeta - z)(\bar{\zeta} - \bar{w})}\right].$$

We use the convention that $\exp[-\infty] = 0$.

Some elementary properties of the exponential transform are

- $E(z, w)$ is analytic in z for $z \notin \overline{\Omega}$, antianalytic in w for $w \notin \overline{\Omega}$.

- $E(z, w) = 1 - \frac{|\Omega|/\pi}{z\bar{w}} +$ smaller terms, as $|z|, |w| \to \infty$.

- $E(z, w) = 1 - \frac{\hat{\Omega}(z)}{\bar{w}} +$ smaller terms, as $|w| \to \infty$ for fixed z. Thus $\hat{\Omega}(z)$ is a derivative at infinity with respect to w of $E(z, w)$.

- $E_{a\Omega+b}(az + b, aw + b) = E_\Omega(z, w)$ for any $a, b \in \mathbb{C}$, $a \neq 0$.

- $|E(z, w)| \leq 2$. Equality is attained only if Ω is a disc (up to null sets) and z, w diametrically opposite points on the boundary.

Example: For the unit disc $B = B_1(0)$ we have

$$E(z, w) = \begin{cases} 1 - \frac{1}{z\bar{w}} & , \quad z, w \notin \overline{B}, \\ 1 - \frac{z}{\bar{w}} & , \quad z \in B, w \notin \overline{B}, \\ 1 - \frac{w}{z} & , \quad z \notin \overline{B}, w \in B, \\ \frac{|z-w|^2}{1-z\bar{w}} & , \quad z, w \in B. \end{cases} \tag{2.2}$$

Investigation of the above expressions shows that $E(z, w)$, for the unit disc, is continuous everywhere. This is almost true in general: for any bounded open $\Omega \subset \mathbb{C}$, $E(z, w)$ is

- continuous in each variable separately,

- jointly continuous except at points (z, z) with z in

$$Z = \left\{z \in \partial\Omega : \int_\Omega \frac{dA(\zeta)}{|\zeta - z|^2} < \infty\right\}. \tag{2.3}$$

Clearly, Z consists of those points (if any) on $\partial\Omega$ at which Ω is "thin", e.g., has a sharp outward cusp.

The general structure of $E(z,w)$ as to analyticity–antianalyticity is the same as that for the disc, namely,

$$
E(z,w) = \begin{cases}
F_1(z,\bar{w}) & \text{in} & \overline{\Omega}^c \times \overline{\Omega}^c, \\
(\bar{z} - \bar{w})F_2(z,\bar{w}) & \text{in} & \Omega \times \overline{\Omega}^c, \\
(z - w)F_3(z,\bar{w}) & \text{in} & \overline{\Omega}^c \times \Omega, \\
|z - w|^2 F_4(z,\bar{w}) & \text{in} & \Omega \times \Omega,
\end{cases}
$$

where F_1, F_2, F_3, F_4 are functions which are analytic in their arguments.

3 Analytic continuation properties

Let $\Omega \subset \mathbb{C}$ be open and bounded. Our main result has a very simple formulation: if $\widehat{\Omega}$ has an analytic continuation from Ω^c across $\partial\Omega$, then the same is true for E in both variables (analytic-antianalytic continuation). Precisely

Theorem 3.1 *Let $z_0 \in \partial\Omega$, $r > 0$ and assume there exists $f(z)$ analytic in $B = B_r(z_0)$ such that $\widehat{\Omega}(z) = f(z)$ for $z \in B\setminus\Omega$. Then there exists $F(z,\bar{w})$, defined in $B \times B$, analytic in z and antianalytic in w such that*

$$
E(z,w) = F(z,\bar{w}) \qquad \text{for} \qquad z, w \in B\setminus\Omega.
$$

In addition,

$$
\partial\Omega \cap B \subset \{z \in B : F(z,\bar{z}) = 0\}. \tag{3.1}
$$

The important point for free boundaries is the last issue. It says that under the assumptions of the theorem, the free boundary is contained in a real-analytic variety. With this regularity at hand the more detailed analysis of possible singular points on the boundary is relatively easy.

Idea of proof of the theorem: One needs to perform an analytic continuation. Formally this is rather easy. Set

$$
S(z) = \bar{z} - \widehat{\Omega}(z) + f(z) \qquad \text{for} \qquad z \in B. \tag{3.2}
$$

Then $S(z)$ is continuous in B, analytic in $\Omega \cap B$ (by (2.1)) and

$$
S(z) = \bar{z} \qquad \text{for} \qquad z \in B\setminus\Omega. \tag{3.3}
$$

Thus $S(z)$ is the *Schwarz function* [D, Sh] for $\partial\Omega$ (here known to be analytic only on one side of $\partial\Omega$ and for convenience set equal to $\bar{z}$ on all $B\setminus\Omega$).

Assuming temporarily some smoothness of $\partial\Omega$, Stokes' formula can be applied to give, for $z, w \in B \setminus \overline{\Omega}$,

$$
E(z, w) = \exp\left[-\frac{1}{2\pi i} \int_\Omega \frac{d\bar\zeta\, d\zeta}{(\bar\zeta - \bar w)(\zeta - z)} \right] \tag{3.4}
$$

$$
= \exp\left[-\frac{1}{2\pi i} \int_{\partial\Omega} \log(\bar\zeta - \bar w)\, \frac{d\zeta}{\zeta - z} \right]
$$

$$
= \exp\left[-\frac{1}{2\pi i} \int_{\partial\Omega} \log(S(\zeta) - \bar w)\, \frac{d\zeta}{\zeta - z} \right].
$$

In the last member an analytic function (in $\Omega \cap B$) is integrated along $\partial\Omega$. Hence the part of the contour $\partial\Omega$ which is inside B can be moved inward a little bit. This shows that as a function of z for fixed $w \notin \overline{\Omega}$, $E(z, w)$ extends analytically to $\overline{\Omega}^c \cup B$.

This is the start of the analytic continuation process. Proceeding in similar ways and also invoking Hartog's theorem one gets a continuation to all of $B \times B$, and one even gets a formula for the continuation, namely,

$$
F(z, \bar w) = \frac{z - \overline{S(w)}}{z - w} \cdot \frac{S(z) - \bar w}{\bar z - \bar w} \cdot E(z, w), \tag{3.5}
$$

suitably interpreted on the diagonal $(z = w)$.

To do everything stringent a couple of technical tools are needed.

Lemma 1 (Caffarelli [Ca1], Karp, Margulis [K-M], Shahgholian) *Under the assumptions of the theorem, $\partial\Omega \cap B$ has area measure zero. Moreover, $Z = \varnothing$ (see (2.3)).*

This lemma enables us to assume that $\Omega = \operatorname{int}(\overline{\Omega})$, i.e., to assume that each point of $\partial\Omega$ is accessible from $\overline{\Omega}^c$. Moreover, it shows (last statement) that $E(z, z) = 0$ for all $z \in \partial\Omega$.

Lemma 2 (Ahlfors [Af], Bers [Be]) *For any open set $\Omega \subset \mathbf{R}^2$ (or $\mathbf{R}^N$) there exist $\psi_n \in C_c^\infty(\Omega)$ with $0 \le \psi_n \nearrow \chi_\Omega$ pointwise as $n \to \infty$ and satisfying*

$$
|\nabla\psi_n(z)| \le \frac{C}{n d(x) \log \frac{1}{d(x)}}
$$

for $0 < d(z) < e^{-2}$, $d(z)$ denoting the distance from $z \in \Omega$ to Ω^c.

Under the hypotheses of the theorem and with $S(z)$ defined by (3.2), Lemma 2 together with standard estimates for the Cauchy transform shows that

$$
\left\| (S(z) - \bar z)\, \bar\partial\psi_n(z) \right\|_\infty \le \frac{C}{n}. \tag{3.6}
$$

From this the use of Stokes' formula in (3.4) can be justified (or avoided) with no smoothness assumption on $\partial\Omega$ whatsoever. One simply approximates χ_Ω by ψ_n and uses partial integration in full space.

Alternatively, one may use (3.6) to check directly the analyticity of the function F in (3.5) by computing the $\partial/\partial\bar{z}$ and $\partial/\partial w$ derivatives. The reason (3.6) is needed is that the product rule for executing the derivatives is not directly applicable to (3.5). Therefore (3.5) has to be approximated and rewritten in such a way that the different factors have disjoint supports.

The above indicates how the existence and analyticity of F follow using Lemma 2. Then (3.1) follows by remarks after Lemma 1, since both E and F are continuous and agree outside $\overline{\Omega}$. This finishes the outline of the proof of the theorem. □

4 Final remarks

We first mention a couple of open questions.

(i) Generalization to higher dimensions: is there any analogue of the exponential transform which is useful for free boundary problems in $\mathbb{R}^N$, $N \geq 3$?

(ii) Can the exponential transform be adapted to problems with nonanalytic data, or to problems governed by more general elliptic operators than the Laplace operator?

(iii) Is the exponential transform useful for evolution problems ($\Omega = \Omega(t)$), e.g., for Hele Shaw flows or Stokes flow?

So far, the exponential transform has been explicitly computed only for a few classes of domains. These include ellipses and quadrature domains. For $\Omega = \{z \in \mathbb{C} : p(z, \bar{z}) > 0\}$ an ellipse, where p is a second degree polynomial, we have

$$E(z, w) = \text{const.} \frac{|z - w|^2}{p(z, \bar{w})},$$

for $z, w \in \Omega$ (cf. (2.2)).

A quadrature domain [Sh] may be defined as a domain $\Omega \subset \mathbb{C}$ such that $\widehat{\Omega}$ on Ω^c agrees with a rational function. Such domains are known to have algebraic boundaries. Writing thus $\partial\Omega = \{z \in \mathbb{C} : Q(z, \bar{z}) = 0\}$ and $\widehat{\Omega}(z) = R(z)/P(z)$, $z \in \Omega^c$, where $Q(z, w)$, $R(z)$, $P(z)$ are polynomials of smallest possible degrees and with leading coefficients one, we have [P1]

$$E(z, w) = \frac{Q(z, \bar{w})}{P(z)\overline{P(w)}}.$$

144

for $z, w \in \Omega^c$. Note that this agrees well with our main theorem: $E(z, w)$ extends analytically into Ω as far as $\widehat{\Omega}$ does, namely, to the zeros of P in the present case.

It would be interesting to compute $E(z, w)$ for some domain which has corners on the boundary, e.g., for a polygon. Preliminary investigations, however, indicate that there is no simple expression for $E(z, w)$ in such cases.

References

[Af]   L.V. AHLFORS. *Finitely generated Kleinian groups*, Amer. J. Math. **86** (1964), 413–429.

[Be]   L. BERS. *An approximation theorem*, J. Analyse Math. **14** (1965), 1–4.

[Ca1] L. CAFFARELLI. *Compactness methods in free boundary problems*, Comm. Partial Diff. Eq. **5** (1980), 427–448.

[Ca2] L.A. CAFFARELLI. Free boundary problems, a survey, in *Topics in calculus of variations*, Lect. Notes Math. **1365** (1989), 31–61.

[C-Pi] R.W. CAREY AND J.D. PINCUS. *An exponential formula for determining functions*, Indiana Math. Univ. J. **23** (1974), 1031–1042.

[D]   P.J. DAVIS. *The Schwarz Function and its Applications*, Carus Mathematical Monographs, Math. Assoc. Amer., 1974.

[F]   A. FRIEDMAN. *Variational Principles and Free Boundary Problems*, Wiley, New York, 1982.

[G-P] B. GUSTAFSSON AND M. PUTINAR. *An exponential transform and regularity of free boundaries in two dimensions*, Ann. Scuola Norm. Sup. Pisa Cl. Sci. (to appear).

[K-M] L. KARP AND A. MARGULIS. *On the Newtonian potential theory for unbounded sources and its applications to free boundary problems*, J. Analyse Math. **70** (1996), 1–63.

[M-P] M. MARTIN AND M. PUTINAR. *Lectures on Hyponormal Operators*, Birkhäuser, Basel, 1989.

[P1]   M. PUTINAR. *Extremal solutions of the two-dimensional L-problem of moments*, J. Funct. Anal. **136** (1996), 331–364.

[P2]   M. PUTINAR. *Linear analysis of quadrature domains*, Ark. Mat. **33** (1995), 357–376.

[P3]   M. PUTINAR. *Extremal solutions of the two-dimensional L-problem of moments II*, J. Approx. Theory (to appear).

[Ro]   J.F. RODRIGUES. *Obstacle problems in Mathematical Physics*, North-Holland, Amsterdam 1987.

[Sa1] M. SAKAI. *Regularity of boundary having a Schwarz function*, Acta Math. **166** (1991), 263–297.

[Sa2] M. SAKAI. *Regularity of free boundaries in two dimensions*, Ann. Scuola Norm Sup. Pisa Cl. Sci. (4) **20** (1993), 323–339.

[Sa3] M. SAKAI. *Regularity of boundaries of quadrature domains in two dimensions*, SIAM J. Math. Anal. **24** (1994), 341–364.

[Sch] D.G. SCHAEFFER. *Some examples of singularities in a free boundary*, Ann. Scuola Norm. Sup. Pisa **4** (1977), 131–144.

[Sh]   H.S. SHAPIRO. *The Schwarz Function and Its Generalization to Higher Dimensions*, Wiley, New York, University of Arkansas Lecture Notes, 1992.

Björn Gustafsson
Department of Mathematics
Royal Institute of Technology
S-100 44 Stockholm, Sweden

N. Kenmochi and M. Kubo

Well-Posedness for a Phase-Field Model with Constraint

Abstract

A phase-field model with constraint of the Penrose-Fife type is considered. The model is a system of nonlinear parabolic PDEs which governs the (absolute) temperature and order parameter with bounded constraint. In this paper, the well-posedness of the system is shown in the product space of the dual of $H^1(\Omega)$ and $L^2(\Omega)$.

1   Introduction

In this paper we discuss the well-posedness of the following phase-field model with constraint:

$$[\theta + \lambda(w)]_t - \Delta\tilde{\alpha} = f(x,t), \quad \tilde{\alpha} \in \alpha(\theta), \quad \text{in } Q_s := \Omega \times (s,T), \tag{1.1}$$

$$w_t - \kappa\Delta w + \tilde{\beta} + g(w) - \tilde{\alpha}\lambda'(w) = 0, \quad \tilde{\beta} \in \beta(w), \quad \text{in } Q_s, \tag{1.2}$$

$$\frac{\partial\tilde{\alpha}}{\partial n} + n_o\tilde{\alpha} = h(x,t), \quad \frac{\partial w}{\partial n} = 0 \quad \text{on } \Sigma_s := \Gamma \times (s,T), \tag{1.3}$$

$$\theta(\cdot,s) = \theta_o, \quad w(\cdot,s) = w_o \quad \text{in } \Omega. \tag{1.4}$$

Here $0 < T < +\infty$ is fixed, s is any number in $[0,T)$ and Ω is a bounded domain in $\mathbf{R}^N$, $1 \leq N \leq 3$, with smooth boundary $\Gamma := \partial\Omega$; α and β are maximal monotone graphs in $\mathbf{R} \times \mathbf{R}$; λ and g are smooth functions on $\mathbf{R}$; κ and n_o are positive constants; f, h, θ_o and w_o are prescribed as data.

This system generalizes various models treated so far; for instance, the models in [1-3, 9, 10], which are derived from non-differentiable free energy functionals. In these papers the existence and uniqueness of a solution were proved only for a restricted class of α and for good initial data θ_o, w_o. But, in the discussion on the large-time behaviour of the evolution operator $E(t,s)$ associated to system (1.1)-(1.3) from the viewpoint of global attractors (cf. [7, 11]), the structure of the free energy functionals suggests us that the operator $E(t,s)$ should be defined on a much bigger set than the class of initial data in which the existence and uniqueness of a solution are guaranteed and it should still have the continuity property in a weak sense.

In this paper we represent the well-posedness of system (1.1)-(1.4) in the product space of the dual of $H^1(\Omega)$ and $L^2(\Omega)$. To this end let us introduce some operators and function spaces.

Let V be the Sobolev space $H^1(\Omega)$ with norm

$$|v|_V := \left\{ \int_\Omega |\nabla v|^2 dx + n_0 \int_\Gamma |v|^2 d\Gamma \right\}^{\frac{1}{2}}, \quad \forall v \in V,$$

V^* be the dual space of V and F be the duality mapping from V onto V^*, namely,

$$\langle Fv, z \rangle := \int_\Omega \nabla v \cdot \nabla z \, dx + n_0 \int_\Gamma vz \, d\Gamma, \quad \forall v, \forall z \in V,$$

where $\langle \cdot, \cdot \rangle$ denotes the duality pairing between V^* and V.

Given $f \in L^2(\Omega)$ and $h \in L^2(\Gamma)$, an element $f^* \in V^*$ is uniquely determined by

$$\langle f^*, z \rangle := \int_\Omega fz \, dx + \int_\Gamma hz \, d\Gamma, \quad \forall z \in V, \tag{1.5}$$

and it is easy to check that $Fv = f^*$ is formally equivalent to

$$-\Delta v = f \text{ in } \Omega, \quad \frac{\partial v}{\partial n} + n_0 v = h \text{ on } \Gamma; \tag{1.6}$$

in fact, (1.6) is satisfied in the variational sense that

$$\int_\Omega \nabla v \cdot \nabla z \, dx + n_0 \int_\Gamma vz \, d\Gamma = \int_\Omega fz \, dx + \int_\Gamma hz \, d\Gamma \ (= \langle f^*, z \rangle), \quad \forall z \in V.$$

By notation Δ_N we denote the Laplacian, with homogeneous Neumann boundary condition, in $L^2(\Omega)$ more precisely,

$$D(\Delta_N) = \left\{ z \in H^2(\Omega) \ \middle| \ \frac{\partial z}{\partial n} = 0 \text{ in } H^{\frac{1}{2}}(\Gamma) \right\}$$

and

$$\Delta_N z = \Delta z \text{ a.e. in } \Omega \text{ for any } z \in D(\Delta_N).$$

It is well-known that $-\Delta_N$ is single valued and maximal monotone in $L^2(\Omega)$.

2 Weak formulation

We give the precise assumptions on the data α, β, λ, g, f and h:

- α is a maximal monotone graph in $\mathbf{R} \times \mathbf{R}$; we fix a proper l.s.c. convex function $\hat{\alpha}$ on $\mathbf{R}$ whose subdifferential $\partial\hat{\alpha}$ coincides with α in $\mathbf{R}$.

- β is a maximal monotone graph in $\mathbf{R} \times \mathbf{R}$ such that the domain $D(\beta)$ is bounded in $\mathbf{R}$, having non-empty interior, say $\overline{D(\beta)} = [\sigma_*, \sigma^*]$ for some constants σ_* and σ^* with $\sigma_* < \sigma^*$; we fix a non-negative proper l.s.c. convex function $\hat{\beta}$ on $\mathbf{R}$ whose subdifferential $\partial\hat{\beta}$ coincides with β.

- λ is of C^1-class on $\mathbf{R}$ and its derivative λ' is locally Lipschitz continuous on $\mathbf{R}$.

- g is a locally Lipschitz continuous function on $\mathbf{R}$.

148

- $f \in L^2(0,T; L^2(\Omega))$ and $h \in L^2(0,T; L^2(\Gamma))$.

Now, we easily see that with a new variable $e := \theta + \lambda(w)$, system (1.1)-(1.3) can be written in the form:

$$e'(t) + F\tilde{\alpha}(t) = f^*(t) \quad \text{in } V^*, \text{ a.e. } t \in [s,T], \tag{2.1}$$

$$w'(t) - \kappa\Delta_N w(t) + \tilde{\beta}(t) + g(w(t)) - \tilde{\alpha}(t)\lambda'(w(t)) = 0 \quad \text{in } L^2(\Omega), \text{ a.e. } t \in [s,T], \tag{2.2}$$

where $f^*(t)$ is the element of V^* determined by (1.5) from $f(t) \in L^2(\Omega)$ and $h(t) \in L^2(\Gamma)$.

Definition 2.1. A couple of functions $\{\theta, w\}$ is called a weak solution of (1.1)-(1.3), if the following conditions (w1)-(w3) are fulfilled:

(w1) $e := \theta + \lambda(w) \in C_w([s,T]; L^2(\Omega))$ (= the space of all weakly continuous functions from $[s,T]$ into $L^2(\Omega)$ with e' $(:= \frac{de}{dt}) \in L^2(s,T; V^*)$, and $w \in W^{1,2}(s,T; L^2(\Omega)) \cap C_w([s,T]; H^1(\Omega)) \cap L^2(s,T; H^2(\Omega))$ with $w(t) \in D(\Delta_N)$ for a.e. $t \in [s,T]$, $w \in D(\beta)$ a.e. on Q_s.

(w2) There exists $\tilde{\alpha} \in L^2(s,T; V)$ with $\tilde{\alpha} \in \alpha(e - \lambda(w))$ a.e. in Q_s such that (2.1) holds.

(w3) There exists $\tilde{\beta} \in L^2(s,T; L^2(\Omega))$ with $\tilde{\beta} \in \beta(w)$ a.e. on Q_s such that (2.2) holds.

A couple of functions $\{\theta, w\}$ is called a weak solution of system (1.1)-(1.4), if it is a weak solution of (1.1)-(1.3), $\theta(s) = \theta_o$ and $w(s) = w_o$, when θ_o and w_o are given in $L^2(\Omega)$.

We are going to establish the well-posedness of system (1.1)-(1.4) in the product space $V^* \times L^2(\Omega)$, paying attention to the dynamics of $\{e, w\}$ rather than $\{\theta, w\}$; physically, $e := \theta + \lambda(w)$ is the internal energy density of the system.

Our main result of this paper is stated as follows:

Theorem 2.1. *Let $\{\theta_i, w_i\}$, $i = 1,2$, be two weak solutions of (1.1)-(1.3). Put $e_i := \theta_i + \lambda(w_i)$ and denote by $\tilde{\alpha}_i$ the function $\tilde{\alpha}$ in (2.1) corresponding to $i = 1,2$. Then, for any $s \leq t_1 \leq t_2 \leq T$,*

$$|e_1(t_2) - e_2(t_2)|_{V^*}^2 + |w_1(t_2) - w_2(t_2)|_{L^2(\Omega)}^2 + C_o \int_{t_1}^{t_2} |\nabla(w_1(\tau) - w_2(\tau))|_{L^2(\Omega)}^2 d\tau$$

$$\leq \exp\left\{ C_1 \int_{t_1}^{t_2} (1 + |\tilde{\alpha}_1(\tau)|_V^{\frac{4}{3}} + |\tilde{\alpha}_2(\tau)|_V^{\frac{4}{3}}) d\tau \right\} \times \tag{2.3}$$

$$\times \left\{ |e_1(t_1) - e_2(t_1)|_{V^*}^2 + |w_1(t_1) - w_2(t_1)|_{L^2(\Omega)}^2 \right\},$$

where C_o and C_1 are positive constants depending only on κ, λ and g.

Clearly, by the above theorem, the weak solution of (1.1)-(1.4) is unique.

The key for our proof of Theorem 2.1 is the subdifferential representation for the term $F\tilde{\alpha}$ of (2.1) in V^*. To this end we consider a function j_o on $L^2(\Omega)$ defined by

$$j_o(z) := \int_\Omega \hat{\alpha}(z)dx \text{ if } \hat{\alpha}(z) \in L^1(\Omega), \text{ and } +\infty \text{ otherwise.}$$

It is easy to see that j_o is proper, l.s.c. and convex on $L^2(\Omega)$, but in general the extension of j_o onto V^* by $+\infty$, denoted by the same notation j_o, is not l.s.c. on V^* for lack of coerciveness of $\hat{\alpha}$. Therefore, let us now consider the Γ-regularization, denoted by j, of j_o onto V^*, namely, j is given by

$$j(z) = \inf\left\{\liminf_{z_n \to z} j_o(z_n); z_n \in L^2(\Omega), z_n \to z \text{ in } V^*\right\}, \quad \forall z \in V^*.$$

By the general theory on Γ-regularization (cf. [6]), j is proper, l.s.c. and convex on V^*, and $j_o \geq j$ on V^*. We denote by ∂j the subdifferential of j as a multivalued mapping from V^* into V, and by $\partial_* j$ that of j as a multivalued mapping from V^* into itself; note here that in the latter case V^* is regarded as a Hilbert space endowed with inner product $(z_1, z_2)_* := \langle z_1, F^{-1}z_2\rangle$, $\forall z_1, z_2 \in V^*$, and that $\partial_* j = F\partial j$.

In our case, we further have:

Theorem 2.2. (cf. [4]) *(i)* $j = j_o$ *on* $L^2(\Omega)$.
(ii) If $z \in D(\partial_* j) \cap L^2(\Omega)$, *then* $\partial_* j(z) = \{F\tilde{\alpha}; \tilde{\alpha} \in V, \tilde{\alpha} \in \alpha(z) \text{ a.e. on } \Omega\}$.

On account of Theorem 2.2 (ii), equation (2.1) can be reformulated in the following more general form:

$$e'(t) + \partial_* j(e(t) - \lambda(w(t))) \ni f^*(t) \text{ in } V^*, \text{ a.e. } t \in [s, T]. \tag{2.4}$$

3 Uniqueness of solutions

For simplicity, the usual inner product in $L^2(\Omega)$ is denoted by $(\cdot, \cdot)_{L^2(\Omega)}$. With the same notation as in Theorem 2.1 we prove:

Lemma 3.1. *For a.e.* $t \in [s, T]$ *it holds that*

$$\frac{1}{2}\frac{d}{dt}|e_1(t) - e_2(t)|_{V^*}^2 + \frac{1}{2}\frac{d}{dt}|w_1(t) - w_2(t)|_{L^2(\Omega)}^2 + \kappa|\nabla(w_1(t) - w_2(t))|_{L^2(\Omega)}^2$$

$$+(\tilde{\beta}_1(t) - \tilde{\beta}_2(t), w_1(t) - w_2(t))_{L^2(\Omega)} + (g(w_1(t)) - g(w_2(t)), w_1(t) - w_2(t))_{L^2(\Omega)} \quad (3.1)$$

$$\leq \int_\Omega \tilde{\alpha}_1(t)\{\lambda(w_2(t)) - \lambda(w_1(t)) - \lambda'(w_1(t))(w_2(t) - w_1(t))\}dx$$

$$+ \int_\Omega \tilde{\alpha}_2(t)\{\lambda(w_1(t)) - \lambda(w_2(t)) - \lambda'(w_2(t))(w_1(t) - w_2(t))\}dx,$$

where $\tilde{\beta}_i$ is the function $\tilde{\beta}$ as in (2.2) for each $i = 1, 2$.

Proof. By (2.4) and Theorem 2.2 (ii), $f^* - e_1' \in \partial_* j(e_1 - \lambda(w_1))$. Hence, a.e. on $[s, T]$,

$$
\begin{aligned}
&(f^* - e_1', e_2 - e_1)_* + (-\tilde{\alpha}_1\lambda'(w_1), w_2 - w_1)_{L^2(\Omega)} \\
= \ &(f^* - e_1', (e_2 - \lambda(w_2)) - (e_1 - \lambda(w_1)))_* \\
&\qquad + (f^* - e_1', \lambda(w_2) - \lambda(w_1))_* + (-\tilde{\alpha}_1\lambda'(w_1), w_2 - w_1)_{L^2(\Omega)} \\
\leq \ &j(e_2 - \lambda(w_2)) - j(e_1 - \lambda(w_1)) \\
&\qquad + (\tilde{\alpha}_1, \lambda(w_2) - \lambda(w_1))_{L^2(\Omega)} + (-\tilde{\alpha}_1\lambda'(w_1), w_2 - w_1)_{L^2(\Omega)};
\end{aligned}
$$

we used above $(f^* - e_1', \lambda(w_2) - \lambda(w_1))_* = (F\tilde{\alpha}_1, \lambda(w_2) - \lambda(w_1))_* = (\tilde{\alpha}, \lambda(w_2) - \lambda(w_1))_{L^2(\Omega)}$. Also, by exchanging e_1 and w_1 for e_2 and w_2 in the above relations, respectively, we have another inequality. Adding these two inequalities and substituting there the expression $-\tilde{\alpha}_i\lambda'(w_i) = -w_i' + \kappa\Delta_N w_i - \tilde{\beta}_i - g(w_i)$, we obtain (3.1).$\Diamond$

Lemma 3.2. For each $\varepsilon > 0$ there is a positive constant C_ε', depending only on ε, such that

$$\int_\Omega (|\tilde{\alpha}_1(t)| + |\tilde{\alpha}_2(t)|)|w_1(t) - w_2(t)|^2 dx \tag{3.2}$$

$$\leq \varepsilon|\nabla(w_1(t) - w_2(t))|_{L^2(\Omega)}^2 + C_\varepsilon'(1 + |\tilde{\alpha}_1(t)|_V^{\frac{4}{3}} + |\tilde{\alpha}_2(t)|_V^{\frac{4}{3}})|w_1(t) - w_2(t)|_{L^2(\Omega)}^2$$

for a.e. $t \in [s, T]$.

Proof. We use an inequality (Gagliardo-Nirenberg inequality (cf. [12]))

$$|w_1 - w_2|_{L^{\frac{12}{5}}(\Omega)}^2 \leq \text{const.}\left\{|\nabla(w_1 - w_2)|_{L^2(\Omega)}^{\frac{1}{2}}|w_1 - w_2|_{L^2(\Omega)}^{\frac{3}{2}} + |w_1 - w_2|_{L^2(\Omega)}^2\right\}. \tag{3.3}$$

Applying the Hölder's inequality, Sobolev's inequality and then (3.3), we have

$$
\begin{aligned}
&\int_\Omega (|\tilde{\alpha}_1| + |\tilde{\alpha}_2|)|w_1 - w_2|^2 dx \\
\leq \ &\text{const.}(|\tilde{\alpha}_1|_{L^6(\Omega)} + |\tilde{\alpha}_2|_{L^6(\Omega)})|w_1 - w_2|_{L^{\frac{12}{5}}(\Omega)}^2 \\
\leq \ &\text{const.}(|\tilde{\alpha}_1|_V + |\tilde{\alpha}_2|_V)\left\{|\nabla(w_1 - w_2)|_{L^2(\Omega)}^{\frac{1}{2}}|w_1 - w_2|_{L^2(\Omega)}^{\frac{3}{2}} + |w_1 - w_2|_{L^2(\Omega)}^2\right\} \\
= \ &\text{const.}|\nabla(w_1 - w_2)|_{L^2(\Omega)}^{\frac{1}{2}} \cdot (|\tilde{\alpha}_1|_V + |\tilde{\alpha}_2|_V)|w_1 - w_2|_{L^2(\Omega)}^{\frac{3}{2}} \\
&\qquad + \text{const.}(|\tilde{\alpha}_1|_V + |\tilde{\alpha}_2|_V)|w_1 - w_2|_{L^2(\Omega)}^2
\end{aligned}
$$

Now, use the Young's inequality with coefficients ε and a suitable C_ε' in the first term of the last side. Then (3.2) is obtained. $\Diamond$

151

Proof of Theorem 2.1: First we observe from the mean value theorem that

$$
\begin{aligned}
&|(\lambda(w_2) - \lambda(w_1)) - \lambda'(w_1)(w_2 - w_1)| \\
&= |\lambda'(w_1 + r(w_2 - w_1))(w_2 - w_1) - \lambda'(w_1)(w_2 - w_1)| \\
&= |\lambda'(w_1 + r(w_2 - w_1)) - \lambda'(w_1)||w_2 - w_1| \\
&\leq L(\lambda')|w_2 - w_1|^2 \qquad \text{a.e. on } Q_s,
\end{aligned}
\tag{3.4}
$$

where $r := r(x, t)$ is a number with $0 < r < 1$ depending on $(x, t) \in Q_s$ and $L(\lambda')$ is the Lipschitz constant of λ' on $[\sigma_*, \sigma^*]$. Similarly we have

$$
|(\lambda(w_1) - \lambda(w_2)) - \lambda'(w_2)(w_1 - w_2)| \leq L(\lambda')|w_1 - w_2|^2 \quad \text{a.e. on } Q_s.
\tag{3.5}
$$

Next, consider the right-hand side of (3.1) in Lemma 3.1. Then, by inequalities (3.4), (3.5) and (3.2), we have a.e. on $[s, T]$

the right-hand side of (3.1)

$$
\leq L(\lambda') \int_\Omega (|\tilde{\alpha}_1| + |\tilde{\alpha}_2|)|w_1 - w_2|^2 dx
$$

$$
\leq L(\lambda')\varepsilon |\nabla(w_1 - w_2)|^2_{L^2(\Omega)} + L(\lambda')C'_\varepsilon(1 + |\tilde{\alpha}_1|^{\frac{4}{3}}_V + |\tilde{\alpha}_2|^{\frac{4}{3}}_V)|w_1 - w_2|^2_{L^2(\Omega)}
$$

By these inequalities and the monotonicity of β it follows from (3.1) that

$$
\frac{1}{2}\frac{d}{dt}|e_1(t) - e_2(t)|^2_{V_*} + \frac{1}{2}\frac{d}{dt}|w_1(t) - w_2(t)|^2_{L^2(\Omega)}
$$

$$
+ (\kappa - L(\lambda')\varepsilon)|\nabla(w_1(t) - w_2(t))|^2_{L^2(\Omega)}
$$

$$
\leq (L(\lambda')C'_\varepsilon + L(g))(1 + |\tilde{\alpha}_1(t)|^{\frac{4}{3}}_V + |\tilde{\alpha}_2(t)|^{\frac{4}{3}}_V)|w_1(t) - w_2(t)|^2_{L^2(\Omega)}
$$

for a.e. $t \in [s, T]$, where $L(g)$ is the Lipschitz constant of g on $[\sigma_*, \sigma^*]$. Now fix a positive number ε so that $\kappa - L(\lambda')\varepsilon > 0$, and then apply the Gronwall's inequality to the above inequality to get (2.3); in fact, we can take $2(\kappa - L(\lambda')\varepsilon)$ and $2(L(\lambda')C'_\varepsilon + L(g))$ as constants C_o and C_1, respectively. $\Diamond$

4 Existence of solutions

In this section we give a result on the existence of a weak solution and some comments on the associated evolution operator.

Let us consider a class of initial data D defined by

$$
D := \{[\theta_o, w_o] \in L^2(\Omega) \times H^1(\Omega); \hat{\alpha}(\theta_o) \in L^1(\Omega), \hat{\beta}(w_o) \in L^1(\Omega)\}.
$$

Then we have:

Theorem 4.1. *Further suppose that*

$$
\frac{h}{n_o} \in \alpha(\tilde{h}) \quad \text{for some } \tilde{h} \in L^2(0, T; L^2(\Gamma)).
\tag{4.1}
$$

Then, for each $[\theta_o, w_o] \in D$, system (1.1)-(1.4) admits a weak solution $\{\theta, w\}$ such that $[\theta(t), w(t)] \in D$ for all $t \in [s, T]$.

In fact, if $[\theta_o, w_o] \in D$, then the weak solution is constructed as a limit of solutions of the regular approximate problems with α and β replaced by their Yosida-regularizations, and under compatibility condition (4.1) we can show the convergence of approximate solutions. We refer for the details to a forthcoming paper [8].

Based on the above existence theorem, we can define the evolution operator $E(t, s)$: $D \to D$, $0 \le s \le t \le T$, which assigns to each $[\theta_o, w_o] \in D$ the element $[\theta(t), w(t)] \in D$, where $\{\theta, w\}$ is the weak solution of (1.1)-(1.4). Also, for each constant $M > 0$ we put

$$D_M := \left\{ [\theta_o, w_o] \in D; \begin{array}{l} j(\theta_o) \le M, |\theta_o|_{V^*} \le M, \\ |\hat{\beta}(w_o)|_{L^1(\Omega)} \le M, |w_o|_{H^1(\Omega)} \le M \end{array} \right\}$$

and denote by D_M^* its closure in $V^* \times L^2(\Omega)$. With these notations, we see by the usual energy estimate that for any number $M > 0$ there is a constant $M' > 0$ such that any weak solution $\{\theta, w\}$ of (1.1)-(1.3) satisfies

$$|\tilde{\alpha}|_{L^2(s,T;V)} \le M', \quad [\theta(t), w(t)] \in D_{M'}, \quad \forall t \in [s, T],$$

as long as the initial datum $[\theta_o, w_o]$ is given in D_M. Therefore, we observe from (2.3) of Theorem 2.1 that for each $M > 0$, $E(t, s)$ is a continuous mapping from D_M into $D_{M'}$ with respect to the topology of $V^* \times L^2(\Omega)$; hence $E(t, s)$ has a unique extension $E^*(t, s): D_M^* \to D_{M'}^*$. Consequently, we have a unique extension $E^*(t, s) : D^* := \bigcup_{M>0} D_M^* \to D^*$ of $E(t, s)$. It is naturally expected that for each $[\theta_o, w_o] \in D^*$ the couple of functions $\{\theta, w\}$ given by $[\theta(t), w(t)] := E^*(t, s)[\theta_o, w_o]$ is a solution of the Cauchy problem for system (2.2),(2.4). See [8] in detail.

Remark 4.1. In [1, 2] the existence of a weak solution was proved under some growth conditions of α, and the uniqueness as well in the case of $\alpha(\theta) = -\frac{1}{\theta} + \theta$. But there was no uniqueness result in the general case. Therefore, our Theorems 2.1 and 4.1 give some mathematical improvements.

Remark 4.2. According to a recent result [5], when λ is convex and the range of α is included in $(-\infty, 0)$, our system can be described as a single evolution equation of the form $u' + \partial\varphi(u) + G(u) \ni \ell$ in $H := V^* \times L^2(\Omega)$, where $\partial\varphi$ is the subdifferential of a proper l.s.c. convex function φ on H and G is a Lipschitz perturbation in H. Therefore, the system is well-posed in H.

References

[1] P. COLLI AND PH. LAURENÇOT, Weak solutions to the Penrose-Fife phase field model for a class of admissible heat flux laws, *Physica D*, **111** (1998), 311-334.

[2] P. COLLI, PH. LAURENÇOT AND J. SPREKELS, Global solution to the Penrose-Fife phase field model with special heat flux laws, pp. 181-188 in the Proceedings

of IUTAM Symposium of *Variations of Domains and Free-Boundary Problems in Solid Mechanics*, ed. by P. Argoul, M. Frmond and Q. S. Nguyen, Kluwer Acad. Pub., Dordrecht-Boston-London, 1999.

[3] P. COLLI AND J. SPREKELS, On a Penrose-Fife model with zero interfacial energy leading to a phase-field system of relaxed Stefan type, *Ann. Mat. Pura. Appl.* (4), **169** (1995), 269-289.

[4] A. DAMLAMIAN AND N. KENMOCHI, Evolution equations generated by sub-differentials in the dual space of $H^1(\Omega)$, Discrete and Continuous Dynamical Systems, **5** (1999), 269-278.

[5] A. DAMLAMIAN AND N. KENMOCHI, Evolution equations associated with non-isothermal phase transitions, pp. 62-77, in Functional Analysis and Global Analysis (the proceedings of the Manila conference), eds. T. Sunada and P. W. Sy, Springer-Verlag, Singapore, 1997.

[6] I. EKELAND AND R. TEMAM, Convex Analysis and Variational Problems, *Studies Math. Appl.* **Vo. 1**, North-Holland, Amsterdam, 1976.

[7] J. K. HALE, Asymptotic Behavior of Dissipative Systems, *Mathematical Surveys and Monographs* **25**, Amer. Math. Soc., Providence, R. I., 1988.

[8] N. KENMOCHI AND M. KUBO, Weak solutions of nonlinear systems for non-isothermal phase change problems, *Adv. Math. Sci. Appl.*, **9** (1999), 499-521.

[9] N. KENMOCHI AND M. NIEZGÓDKA, Systems of nonlinear parabolic equations for phase change problems, *Adv. Math. Sci. Appl.*, **3**(1993/94), 89-117.

[10] PH. LAURENÇOT, Solutions to a Penrose-Fife model of phase-field type, *J. Math. Anal. Appl.*, **185**(1994), 262-274.

[11] R. TEMAM, *Infinite Dimensional Dynamical Systems in Mechanics and Physics*, Springer-Verlag, Berlin, 1988.

[12] E. ZEIDLER, *Nonlinear Functional Analysis and its Applications* II/B, Springer-Verlag, New York-Berlin-Heidelberg-London-Paris-Tokyo, 1990.

Nobuyuki Kenmochi
Department of Mathematics
Faculty of Education, Chiba University
1-33 Yayoi-chō, Inage-ku, Chiba 263, Japan

Masahiro Kubo
Department of Mathematics
Faculty of Science and Engineering, Saga University
1 Honjo-machi, Saga 840, Japan

M. K. Korten*

On a Structure Theorem for Some Free Boundary Problems for the Heat Equation

1 Introduction

In one space dimension and for a given initial datum $u_I(x) \in C_0^\infty$ (say such that $u_I(x) > 1$ in some interval), the equation

$$u_t = \Delta(u - 1)_+ \tag{1.1}$$

can be thought of as describing the energy per unit volume in a Stefan-type problem where the latent heat of the phase change is given by $(1 - u_I(x))_+$. Note that discontinuous solutions should be expected for (1.1) (see [BKM]).

In previous papers ([AK], [K]) a-priori regularity of nonnegative solutions in the sense of distributions $u \in L^1_{\text{loc}}(\mathbb{R}^n \times (0, T))$ was found, mainly u, $\nabla_x(u - 1)_+$ and $\frac{\partial}{\partial t}(u - 1)_+ \in L^2_{\text{loc}}(\mathbb{R}^n \times (0, T))$, and continuity of $(u - 1)_+$ in $(\mathbb{R}^n \times (0, T))$. Also $(u - 1)_+$ is a (weak) subsolution to the heat equation. From the weak Harnack inequality

$$\int_{\mathbb{R}^n} u(x, t) \exp(-c|x|^2) \, dx \le M(u, n, T),$$

for some $c = c(T) \in \mathbb{R}_+$ and $0 < t < T/2$ (see [AK]), existence of a unique initial trace $0 \le \mu(x)$ follows, which is a Radon measure satisfying the growth condition at infinity

$$\int_{\mathbb{R}^n} \exp(-c|x|^2) \, d\mu(x) < \infty,$$

and is taken in the appropriate sense. Each such measure gives rise to a solution to (1.1), which is unique (see [K]). In [K1] we prove that for all $\beta, k > 0$ and a.e. $x_0 \in \mathbb{R}^n$, there exists

$$\lim_{\substack{(x,t) \to x_0 \\ (x,t) \in \Gamma_\beta^k(x_0)}} (u(x,t) - 1)_+ = (f(x_0) - 1)_+,$$

where $f = \partial\mu/\partial| \,|$ is the Radon-Nikodym derivative of the initial trace with respect to the Lebesgue measure and

$$\Gamma_\beta^k(x_0) = \{(x, t) : |x - x_0| < \beta\sqrt{t}, \, 0 < t < k\}$$

are the parabolic "nontangential" approach regions.

In this contribution we announce a structure theorem for the free boundary of the one-phase Stefan problem formulated in enthalpic variables (1.1). According to

*Research partially supported by PIDs 3668/92 and 3164/92-CONICET and EX 071-UBA.

the preceding paragraph, our results hold for nonnegative $u \in L^1_{loc}(\mathbb{R}^n \times (0,T))$ which solve (1.1) in the sense of distributions, however, still under a hypothesis on $\nabla(u-1)_+$ which we hope to remove in forthcoming work. We also show (and this will be the content of Section 2) how the combustion problem can be seen as a particular case of the foregoing theory. This problem has been studied recently by many authors, see the references in [CV], [V]. It was believed until now that this is a *different* free boundary problem for the heat equation.

From the broad literature on the Stefan problem, which was treated extensively by J. R. Cannon, C. D. Hill, A. Fasano and M. Primicerio, among others, our work is close in its scope to the papers [ACS1] and [ACS2]. There, following the steps in the study of the regularity of minimal surfaces, they propose the following program for the study of the regularity of free boundaries:

a) Lipschitz free boundaries are smooth.

b) "Flat" free boundaries (in some Lebesgue differentiability or measure theoretical sense) are Lipschitz (but have some "waiting time" points).

c) Generalized free boundaries are "flat".

Here we study c) for the free boundary of the 1-phase Stefan problem (1.1). Specifically we will show that the free boundary is but a null set for a measure supported on the free boundary, which takes into account the loss of energy at the change of phase, and which we will call the "boundary measure" (see Section 3) n-rectifiable.

Our approach is very different from the one in [ACS1] and [ACS2]: Since our solution is understood in the distribution sense in $\mathbb{R}^n \times (0,T)$, the regularity information we have holds across the free boundary. Also such solutions can satisfy *only* the Stefan condition on the interphase (see [B] and Sections 2 and 3 below), while in [ACS1] and [ACS2] solutions are viscosity solutions and the free boundary condition is satisfied in a generalized sense. Also our setting includes in a natural way the presence of a "mushy" region (the set where the initial datum is between 0 and 1), where the energy is positive but not enough to start the phase change. We recover for H^n a. e. $x \in \mathbb{R}^n$ the pointwise jump condition well-known in the classical Stefan problem, while in [ACS1] and [ACS2] this condition is replaced by an asymptotic expansion of the solution(s) holding at points of the free boundary.

The method of proof is simple, basically it reduces to careful handling of a measure μ (the boundary measure) which is well identified as a consequence of the a-priori information we have about u and $(u-1)_+$, and the use of results from geometric measure theory and functions of bounded variation taken from [EG], [M] and [Z]. The most delicate part is solved using the powerful result of Preiss ([P], see also [M]) on rectifiability of measures. We owe to [BGO] the crucial idea to use the right-hand side of (1.1) to obtain an alternative decomposition of the boundary measure.

The awareness that for a solution u of a parabolic problem u_t may naturally belong to $BV(t)$ can be traced back to [V]. Recently this fact has been used in [DP] and [GL]. In [CV] the authors are also aware of the existence of a boundary measure, absolutely continuous with respect to surface measure. In fact, more can be said (c.f. Section 3

below). However it should be said that our approach does not apply its full generality
to the two-phase Stefan problem: it relies heavily on a-priori regularity, and the
main steps to obtain this (in particular, the crucial result $L^1_{\text{loc}} \Rightarrow L^2_{\text{loc}}$, see [K], also
[DK]) require nonnegative (or lower bounded) solutions ... and this is not a natural
requirement for solutions to the two-phase problem. Whether this requirement is of
technical nature or has deeper roots is not clear up to now.

Most of these results where obtained during a 7 month visit to the research group
of Prof. Bogdan Bojarski at the Mathematical Institute of the University of Warsaw.
The author wishes to thank the members of the group for making the stage possible,
and the University of Buenos Aires and CONICET for the corresponding leave of
absence. I am indebted to Paweł Strzelecki and Piotr Hajłlasz for uncountable helpful
discussions on the subject of this contribution.

2   The combustion problem

Let us recall that usually the combustion problem is stated as follows: find u, Ω such
that

$$u_t = \Delta u \qquad (x, t) \text{ in } \Omega, \tag{2.1a}$$

$$u = 0 \text{ and } \frac{\partial u}{\partial n} = -1 \qquad \text{on } \Gamma. \tag{2.1b}$$

$$u(x, 0) = u_I \text{ on } \Omega_I, \tag{2.1c}$$

where Γ is the lateral boundary of Ω, $\Gamma \cap \{t = 0\} = \Gamma_0 = \partial\Omega_I$, and u_I, Ω_I are
given. Next, since this part of [B] remained unpublished after the author's death,
we will state the result on the pointwise jump condition for smooth free boundaries
for degenerate parabolic equations and its consequence, the shrinking of the "mushy"
region. The proof follows using the Green's identity

$$\Delta\phi\alpha(u) - \phi\Delta\alpha(u) = \nabla.(\alpha(u)\nabla\phi) - \nabla.(\phi\nabla\alpha(u))$$

and Gauss' theorem in Ω^+ and Ω^- to compute $\int \int_{\Omega^+}$ and $\int \int_{\Omega^-}$ separately. Then we
will show how this applies to the combustion problem.

Theorem (Bouillet [B]): If $u \in L^\infty(\Omega)$, the solution in the sense of distributions
is an equation of the type
$$u_t = \Delta\alpha(u)$$
in $\Omega \subset \mathbf{R} \times (t_1, t_2)$, $\alpha(0) = 0$, α monotone nondecreasing uniformly Lipschitz, u
piecewise smooth supposed to have side limits together with its spatial gradient at
each side of a smooth surface of discontinuity S, then at any point of the free boundary
(the surface of discontinuity) the Rankine-Hugoniot condition

$$[\nabla_x\alpha(u^+) - \nabla_x\alpha(u^-)].\bar\nu = (u^+ - u^-)\nu_t = 0. \tag{2.2}$$

157

is satisfied, where (ν, ν_t) is the unit normal vector pointing into $\{u > u^+\}$.

Remark (Bouillet [B]): If $I = [u_*, u^*]$ is an interval in which $\alpha' = 0$, and $u^-(x, t) < u^*$ for $(x, t) \in \Omega^-, u^* \leq u(x, t)$ for $(x, t) \in \Omega^+$, and $u^+ = u^*$, then the assumption of piecewise smoothness of $u(x, t)$ would give

$$u_t = u_t^-(x, t) = 0 \qquad \text{in } \omega^-, \text{ i.e.,}$$

$$u^-(x, t) = u_I(x) \qquad \text{whenever } (x, t) \in \Omega^-,$$

and

$$-(u^* - u_I(x)) = \nabla_x \alpha(u^+(x, t)).\nabla \Psi(x, t) \tag{2.2'}$$

on S. Now as u^* is the lowest value of $u(x, t)$ in $\Omega^+ \cap \{t\}, \nabla_x \alpha(u^+(x, t))$ points into $\Omega^+ \cap \{t\} : u^* - u_I(x) > 0$, and therefore $\nabla_x \Psi$ points out of Ω^+ and into $\Omega^- \cap \{t\}$. hinting at the shrinking behaviour of the "hyperbolic" region Ω^- (where $\alpha(u) = $ constant), as time elapses.

Let us come back to the combustion problem. From the a-priori regularity ([AK], [K]) and the result upon Lipschitz continuity of the solution u up to the free boundary if the latter is Lipschitz ([ACS1]), we know that nonnegative distributional solutions $u \in L^1_{loc}$ meet the above hypotheses if u is continuous on the "mushy" zone Ω^- Now, if we require the initial datum u_I of the combustion problem (2.1a) to avoid the interval $(0, 1)$ and take $\alpha(u) = (u - 1)_+$, (2.2') is exactly the free boundary condition for the combustion problem, which we recover as a particular case of the one-phase Stefan problem in enthalpic variables, therefore inheriting the regularity results, and existence and uniqueness even with (nonnegative) measures as data under optimal conditions of growth and the Fatou theory for the Cauchy problem, and the results in [B] quoted above hold. To see that $u(x, t) = u_I(x)$ (the absolutely continuous part of the initial trace) in Ω^-, we refer to lemma 3.1 below.

We included this paragraph to compare it with [CV]. After we prove the structure result in the next section the classical jump condition will be shown to hold in the above form (2.2') at μ a. e. point of the free boundary.

3. Rectifiability of the free boundary

This section will be devoted to our main result, theorem 3.5. After some definitions we will identify the measure supported on the free boundary. The rectifiability of this measure will follow from Preiss' result (theorem 3.4 below). We begin with some general notations and definitions (for both of them we will follow Mattila [M]).

Throughout this contribution we will write $B^{n+1}((x_0, t_0), r)$ for the $n+1$-dimensional euclidean ball centered at (x_0, t_0) whose radius is r, and $B^n(x_0, r)$ for the n-dimensional euclidean ball centered at x_0 whose radius is r. For $n \in I\!N$, $\mathcal{L}^n$ will stand for n-dimensional Lebesgue measure and $\mathcal{H}^n$ for n-dimensional Hausdorff measure. For

$0 \leq m < \infty$ we define *m-dimensional upper and lower densities* of a Radon measure μ at a point $a \in \mathbf{R}$ as

$$\Theta^{*m}(\mu, a) = \limsup_{r \to 0} (2r)^{-m} \mu(B(a, r)),$$

$$\Theta_{*m}(\mu, a) = \liminf_{r \to 0} (2r)^{-m} \mu(B(a, r)).$$

If they agree, their common value

$$\Theta^m(\mu, a) = \Theta^{*m}(\mu, a) = \Theta_{*m}(\mu, a)$$

will be called the *m-density* of μ at a. A set $E \subset \mathbf{R}^n$ is called *m-rectifiable* if there exist Lipschitz maps $f_i : \mathbf{R}^m \to \mathbf{R}^n$, $i = 1, 2, \ldots$, such that

$$\mathcal{H}^m \left(E \setminus \cup_{i=1}^\infty f_i(\mathbf{R}^n) \right) = 0.$$

We will say that a Radon measure μ on $\mathbf{R}$ is *m-rectifiable* if μ is absolutely continuous with respect to $\mathcal{H}^m$ measure and there exists an m-rectifiable Borel set E such that $\mu(\mathbf{R}^n \setminus E) = 0$. Once shown that E is an m-rectifiable subset of $\mathbf{R}^n$ we have that ([F], Chapter 3) $\mathcal{H}^m(E \setminus \cup_{i=1}^\infty M_i)$, where M_i are m-dimensional inmersed submanifolds of $\mathbf{R}$. Next we define the function

$$t(x) := \inf\{t > 0 : u(x, t) > 1\},$$

$t(x)$ may be zero (for example, on sets where the initial trace is a.e. > 1). We will by convention write $t(x) = T$ when $t(x) \notin [0, T)$. This may happen on sets where the initial trace is a.e. smaller than 1 which are never reached by the diffusion. We define the free boundary as the graph F of the function $t(x)$:

$$F := \{(x, t) \in \mathbf{R} \times (0, T) : t = t(x)\}.$$

Whenever we refer to a function $f \in L^1_{\text{loc}}$ we will understand the precise representative of f, namely, $\lim_{r \to 0} |B_r|^{-1} \int_{B_r} f$.

3.1. Lemma: The function $t(x)$ is upper semicontinuous, hence borelian and measurable.

3.2. Lemma: If for some $\bar{t} > 0$, $u(x, \bar{t}) < 1$ for $x \in E$, where E is measurable and $|E| > 0$, then $u(x, t) = u(x, \bar{t})$ for a. e. $x \in E$ and $0 < t < \bar{t}$.

Let us recall that $(u - 1)_+$ is a (weak) subsolution to the heat equation, that is, $\Delta(u - 1)_+ - \frac{\partial}{\partial t}(u - 1)_+ \geq 0$ in $D'(\mathbf{R}^n \times (0, T))$. This defines a (nonnegative) measure

$$\mu = \Delta(u - 1)_+ - \frac{\partial}{\partial t}(u - 1)_+ = u_t - \frac{\partial}{\partial t}(u - 1)_+. \tag{3.2}$$

159

3.3. Lemma: The following holds for μ:

i) $\mu = (1 - u_I(x))_+ \, \delta_{t=t(x)}$, where $u_I(x)$ is the initial trace of u (see [AK]), which due to lemma 3.1 is absolutely continuous with respect to $\mathcal{L}^n$ on the set $\{(x,t) : u(x,t) < 1\}$.

ii) μ is a positive Radon measure, and a Carleson measure.

iii) The measure μ is carried by F and $\mathcal{L}_{n+1}(F) = 0$.

iv) $\mu(B^{n+1}((x_0, t_0), r)) \leq \mathcal{L}^n(B_r^n) = cr^n$, whence μ defines a continuous linear functional on $BV(\mathbb{R}^{n+1})$, and $\mu(B) = 0$ on all Borel sets B such that $\mathcal{H}^n = 0$ (see [Z], Thm. 5.2.4). Moreover,

$$\limsup_{r \to 0} \mu(B^{n+1}((x_0, t_0), r)) \leq (1 - u_I(x_0))_+ dx < \infty$$

for $x \in \mathbb{R}$, i.e., the upper density of μ, $\Theta^{*m}(\mu, (x,t)) < \infty$, $\forall (x,t) \in \mathbb{R}^n \times (0,T)$.

v) $\mu = \mathrm{div}(V)$, $V \in H^1_{loc}(\mathbb{R}^n \times (0,T))$; moreover, $\mu \in H_0^{1*}(\mathbb{R} \times (0,T))$, and for $f \in H_0^1$, $(\mu, f) = -(V, \nabla_{x,t} f)$.

Next let us recall Preiss' theorem ([P], also thm. 17.8 in [M]).

3.4. Theorem (Preiss [P]):
Let μ be a Radon measure on $\mathbb{R}$ such that the density $\Theta^m(\mu, (x,t))$ exists and is positive and finite for μ almost all $(x,t) \in \mathbb{R}$. Then μ is m-rectifiable.

Hypothesis H: $|\nabla(u - 1)_+|_\infty < \infty$.

If **H** holds in a ball the results below hold in the same ball. Heuristically, **H** concerns the points (x,t) where the diffusive region touches for the first time an isolated region Ω where the initial datum is identically 1. In view of lemma 3.3, iv) we conjecture that μ does not charge the set of points where $\nabla(u - 1)_+$ blows up.

Theorem 3.5:
The boundary measure μ is n-rectifiable.

In order to fit into Preiss' theorem's hypotheses we need to show that the density $\Theta_n(\mu, (x,t))$ exists and is positive μ-a.e. in $\mathbb{R} \times (0,T)$. The following results ensure this.

3.4. Lemma: Let

$$E = \{(x_0, t_0) \in \mathbb{R} \times (0,T) : \lim_{r \to 0} \mu(B^n(x_0, r) \times (t_0 - r, t_0 + r)) = 0\}.$$

Then $\mu(E) = 0$.

The proof uses the fact that $g(x) = (x, t(x))$ and $\pi_{|F}(x,y) = x$ are each other's inverse function, Besicovich's covering lemma (for π of a fine covering of E by cylinders C_λ with $\mu(C_\lambda) < \epsilon$, $0 < \epsilon$ fixed arbitrarily, and the decomposition of μ given by $\mu(S) = \int_{\pi(S \cap F)} (1 - u_I(x))_+ dx$.

Now recall that $(u-1)_+(x,t) \in C^\infty(\{(x,t) : (u-1)_+(x,t) > 0\}$. Then we can extend $\nabla_{x,t}(u-1)_+$ to the closure of $\{(u-1)_+ > 0\}$. This defines

$$\lim_{\substack{(x,t)\to(x_0,t_0) \\ (x,t)\in\{(u-1)_+>0\}}} \nabla_{x,t}(u-1)_+ = \nabla^+_{x,t}(u-1)_+(x_0,t_0).$$

If it is zero at (x_0,t_0), then $\Theta^n(\mu, (x_0,t_0)) = 0$. Suppose some component of

$$\nabla^+_{x,t}(u-1)_+(x_0,t_0) \quad (\text{say}, \ \frac{\partial(u-1)^+_+(x_0,t_0)}{\partial t}\ ) \neq 0.$$

Then the inverse function theorem yields a smooth surface $t_\epsilon(x)$ in a ball $B = B^{n+1}_r(x_0, t_0)$ such that $(u-1)_+(x, t_\epsilon(x)) = \epsilon$ for sufficiently small ϵ. From the decomposition of

$$\mu(B) = -\int_{\partial B} (\nabla(u-1)_+, -(u-1)_+).\nu \, dH^n$$

and Gauss-Green's theorem we have

$$\mu(B) = -\int_{\partial\{(u-1)_+>\epsilon\}\cup\{\partial B\cap\{0<(u-1)_+<\epsilon\}} (\nabla(u-1)_+, -(u-1)_+).\nu \, dH^n.$$

This and **H** allow to application of the compactness theorem for BV to obtain $\chi_{\{(u-1)_+>0\}} \in BV$, which in turn yields

$$\mu(B) = \int_{B\cap\{(u-1)_+>0\}} V^+.\nu \, dH^n = \int_{\pi(B\cap F)} (1 - u_I)_+ \, dL^n.$$

Now $H^n \lfloor (B \cap \{(u-1)_+ > 0\})$ is a Radon measure, and the Radon-Nykodym theorem gives

$$\Theta^n(\mu, (x_0,t_0)) = \nabla^+(u-1)_+(x_0,t_0).\nu = (1 - u_I)_+(x_0),$$

which is the classical Stefan condition.

References

[ACS1] I. ATHANASOPOULOS, L. CAFFARELLI, S. SALSA, *Regularity of the free boundary in parabolic phase transition problems*, preprint no. 95-4/1995, Dept. of Mathematics, University of Crete.

[ACS1] I. ATHANASOPOULOS, L. CAFFARELLI, S. SALSA, *Caloric functions in Lipschitz domains and the regularity of solutions of phase transition problems*, preprint IAS.

[AK]   D. ANDREUCCI, M.K. KORTEN, *Initial traces of solutions to a one-phase Stefan problem in an infinite strip*, Rev. Mat. Iberoamericana, Vol. 9, No. 2 (1993), 315-332.

[BGO]   L. BOCCARDO, T. GALLOUËT, L. ORSINA, *Existence and uniqueness of entropy solutions for nonlinear elliptic equations with measure data*, Ann. Inst. H. Poincaré Anal. Non Linéaire, to appear.

[B]   J. E. BOUILLET, *Signed solutions to diffusion-heat conduction equations*, Trab. de Matemática, IAM - CONICET, preprint 101, Dec. 1986.

[BKM]   J. E. BOUILLET, M. K. KORTEN AND V. MÁRQUEZ, *Singular limits and the "Mesa" problem*, Trabajos de Matemática (IAM), preprint 133, June 1988, to appear in Revista de la Unión Matemática Argentina.

[CV]   L. A. CAFFARELLI, J. L. VÁZQUEZ, *A free-boundary problem for the heat equation arising in flame propagation*, Trans. Am. Math. Soc. 347, No. 2 (1995), 411-441.

[DK]   B. E. J. DAHLBERG, C. E. KENIG, *Weak solutions of the porous medium equation*, Trans. Am. Math. Soc. 336, No. 2, 711-725 (1993).

[DP]   J. I. DÍAZ, J. F. PADIAL, *Uniqueness and existence of solutions in the $BV_t(Q)$ space to a doubly nonlinear parabolic problem*, Publications Matemàtiques 40 (1996), 527-560.

[EG]   C. EVANS, R. GARIEPY, *Measure theory and fine properties of functions*, CRC Press, Boca Raton 1992.

[F]   H. FEDERER, *Geometric measure theory*, Springer-Verlag Berlin-Heidelberg 1969.

[GL]   G. GILARDI, S. LUCKHAUS, *A regularity result for the solution of the dam problem*, Nonl. An., Th., Meth. and Appl, Vol. 26, No. 1, 113-138 (1996).

[GS]   B. GUSTAFSSON, H. SHAHGHOLIAN, *Existence and geometric properties of solutions of a free boundary problem in potential theory*, J. reine angew. Math. 473, 137-179 (1996).

[K]   M. K. KORTEN, *Non-negative solutions of $u_t = \Delta(u - 1)_+$: Regularity and uniqueness for the Cauchy problem*, Nonl. Anal., Th., Meth. and Appl, Vol. 27, No. 5, 589-603 (1996).

[K1]   M. K. KORTEN, *A Fatou theorem for the equation $u_t = \Delta(u-1)_+$*, submitted.

[M]   P. MATTILA, *Geometry of sets and measures in euclidean spaces*, Cambridge Studies in Advanced Mathematics 44, Cambridge University Press 1995.

[P]   D. PREISS, *Geometry of measures in* ℝ : *distribution, rectifiability and densities*, Ann. of Math. 125 (1987), 537-643.

[Va] J. L. VÁZQUEZ, *The free boundary problem for the heat equation with fixed gradient condition*, in Free Boundary Problems, Theory and Applications. M. Niezgódka, P. Strzelecki (eds.), Pitman Res. Notes in Maths. Series 363, 277-302 (1996).

[V]   A. I. VOL'PERT, *The spaces BV and quasilinear equations*, Math. Sb. 73(115), No. 2, 255-302 (1967).

[Z]   W. P. ZIEMER, *Weakly differentiable functions*, Springer-Verlag New York Inc. 1989.

Marianne Korten
Departamento de Matemática
Facultad de Ciencias Exactas y Naturales
Universidad de Buenos Aires
Pab. No. 1, Ciudad Universitaria
1428 Buenos Aires
Argentina

C. LEDERMAN*AND N. WOLANSKI

Limit, Pointwise, Viscosity and Classical Solutions to a Free Boundary Problem in Combustion

1 Introduction

In these notes we will present a survey of mathematical results on a free boundary problem in combustion which we have recently obtained in [CLW1] and [CLW2] (joint work with L. A. Caffarelli), in [LW], and in [LVW] (joint work with J. L. Vázquez). These papers deal with models of propagation of premixed flames of the following types:

$$\Delta u^\varepsilon - u_t^\varepsilon = \beta_\varepsilon(u^\varepsilon) \quad \text{or} \quad \Delta u^\varepsilon = \beta_\varepsilon(u^\varepsilon),$$

where $1/\varepsilon$ stands for the large activation energy, u^ε is a normalized temperature and $\beta_\varepsilon(s) = \frac{1}{\varepsilon}\beta(\frac{s}{\varepsilon})$. Convergence results have been obtained as $\varepsilon \to 0$ to the following FBPs:

$$\Delta u - u_t = 0 \quad \text{or} \quad \Delta u = 0,$$

in the complement of $\partial\{u > 0\}$, with free boundary conditions

$$u = 0, \quad u_\nu = \sqrt{2M} \quad \text{on } \partial\{u > 0\}$$

in the one-phase case and

$$u = 0, \quad (u_\nu^+)^2 - (u_\nu^-)^2 = 2M \quad \text{on } \partial\{u > 0\}$$

in the two-phase case. Here $M = \int \beta(s)\, ds$ and ν is the inward unit spatial normal to the free boundary $\partial\{u > 0\}$.

In the two-phase parabolic case, limits are shown to be solutions to the FBP in pointwise and viscosity senses (see [CLW1] and [CLW2]). In the elliptic case, among other results, two-phase nondegenerate limits are shown to be classical solutions (see [LW]). In the one-phase parabolic case, different notions of solution to the FBP are proved to coincide and produce a unique solution (see [LVW]).

2 Pointwise and viscosity solutions for the limit of a two-phase parabolic singular perturbation problem

In [CLW1] and [CLW2] we are concerned with the following problem: Study the uniform properties, and the limit as $\varepsilon \to 0$, of solutions $u^\varepsilon(x, t)$ of the equation:

$$\Delta u^\varepsilon - u_t^\varepsilon = \beta_\varepsilon(u^\varepsilon), \tag{P_ε}$$

*Partially supported by UBA grants EX071, EX197 & CONICET grant PID3668/92

where $\varepsilon > 0$, $\beta_\varepsilon \geq 0$, $\beta_\varepsilon(s) = \frac{1}{\varepsilon}\beta(\frac{s}{\varepsilon})$, support $\beta=[0,1]$, with β a Lipschitz continuous function and $\int \beta(s)ds = M$. Here M is a positive constant, and the functions $u^\varepsilon(x,t)$ are defined in $\mathbb{R}^N \times \mathbb{R}$, or in a subset of it.

This problem has been studied in [CV] for the one-phase case, i.e., restricted to the case where $u^\varepsilon \geq 0$. It appears in combustion, in the description of laminar flames as an asymptotic limit for high activation energy. In [CV] the authors studied the initial value problem associated to P_ε. Previously, the elliptic version of this problem in the one-phase case and, in particular, the convergence problem for travelling waves had been studied in [BCN].

For the two-phase case, it was shown in [C1] (and in the more general work [C2]) that uniformly bounded solutions to P_ε are locally uniformly Lipschitz in space.

The purpose of [CLW1] and [CLW2] is to continue with the local study of problem P_ε: We prove further uniform estimates for uniformly bounded solutions to P_ε, we pass to the limit and we study the limit function u. As in [C1] and [C2], we are concerned with the two-phase version of problem P_ε; that is, our solutions are allowed to change sign and become negative. The approach in our papers is local, also in the spirit of [C1] and [C2], since we do not force the solutions u^ε to be globally defined, nor to take on prescribed initial or boundary values.

We consider a family u^ε of solutions to P_ε in a domain $\mathcal{D} \subset \mathbb{R}^{N+1}$ which are uniformly bounded in L^∞ norm in $\mathcal{D}$. We get uniform estimates that allow us to pass to the limit as $\varepsilon \to 0$, we find properties of the limit function u in general situations, and we prove that u is a solution to the free boundary problem

$$\begin{aligned} \Delta u - u_t &= 0 \quad\quad \text{in } \mathcal{D} \setminus \partial\{u > 0\} \\ u &= 0 \;\;,\;\; (u_\nu^+)^2 - (u_\nu^-)^2 = 2M \quad \text{on } \mathcal{D} \cap \partial\{u > 0\} \end{aligned} \quad\quad (P)$$

in a pointwise sense at regular free boundary points and in a viscosity sense when $\{u \equiv 0\}^\circ = \emptyset$. (Here ν is the inward unit spatial normal to the free boundary $\mathcal{D} \cap \partial\{u > 0\}$, $u^+ = \max(u,0)$ and $u^- = \max(-u,0)$).

First, we prove that the functions u^ε are locally uniformly Hölder continuous with exponent $\frac{1}{2}$ in time in $\mathcal{D}$. Next, we get further uniform estimates and we start our study of the limit function u. We show that u satisfies

$$\Delta u - u_t = \mu$$

where μ is a nonnegative measure which is supported on $\mathcal{D} \cap \partial\{u > 0\}$.

In order to understand the nature of the measure μ, we study the case in which the limit function u is the difference of two hyperplanes. In the case of a two-phase limit we show that u satisfies the free boundary condition

$$(u_\nu^+)^2 - (u_\nu^-)^2 = 2M \quad \text{on } \partial\{u > 0\} \quad\quad (1)$$

which is the extension to the two-phase case of the free boundary condition obtained in [BCN] and [CV]. On the other hand, we show that for a range of values of α with $0 < \alpha < \sqrt{2M}$ there are locally uniformly bounded solutions to P_ε converging to $\alpha x_1^+ + \alpha x_1^-$. This indicates that some extra hypotheses should be made if one intends to prove that a limit function satisfies the free boundary condition (1) in a classical sense.

Then, we study the behavior of a limit function near an arbitrary free boundary point $(x_0, t_0) \in \mathcal{D} \cap \partial\{u > 0\}$. We prove that if $\limsup_{(x,t) \to (x_0, t_0)} |\nabla u^-| \le \gamma$, then we have $\limsup_{(x,t) \to (x_0, t_0)} |\nabla u^+| \le \sqrt{2M + \gamma^2}$.

Next, we introduce the concepts of pointwise and viscosity solutions to problem P. On one hand, in Theorem 2.1 we prove that in the two-phase case, the free boundary condition (1) is satisfied in a pointwise sense at any "regular" free boundary point (x_0, t_0). We point out that we only make assumptions on u at (x_0, t_0).

On the other hand, we prove, when $\{u \equiv 0\}^\circ = \emptyset$, that the limit function u is a viscosity solution to the free boundary problem P (Theorem 2.2). By a viscosity solution, we mean a weak solution of the free boundary problem in the sense that it is a continuous function for which comparison principles with classical supersolutions and subsolutions hold.

We point out that any limit function u is a viscosity supersolution (Theorem 4.1 in [CLW2]) (that is, we do not need the assumption $\{u \equiv 0\}^\circ = \emptyset$).

To conclude this section we state here the main results in [CLW1] and [CLW2].

Theorem 2.1 *(Theorem 3.1 in [CLW2]) Let u^{ε_j} be solutions to P_{ε_j} in a domain $\mathcal{D} \subset \mathbb{R}^{N+1}$ such that $u^{\varepsilon_j} \to u$ uniformly on compact subsets of $\mathcal{D}$ and $\varepsilon_j \to 0$. Let $(x_0, t_0) \in \mathcal{D} \cap \partial\{u > 0\}$ be such that there exists an inward unit spatial normal to the free boundary $\mathcal{D} \cap \partial\{u > 0\}$ in a parabolic measure theoretic sense (see Def. 2.1) and such that $\{u \equiv 0\}$ has zero "parabolic density" (see Def. 2.2) at (x_0, t_0). Then, there exist $\alpha > 0, \gamma \ge 0$ such that*

$$u(x, t) = \alpha \langle x - x_0, \nu \rangle^+ - \gamma \langle x - x_0, \nu \rangle^- + o(|x - x_0| + |t - t_0|^{\frac{1}{2}})$$

with

$$\alpha^2 - \gamma^2 = 2M.$$

Here ν is the inward unit spatial normal to the free boundary at (x_0, t_0).

Theorem 2.2 *(Theorems 4.1, 4.2 in [CLW2]) Let u^{ε_j} be a family of solutions to P_{ε_j} in a domain $\mathcal{D} \subset \mathbb{R}^{N+1}$ such that $u^{\varepsilon_j} \to u$ uniformly on compact subsets of $\mathcal{D}$ and $\varepsilon_j \to 0$. If $\{u \equiv 0\}^\circ \cap \mathcal{D} = \emptyset$, then u is a viscosity solution of the free boundary problem P.*

166

Definition 2.1 We say that ν is the inward unit spatial normal to the free boundary $\partial\{u > 0\}$ at a point $(x_0, t_0) \in \partial\{u > 0\}$ in the parabolic measure theoretic sense, if $\nu \in \mathbb{R}^N$, $|\nu| = 1$ and

$$\lim_{r \to 0} \frac{1}{r^{N+2}} \iint_{Q_r(x_0, t_0)} |\chi_{\{u>0\}} - \chi_{\{(x,t) / \langle x - x_0, \nu \rangle > 0\}}| \, dx \, dt = 0.$$

(We denote $Q_r(x_0, t_0) = \{(x, t) / |x - x_0| < r, |t - t_0| < r^2\}$.)

Definition 2.2 We say that $\{u \equiv 0\}$ has zero "parabolic density" at $(x_0, t_0) \in \partial\{u > 0\}$ if it holds that

$$\lim_{r \to 0} \frac{|\{u \equiv 0\} \cap Q_r(x_0, t_0)|}{|Q_r(x_0, t_0)|} = 0.$$

3 Viscosity solutions and regularity of the free boundary for the limit of an elliptic two-phase singular perturbation problem

In [LW] we pursue further the study in [CLW1] and [CLW2] in the case where the functions u^ϵ do not depend on t; that is, u^ϵ are solutions to

$$\Delta u^\epsilon = \beta_\epsilon(u^\epsilon). \tag{E_ϵ}$$

In fact, we consider a family u^ϵ of uniformly bounded solutions to E_ϵ in a domain $\Omega \subset \mathbb{R}^N$ and we prove that the limit function u is a solution — in a more appropriate elliptic sense — to the free boundary problem:

$$\begin{aligned} \Delta u &= 0 & \text{in } \Omega \setminus \partial\{u > 0\}, \\ u &= 0, \ (u_\nu^+)^2 - (u_\nu^-)^2 = 2M & \text{on } \Omega \cap \partial\{u > 0\}. \end{aligned} \tag{E}$$

Here ν is the inward unit normal to the free boundary $\Omega \cap \partial\{u > 0\}$. Moreover, under suitable assumptions, we prove that the free boundary is smooth and therefore, the free boundary condition is satisfied in the classical sense.

As in [CLW1] and [CLW2], our study here is of a local nature and we consider the two-phase case.

Our first result is of a pointwise nature. Although the pointwise result described above (Theorem 2.1) applies to the present elliptic situation, we get here — under weaker assumptions at the point $x_0 \in \partial\{u > 0\}$ — the same conclusion: it holds that $(u_\nu^+)^2 - (u_\nu^-)^2 = 2M$ at x_0, where ν is the inward unit normal to $\partial\{u > 0\}$ at x_0 in the measure theoretic sense (Theorem 3.1). This result applies, in particular, to the one-phase case.

Next, we prove that under suitable assumptions, our limit function u is a solution to E in a viscosity (elliptic) sense (Theorem 3.2). A viscosity (elliptic) solution is a continuous function, harmonic away from the free boundary, which satisfies the free boundary condition in terms of proper asymptotic developments. In particular, we prove that u is a viscosity solution if $\{u \equiv 0\}° = \emptyset$ (Corollary 4.1 in [LW]) or if u^+ is nondegenerate on $\partial\{u > 0\}$ (Corollary 4.2 in [LW]).

Finally, we study the regularity of the free boundary. We use results from [C3] — where a regularity theory for viscosity solutions to a class of elliptic free boundary problems which includes E was developed. We want to remark here that there are limit functions u which do not satisfy the free boundary condition in the classical sense on any portion of $\partial\{u > 0\}$ (for instance, $u = \alpha x_1^+ + \alpha x_1^-$ with $0 < \alpha < \sqrt{2M}$, see [CLW1], Remark 5.1). Thus, extra hypotheses have to be made in order to get regularity results.

On one hand, we prove that under suitable assumptions on our limit function u, there is a subset of the free boundary $\partial\{u > 0\}$ which is locally a $C^{1,\alpha}$ surface. As a consequence, on this portion of $\partial\{u > 0\}$, the free boundary condition is satisfied in the classical sense. Moreover, this smooth subset is open and dense in $\partial\{u > 0\}$ and the remainder of the free boundary has $(N - 1)$-dimensional Hausdorff measure zero (Theorem 3.3).

On the other hand, we prove in Theorem 3.4 that if u^- is locally uniformly non-degenerate on $\partial\{u > 0\}$, then the free boundary is locally a $C^{1,\alpha}$ surface. Hence, there are no singularities. We point out that the local uniform nondegeneracy of u^- on $\partial\{u > 0\}$ is a necessary condition for the regularity of the free boundary in the strictly two-phase case, even if we only require the free boundary to be locally Lipschitz continuous (see Remark 5.2 in [LW]).

To conclude this section, we state here the main results in [LW].

Theorem 3.1 *(Theorem 3.1 in [LW]) Let u^{ε_j} be solutions to E_{ε_j} in a domain $\Omega \subset \mathbb{R}^N$ such that u^{ε_j} converge to a function u uniformly on compact subsets of Ω and $\varepsilon_j \to 0$. Let $x_0 \in \Omega \cap \partial\{u > 0\}$ be such that $\partial\{u > 0\}$ has at x_0 an inward unit normal ν in the measure theoretic sense. If $\liminf_{r \to 0} \frac{|\{u<0\} \cap B_r(x_0)|}{|B_r(x_0)|} = 0$, we assume, in addition, that u^+ is nondegenerate at x_0 (see Def. 3.1).*

Then, there exist $\alpha > 0$ and $\gamma \geq 0$, such that

$$u(x) = \alpha\langle x - x_0, \nu\rangle^+ - \gamma\langle x - x_0, \nu\rangle^- + o(|x - x_0|)$$

with

$$\alpha^2 - \gamma^2 = 2M.$$

168

Theorem 3.2 *(Theorems 4.1, 4.2 in [LW]) Let u^{ε_j} be solutions to E_{ε_j} in a domain $\Omega \subset \mathbb{R}^N$ such that $u^{\varepsilon_j} \to u$ uniformly on compact subsets of Ω and $\varepsilon_j \to 0$. Assume, in addition, that at every regular point from the nonpositive side $x_0 \in \Omega \cap \partial\{u > 0\}$ (that is, such that there exists a ball $B_\rho(y) \subset \{u \leq 0\}$ with $x_0 \in \partial B_\rho(y)$), u satisfies one of the following hypotheses:*

H1) u^+ is nondegenerate at x_0 (see Def. 3.1).

   or else

H2) $u < 0$ in $B_\rho(y)$.

   Then u is a viscosity solution in Ω.

Theorem 3.3 *(Theorem 5.1 in [LW]) Let u^{ε_j} be solutions to E_{ε_j} in a domain $\Omega \subset \mathbb{R}^N$ such that $u^{\varepsilon_j} \to u$ uniformly on compact subsets of Ω and $\varepsilon_j \to 0$. Assume, in addition, that*

i) u^+ is locally uniformly nondegenerate on $\Omega \cap \partial\{u > 0\}$ (see Def. 3.1)

ii) The set $\{u \leq 0\}$ has locally uniform positive density on $\Omega \cap \partial\{u > 0\}$ (see Def. 3.2).

Then, there is a subset $\mathcal{R}$ of the free boundary $\Omega \cap \partial\{u > 0\}$ ($\mathcal{R} = \partial_{red}\{u > 0\}$) which is locally a $C^{1,\alpha}$ surface. Moreover, $\mathcal{R}$ is open and dense in $\Omega \cap \partial\{u > 0\}$ and the remainder of the free boundary has $(N-1)$-dimensional Hausdorff measure zero.

Theorem 3.4 *(Theorem 5.2 in [LW]) Let u^{ε_j} be solutions to E_{ε_j} in a domain $\Omega \subset \mathbb{R}^N$ such that $u^{\varepsilon_j} \to u$ uniformly on compact subsets of Ω and $\varepsilon_j \to 0$. Assume, in addition, that there is $\mathcal{U} \subset\subset \Omega$ such that u^- is uniformly nondegenerate on $\mathcal{U} \cap \partial\{u > 0\}$ (see Def. 3.1).*

   Then $\mathcal{U} \cap \partial\{u > 0\}$ is locally a $C^{1,\alpha}$ surface.

Definition 3.1 Let $v \geq 0$ be a continuous function in a domain $\Omega \subset \mathbb{R}^N$. We say that v is nondegenerate at a point $x_0 \in \Omega \cap \{v = 0\}$ if there exist $c > 0$ and $r_0 > 0$ such that

$$\fint_{B_r(x_0)} v \geq cr \qquad \text{for } 0 < r \leq r_0. \tag{2}$$

We say that v is uniformly nondegenerate on $\Gamma \subset \Omega \cap \{v = 0\}$, if there exist $c > 0$ and $r_0 > 0$ such that (2) holds for every $x_0 \in \Gamma$.

169

Definition 3.2 Let v be a continuous function in a domain $\Omega \subset \mathbb{R}^N$. We say that the set $\{v \leq 0\}$ has positive density at a point $x_0 \in \Omega \cap \partial\{v > 0\}$, if there exist $c > 0$ and $r_0 > 0$ such that

$$\frac{|\{v \leq 0\} \cap B_r(x_0)|}{|B_r(x_0)|} \geq c \quad \text{for } 0 < r \leq r_0. \tag{3}$$

We say that the set $\{v \leq 0\}$ has uniform positive density on $\Gamma \subset \Omega \cap \partial\{v > 0\}$, if there exist $c > 0$ and $r_0 > 0$ such that (3) holds for every $x_0 \in \Gamma$.

4 Uniqueness of solution to a free boundary problem from combustion

In [LVW] we consider the initial boundary value problem associated to the free boundary problem (P) described in Section A, in the one-phase case (i.e., assuming $u \geq 0$). We will call this problem (IVP). The purpose of [LVW] is to contribute to the questions of unique characterization of the solution to the free boundary problem (IVP) and the consistency of the different solution concepts.

This free boundary problem admits classical solutions only for good data and for small times. Several generalized concepts of solution have been proposed, among them the concepts of limit solution and viscosity solution. We investigate conditions under which the three concepts agree and produce a unique solution.

A limit solution is a function which is obtained as the limit (as $\varepsilon \to 0$) of solutions u^ε to

$$\Delta u^\varepsilon - u_t^\varepsilon = \beta_\varepsilon(u^\varepsilon) \tag{P_ε}$$

where $\beta_\varepsilon(s) = \frac{1}{\varepsilon}\beta(\frac{s}{\varepsilon})$, with β a Lipschitz continuous function such that $\beta > 0$ in $(0,1)$ and zero otherwise, and $\int \beta(s)\, ds = M$.

A viscosity solution is a continuous function which satisfies comparison principles with classical supersolutions and subsolutions to the free boundary problem.

Theorem 4.1 *(Theorem 6.1 in [LVW]) Let us assume that $\Omega = (0, +\infty) \times \Sigma$, with $\Sigma \subset \mathbb{R}^{N-1}$, a smooth bounded domain. Let u be a classical solution to IVP in $\overline{\Omega} \times [0, T]$ such that $\frac{\partial u}{\partial n} = 0$ on $(0, +\infty) \times \partial\Sigma \times (0, T)$. Assume in addition, that $\partial\{u_0 > 0\}$ is bounded and that $u_0(x) = u(x, 0)$ is a strictly decreasing function on $\overline{\{u_0 > 0\}}$ in the direction e_1. We also require that $u(0, x', t) > c_1 > 0$ and $u_{x_1}(0, x', t) \leq -c_2 < 0$ for $(x', t) \in \Sigma \times (0, T)$.*

Let $\widehat{u} = \lim u^{\varepsilon}$, where u^{ε} is a family of bounded solutions to P_{ε} with $\frac{\partial u^{\varepsilon}}{\partial \eta} = 0$ on $(0, +\infty) \times \partial\Sigma \times (0, T)$, such that $u^{\varepsilon} \to \widehat{u}$ uniformly on compact subsets of $\overline{\Omega} \times [0, T]$.

Assume that $\widehat{u} = u$ on $\partial_p(\Omega \times (0, T)) \cap (\{x_1 = 0\} \cup \{t = 0\})$ and, in addition, $\{x \, / \, u^{\varepsilon}(x, 0) > 0\} \to \{x \, / \, u_0(x) > 0\}$.

Then $\widehat{u} = u$ in $\overline{\Omega} \times [0, T]$.

Remark 4.1 In order for P_{ε} to approximate problem IVP properly, we need to impose some condition on how the initial datum $u_0(x)$ is approximated by $u_0^{\varepsilon}(x)$ so that we get a limit solution with a free boundary starting from the initial free boundary $\partial\{x \, / \, u_0(x) > 0\}$. If, for example, we take u_0^{ε} satisfying $u_0^{\varepsilon} \geq \varepsilon$, we obtain that the limit function $\widehat{u}$ is the solution to the heat equation with initial datum u_0, and therefore, the uniqueness result stated above does not hold. (However, we can relax the assumption that $\{x \, / \, u_0^{\varepsilon}(x) > 0\} \to \{x \, / \, u_0(x) > 0\}$, imposing instead a suitable control on the growth of u_0^{ε} away from $\{x \, / \, u_0(x) > 0\}$.)

Theorem 4.2 (*Theorem 5.1 in [LVW]*) Let $\mathcal{D} = \Omega \times (0, T)$, where $\Omega = \mathbf{R} \times \Sigma$, with $\Sigma \subset \mathbf{R}^{N-1}$ a smooth bounded domain. Let u be a classical solution to IVP in $\overline{\mathcal{D}}$. Assume that $\partial\{u > 0\} \cap \partial_p\mathcal{D}$ is bounded and that, on $\overline{\{u > 0\}} \cap \partial_p\mathcal{D}$, u is a strictly decreasing function in the direction e_1. Let v be a viscosity solution to IVP in $\overline{\mathcal{D}}$ such that

(i) $v = u$ on $\partial_p\mathcal{D}$,

(ii) $\overline{\{v > 0\}} \cap \partial_p\mathcal{D} = \overline{\{u > 0\}} \cap \partial_p\mathcal{D}$.

Then $v = u$ in $\overline{\mathcal{D}}$.

In addition, we can prove the same result if we let $\Omega = (0, +\infty) \times \Sigma$. In this case, we require that $u(0, x', t) > c_1 > 0$ and $u_{x_1}(0, x', t) \leq -c_2 < 0$ for $(x', t) \in \Sigma \times (0, T)$.

Moreover, the result also holds if we replace on $(0, +\infty) \times \partial\Sigma \times (0, T)$ (or $\mathbf{R} \times \partial\Sigma \times (0, T)$) the Dirichlet condition by the homogeneous Neumann condition, and we make a monotonicity assumption only on $\{t = 0\}$.

Remark 4.2 Conditions (i) and (ii) imply that the free boundaries of u and v coincide on $\partial_p\mathcal{D}$ and both start from $\partial\{u_0 > 0\}$. A condition like (ii) is necessary in order to get the uniqueness result since, for instance, the solution of the heat equation with Dirichlet datum u is a viscosity solution to IVP.

Proposition 4.1 If u is a classical solution to IVP in $\overline{\mathcal{D}}$, then u is a viscosity solution to IVP in $\overline{\mathcal{D}}$.

Final Comments on Section 4 In conclusion, under suitable assumptions on the domain and on the initial boundary data, *if a classical solution to the free boundary problem exists in a certain time interval, then it is at the same time the unique classical solution, the unique limit solution and also the unique viscosity solution in that time interval. In particular, there is a unique limit solution independent of the choice of the function β.*

References

[BCN] H. BERESTYCKI, L. A. CAFFARELLI, L. NIRENBERG, Uniform estimates for regularization of free boundary problems, *Analysis and Partial Differential Equations*, Cora Sadosky, *Lecture Notes in Pure and Applied Mathematics*, vol. 122, Marcel Dekker, New York, 1988.

[C1] L.A. CAFFARELLI, A monotonicity formula for heat functions in disjoint domains, *Boundary value problems for P.D.E.s and applications*, dedicated to E. Magenes, (J.L. Lions, C. Baiocchi eds.), Masson Paris 1993. pp. 53–60.

[C2] ——, Uniform Lipschitz regularity of a singular perturbation problem, *Diff. and Int. Eqs.* **8(7)** (1995), 1585–1590.

[C3] ——, A Harnack inequality approach to the regularity of free boundaries. Part II: Flat free boundaries are Lipschitz, *Comm. Pure and Appl. Math.* **42** (1989), 55–78.

[CLW1] L.A. CAFFARELLI, C. LEDERMAN, N. WOLANSKI, Uniform estimates and limits for a two phase parabolic singular perturbation problem, to appear in *Indiana Univ. Math. Journal*.

[CLW1] L.A. CAFFARELLI, C. LEDERMAN, N. WOLANSKI, Pointwise and viscosity solutions for the limit of a two phase parabolic singular perturbation problem, to appear in *Indiana Univ. Math. Journal*

[CV] L.A. CAFFARELLI, J.L. VAZQUEZ, A free boundary problem for the heat equation arising in flame propagation, *Trans. Amer. Math. Soc.*, **347** (1995), 411–441.

[LVW] C. LEDERMAN, J.L. VAZQUEZ, N. WOLANSKI, Uniqueness of solution to a free boundary problem from combustion, preprint

[LW] C. LEDERMAN, N. WOLANSKI, Viscosity solutions and regularity of the free boundary for the limit of an elliptic singular perturbation problem, preprint.

Claudia Lederman
Departamento de Matematica
Facultad de Ciencias Exactas
Universidad de Buenos Aires
(1428) Buenos Aires - Argentina

Noemí Wolanski
Departamento de Matematica
Facultad de Ciencias Exactas
Universidad de Buenos Aires
(1428) Buenos Aires - Argentina

Part 3.

Free Boundary Problems in Fluid Mechanics

L. K. Antanovskiĭ and R. H. J. Grimshaw

Hysteresis Behaviour of a Pointed Drop in Taylor's Four-Roller Mill

Abstract

Time-dependent deformation of a two-dimensional inviscid drop subjected to quasi-steady straining flow, such as that created in Taylor's four-roller mill (Taylor[3]) at a slowly varying rotation rate of the cylinders, is addressed. Following the spirit of matched asymptotic expansions, the problem is decomposed into the outer Stokes problem with no drop effect, which is solved by the boundary-element method, and the inner free-boundary problem for the drop behaviour in unbounded fluid. Using complex variable techniques, a broad class of explicit solutions is found, which is described by a rational conformal mapping of the unit disc onto the flow domain, with time-dependent coefficients satisfying ordinary differential equations. Some numerical simulations of the resulting equations are implemented, which suggest that, with increasing rotation rate of the cylinders, a sudden cusping of a sufficiently small drop occurs at a critical strain. This cusped configuration survives up to any strain, but when the strain is being gradually removed, the drop becomes suddenly rounded at a lower critical strain. This proves that, under certain circumstances, a two-dimensional inviscid drop placed in quasi-steady creeping flow exhibits hysteresis behaviour.

1 Introduction

The mathematical modelling of Taylor's experiments on deformation of an inviscid drop subjected to extensional flow in a four-roller mill [3] was presented in [1]. Employing the plane-flow model and using the method of matched asymptotic expansions, the problem was decomposed into the outer problem for the Stokes equations with no drop effect and the inner free-boundary problem for the drop behaviour in unbounded fluid. In particular, the far-field flow was calculated in the form

$$V(z) = G \left\{ x - \mathrm{i}\,y + q \left[x \left(x^2 + 3y^2 \right) - \mathrm{i}\,y \left(3x^2 + y^2 \right) \right] \right\} , \tag{1}$$

$$P(z) = 6\mu G q \left(x^2 - y^2 \right) , \tag{2}$$

where G is some constant proportional to the angular velocity of the rollers, and q is a geometry-dependent positive parameter. These constants determine the following dimensionless groups:

$$\lambda = \frac{2\mu G R}{\sigma} , \quad \varepsilon = q R^2 ,$$

where μ is the fluid viscosity, σ the surface tension, and $R = \sqrt{A/\pi}$ with A being the drop area. The parameter λ is called the capillary number which is the ratio of disruptive elongational stresses and cohesive surface-tension forces.

Using complex variable techniques, the inner problem was solved explicitly in terms of the conformal mapping $z(\zeta)$ of the unit disc $|\zeta| < 1$ onto the flow domain:

$$z(\zeta) = \frac{R\,(\alpha + \beta\,\zeta^2)}{\zeta\,(1 - \varepsilon\,\alpha\,\beta\,\zeta^2)}\ . \tag{3}$$

The coefficients α and β satisfy a system of transcendental equations which can be reduced to the response diagram $\lambda = F(\delta, \varepsilon)$ plotted in Fig. 1, where δ is the non-dimensional deformation parameter

$$\delta = \frac{R_{\max} - R_{\min}}{R_{\max} + R_{\min}}\ .$$

Here $R_{\max}$ and $R_{\min}$ are the major and minor axes of the drop. It is seen that the response diagram can be uniquely resolved in terms of δ provided that $\varepsilon \geq \varepsilon_0 \approx 0.004$. However, for $0 < \varepsilon < \varepsilon_0$, the response diagram does not determine a one-to-one function, and, for some values of λ, there are actually three solutions: $\delta_1 < \delta_2 < \delta_3$. It is conceivable that the solution δ_2 is unstable, and a typical hysteresis loop can occur as shown in Fig. 2. The shapes corresponding to sudden transition from a rounded drop to a cusped one (state 1 to state 2) and from a cusped drop to a rounded one (state 3 to state 4) are depicted in Fig. 3. The structure and the hysteresis behaviour of the steady-state drop for small ε can be readily understood by realising that at $\varepsilon = 0$, there are actually two branches of the response diagram. One is $\lambda = F(\delta, 0)$ and the other is $\delta = 1$, which corresponds to an infinitely long drop. For small ε these distinct branches merge to give the curve shown in Fig. 1. Note that the paper [1] employed steady-state solutions only, and so any stability question was out of scope.

This paper addresses the unsteady formulation of the above problem, which is associated with the quasi-steady deformation of a drop due to a slowly varying rotation rate of the rollers. In this case, the capillary number λ is some function of time t but the geometry-dependent parameter ε is constant. It is shown that, surprisingly, the class of exact solutions (3) persists for the unsteady formulation with time-dependent coefficients α and β. The governing equation takes the form of an ordinary differential equation with transcendental nonlinearity, which is solved numerically. It is shown that, under certain conditions, the predicted hysteresis behaviour of the drop does take place.

The analytical calculations required for this paper were performed using *Mathematica*, and the numerical computations were carried out employing *Matlab*.

2   Problem formulation

Consider a two-dimensional, inviscid, neutrally buoyant drop placed at the centre of Taylor's four-roller mill. Let $\mathcal{D}(t)$ be the flow domain occupied by the exterior fluid, with the free boundary $S(t)$ and rigid walls B, where t is time. The fluid velocity $\mathbf{v}$ and pressure p are governed by the boundary-value problem

$$\nabla \cdot \mathbf{v} = 0 \,, \quad \nabla p = \mu \, \Delta \mathbf{v} \quad \text{in } \mathcal{D}(t) \,, \tag{4}$$

$$V_n = \mathbf{v} \cdot \mathbf{n} \,, \quad p\,\mathbf{n} - \mu\,(\mathbf{n} \cdot \nabla \mathbf{v} + \nabla \mathbf{v} \cdot \mathbf{n}) = p_0\,\mathbf{n} + \sigma \kappa \, \mathbf{n} \quad \text{on } S(t) \,, \tag{5}$$

$$\mathbf{v} = \mathbf{V}(\mathbf{z}, t) \quad \text{on } B \,. \tag{6}$$

Here V_n is the speed of the displacement of $S(t)$ in the direction of its inward normal vector $\mathbf{n}$, κ is the sum of the principal curvatures, $p_0 = p_0(t)$ is the unknown pressure in the drop, and $\mathbf{V}(\mathbf{z}, t)$ is the velocity of the rollers and the mill walls. The problem has to be completed with the initial condition for the drop shape.

The solution of this problem will proceed along the lines described in [1] using the method of matched asymptotic expansions and employing complex variable techniques. In particular, the first step includes the solution of the outer problem with no drop effect, which can be solved by the boundary-element method. It is worth noting that, due to linearity of the Stokes system, the outer problem can be solved just once for the unit angular velocity of the rollers, and then the solution is simply multiplied by the actual angular velocity. Then the local expansion of the flow field is used as the far field for the inner free-boundary problem in the infinite domain exterior to the drop. Since the flow domain becomes simply connected, the use of conformal mappings is advantageous.

Identifying each plane vector with a complex number, let us use the notation $z = x + \mathrm{i}\,y$ for a position point, and $v = v_x + \mathrm{i}\,v_y$ for velocity, where x and y are the Cartesian coordinates of the flow plane. The two-dimensional version of the problem can be reformulated in terms of a bianalytic stress–stream function, $w = \varphi + \mathrm{i}\,\psi$, where φ is Airy's stress function and ψ is the stream function [2]. The bianalyticity condition implies the Goursat representation $w = w_0(z, t) + \bar{z}\,w_1(z, t)$, where $w_0(z, t)$ and $w_1(z, t)$ are time-dependent analytic functions in z. Herein and throughout this paper an overbar is used to denote the complex conjugate. Note that

$$v = \mathrm{i}\,\nabla \psi \,, \quad p = p_0 - \mu\,\Delta\varphi \,, \quad F(\mathrm{d}z) = \mathrm{d}\,(\mathrm{i}\,p_0\,z - 2\mu\,\mathrm{i}\,\nabla\varphi) \,,$$

where

$$F(\mathrm{d}z) = \mathrm{i}\,\left(p\,\mathrm{d}z + 2\mu\,\frac{\partial v}{\partial \bar{z}}\,\mathrm{d}\bar{z}\right)$$

is the stress exerted at the element $\mathrm{d}z$. For irrotational flow, φ becomes the velocity potential.

The inner problem is a particular case of the more general problem: find the evolution of a drop placed in quasi-steady Stokes flow with the given asymptotics at infinity

$$w = W(z, t) \equiv W_0(z, t) + \bar{z}\,W_1(z, t) \,,$$

where $W_0(z,t)$ and $W_1(z,t)$ are some entire analytic functions. It is straightforward to demonstrate that Eqs. (5) and (6) reduce to the problem [1]

$$V_n = \frac{\partial \psi}{\partial s} \ , \quad \varphi = 0 \ , \quad 2\mu \frac{\partial \varphi}{\partial n} = \sigma \quad \text{on } S(t) \ , \tag{7}$$

$$w(z,t) = W(z,t) + \frac{p_0(t)}{4\mu}|z|^2 + O(1) \quad \text{as } z \to \infty \ . \tag{8}$$

Since the stress function φ is uniquely determined by v and p apart from a linear function, integration constants are omitted.

3 Exact solution

By Riemann's theorem, there exists a conformal mapping of the unit disc $\mathcal{G} = \{|\zeta| < 1\}$ onto the flow domain $\mathcal{D}(t)$, given by $z = z(\zeta, t)$. Moreover, this mapping function is unique if we require that

$$\lim_{\zeta \to 0} [\zeta \, z(\zeta, t)] > 0 \ .$$

The partial time derivative of $z(\zeta, t)$ will be denoted by $z_t(\zeta, t)$.

Theorem 1 *The time-dependent conformal mapping $z(\zeta, t)$ satisfies the non-local evolutionary problem*

$$\Re \left\{ \overline{e^{i\vartheta} z'(e^{i\vartheta}, t)} \, z_t \left(e^{i\vartheta}, t \right) - \mathbf{L} \left[\overline{z(e^{i\nu}, t)} \, z_t \left(e^{i\nu}, t \right) + U \left(e^{i\nu}, t \right) \right] \left(e^{i\vartheta} \right) \right\} = 0 \ , \tag{9}$$

where

$$\mathbf{L} \left[f \left(e^{i\nu} \right) \right] \left(e^{i\vartheta} \right) = \frac{\partial}{\partial \vartheta} \left[\frac{1}{4\pi} \int_{-\pi}^{\pi} f \left(e^{i\nu} \right) \cot \frac{\vartheta - \nu}{2} \, d\nu + \frac{i}{2} f \left(e^{i\vartheta} \right) \right] \ ,$$

$$U(\zeta, t) = U_0(\zeta, t) + \overline{z \left(1/\overline{\zeta}, t \right)} \, U_1(\zeta, t) \ ,$$

$$U_0(\zeta, t) = 2W_0[z(\zeta, t), t] \ ,$$

$$U_1(\zeta, t) = 2W_1[z(\zeta, t), t]$$
$$- \frac{\zeta z'(\zeta, t)}{2\pi} \int_{-\pi}^{\pi} \Re \left[\frac{2W_1 \left[z \left(e^{i\vartheta}, t \right), t \right]}{e^{i\vartheta} z' \left(e^{i\vartheta}, t \right)} + \frac{\sigma}{2\mu \left| z' \left(e^{i\vartheta}, t \right) \right|} \right] \frac{e^{i\vartheta} + \zeta}{e^{i\vartheta} - \zeta} \, d\vartheta \ .$$

Proof: Let us change variables $z = z(\zeta, t)$ retaining the same notation for the unknown functions, for example,

$$w(\zeta, t) = w_0(\zeta, t) + \overline{z(\zeta, t)} \, w_1(\zeta, t) \ .$$

180

Then Eqs. (7) and (8) reduce to

$$\Re\left[\overline{e^{i\vartheta} z'\left(e^{i\vartheta}, t\right)} z_t\left(e^{i\vartheta}, t\right)\right]$$
$$+ \frac{\partial}{\partial\vartheta}\Im\left[w_0\left(e^{i\vartheta}, t\right) + \overline{z\left(e^{i\vartheta}, t\right)} w_1\left(e^{i\vartheta}, t\right)\right] = 0 , \tag{10}$$

$$\Re\left[w_0\left(e^{i\vartheta}, t\right) + \overline{z\left(e^{i\vartheta}, t\right)} w_1\left(e^{i\vartheta}, t\right)\right] = 0 , \tag{11}$$

$$\Re\left[\frac{2w_1\left(e^{i\vartheta}, t\right) - z_t\left(e^{i\vartheta}, t\right)}{e^{i\vartheta} z'\left(e^{i\vartheta}, t\right)}\right] + \frac{\sigma}{2\mu\left|z'\left(e^{i\vartheta}, t\right)\right|} = 0 , \tag{12}$$

$$w_0(\zeta, t) = W_0[z(\zeta, t), t] + O(1) , \quad \zeta \to 0 , \tag{13}$$

$$w_1(\zeta, t) = W_1[z(\zeta, t), t] + \frac{p_0(t)}{4\mu} z(\zeta, t) + O(1) , \quad \zeta \to 0 . \tag{14}$$

In terms of the functions $U_0(\zeta, t)$ and $U_1(\zeta, t)$ the unique solution of the problem (12) and (14) takes the explicit form

$$2w_1(\zeta, t) = U_1(\zeta, t) + z_t(\zeta, t) .$$

Likewise, solving the Schwartz problem (11) and (13) for $w_0(\zeta, t)$, we find

$$2w_0(\zeta, t) = U_0(\zeta, t) + i\, C(t)$$
$$- \frac{1}{2\pi} \int\limits_{-\pi}^{\pi} \Re\left\{U_0\left(e^{i\vartheta}, t\right) + \overline{z\left(e^{i\vartheta}, t\right)}\left[U_1\left(e^{i\vartheta}, t\right) + z_t\left(e^{i\vartheta}, t\right)\right]\right\} \frac{e^{i\vartheta} + \zeta}{e^{i\vartheta} - \zeta}\, d\vartheta ,$$

where $C(t)$ is a real function. The proof of Theorem 1 is completed by inserting $w_0(\zeta, t)$ and $w_1(\zeta, t)$ into the right-hand side of Eq. (10). $Q.E.D.$

Let us transform the boundary-value problem (9) to a functional equation assuming that the conformal mapping $z(\zeta, t)$ is a rational function (say, a Padé approximant). It is straightforward to check that

$$\mathbf{L}\left[f\left(e^{i\nu}\right)\right]\left(e^{i\vartheta}\right) = \sum_{k=1}^{\infty} k\, f_{-k} e^{-ik\vartheta} , \quad f_{-k} = \frac{1}{2\pi} \int\limits_{-\pi}^{\pi} f\left(e^{i\vartheta}\right) e^{ik\vartheta}\, d\vartheta .$$

In particular, the operator $\mathbf{L}$ transforms to zero those and only those functions that can be analytically continued from the circumference ∂G onto the whole disc G, and therefore the function $U(\zeta, t)$ can be replaced with its principal part without loss of generality. The operator $\mathbf{L}$ can be expressed in terms of boundary values of an analytic function, namely,

$$\overline{\mathbf{L}\left[f\left(e^{i\nu}\right)\right]\left(e^{i\vartheta}\right)} = e^{i\vartheta}\Lambda'\left(e^{i\vartheta}\right) ,$$

181

where

$$\Lambda(\zeta) = \frac{1}{2\pi i} \int_{\partial \mathcal{G}} \frac{\overline{f(\tau)}\,d\tau}{\tau - \zeta} \ , \quad |\zeta| < 1 \ .$$

Let us compute $\mathbf{L}\left[f\left(e^{i\nu}\right)\right]\left(e^{i\theta}\right)$ for a function $f(\zeta)$ being analytic within $\mathcal{G}$ with a discrete set of singularities. Using the formula

$$f_{-k} = \sum_{\omega \in \mathcal{G}} \operatorname*{res}_{\zeta = \omega}\left[f(\zeta)\,\zeta^{k-1}\right] \ ,$$

one finds

$$\mathbf{L}\left[f\left(e^{i\nu}\right)\right](\tau) = \sum_{\omega \in \mathcal{G}} \sum_{k=1}^{\infty} \operatorname*{res}_{\zeta = \omega}\left[f(\zeta)\,\zeta^{k-1}\right] k\tau^{-k} \ .$$

For instance, if $f(\zeta)$ is a meromorphic function in $\mathcal{G}$,

$$f(\zeta) = \zeta^{-p} F_0(\zeta) + \sum_{j=1}^{m} \frac{F_j(\zeta)}{\zeta - a_j} \ , \quad 0 < |a_j| < 1 \ ,$$

where $F_j(\zeta)$ are regular analytic functions, then

$$\overline{\mathbf{L}\left[f\left(e^{i\nu}\right)\right](\tau)} \equiv \tau \Lambda'(\tau) = \sum_{k=1}^{p} \frac{k\,\overline{F_0^{(p-k)}(0)}}{(p-k)!}\tau^k + \tau \sum_{j=1}^{m} \frac{\overline{F_j(a_j)}}{(1 - \bar{a}_j \tau)^2} \ . \tag{15}$$

Finally, let us effect the decomposition

$$\overline{\tau z'(\tau, t)}\, z_t(\tau, t) = \Phi_+(\tau, t) - \Phi_-(\tau, t) \ , \quad |\tau| = 1 \ ,$$

where $\Phi_\pm(\zeta, t)$ are analytic functions in the interior and exterior of the disc $\mathcal{G}$, respectively, with $\Phi_-(\infty, t) = 0$. These functions are given by the Cauchy-type integral

$$\Phi_\pm(\zeta, t) = \frac{1}{2\pi i} \int_{\partial \mathcal{G}} \frac{\overline{\tau z'(\tau, t)}\, z_t(\tau, t)}{\tau - \zeta}\,d\tau \ ,$$

where the subscript '+' is taken for $|\zeta| < 1$, and '−' for $|\zeta| > 1$. Then Eq. (9) transforms to

$$\Re\left[\Phi_+(\tau, t) - \overline{\Phi_-\left(1/\bar{\tau}, t\right)} - \tau \Lambda'(\tau, t)\right] = 0 \ ,$$

where $\tau \Lambda'(\tau, t)$ is equal to

$$\overline{\mathbf{L}\left[z\left(e^{i\nu}, t\right) z_t\left(e^{i\nu}, t\right) + U\left(e^{i\nu}, t\right)\right](\tau)} \ .$$

This immediately gives the equation

$$\Phi_+(\zeta, t) - \overline{\Phi_-\left(1/\bar{\zeta}, t\right)} - \zeta \Lambda'(\zeta, t) = i\,\Im\left[\Phi_+(0, t)\right] \ , \tag{16}$$

182

which is equivalent to (9).

It is worth noting that integrating Eq. (9) by ϑ (or, equivalently, putting $\zeta = 0$ in Eq. (16)) gives the conservation of the drop area:

$$\Re \sum_{w \in G} \operatorname*{res}_{\zeta = w} \left[z \left(1 / \bar{\zeta}, t \right) z'(\zeta, t) \right] = -\frac{A}{\pi} , \tag{17}$$

which is written in terms of residues, provided that $z(\zeta, t)$ can be analytically continued onto the whole complex plane as a single-valued function with a discrete set of singularities.

Following [1], let us confine ourselves to the polynomial far-field flow (1) and (2) with G being some function of time t but the geometry-dependent parameter q being constant. In this case, rendering all the distances dimensionless by the drop radius R, the stress–stream function assumes the form

$$W(z, t) = -\frac{1}{2} \lambda(t) \left[1 + \varepsilon |z|^2 \right] z^2 ,$$

where the capillary number $\lambda(t)$ is a function of time t but the geometry-dependent parameter ε is constant.

Theorem 2 *The evolutionary problem (9) with*

$$W_0(z, t) = -\frac{1}{2} \lambda(t) z^2 , \quad W_1(z, t) = -\frac{1}{2} \lambda(t) \varepsilon z^3 ,$$

admits the exact solution

$$z(\zeta, t) = \frac{\alpha(t) + \beta(t) \zeta^2}{\zeta [1 - \varepsilon \alpha(t) \beta(t) \zeta^2]} , \tag{18}$$

where $\alpha(t)$ and $\beta(t)$ are time-dependent real coefficients satisfying the equations

$$\alpha = \sqrt{\frac{2 (1 + \beta^2)}{A_\varepsilon(\beta) + \sqrt{A_\varepsilon^2(\beta) - B_\varepsilon(\beta)}}} , \tag{19}$$

$$\frac{d\beta}{dt} = D_\varepsilon(\alpha, \beta) \left[\lambda(t) - C_\varepsilon(\alpha, \beta) \right] , \tag{20}$$

where

$$A_\varepsilon(\beta) = 1 - \varepsilon \beta^2 \left[2 (1 - \varepsilon) + \varepsilon \beta^2 \right] ,$$

$$B_\varepsilon(\beta) = 4\varepsilon^2 \beta^2 \left[3 + (2 + \varepsilon) \varepsilon \beta^2 \right] \left(1 + \beta^2 \right) ,$$

$$C_\varepsilon(\alpha, \beta) = \frac{\beta}{\pi \alpha} \int_0^\pi \frac{d\vartheta}{|\alpha (1 - 3\varepsilon \alpha \beta e^{i\vartheta}) - \beta e^{i\vartheta} (1 + \varepsilon \alpha \beta e^{i\vartheta})|} ,$$

$$D_\varepsilon(\alpha, \beta) = \alpha^3 \left(1 + \varepsilon\,\beta^2\right)\left(1 - \varepsilon\,\beta^2 - 3\varepsilon^2\,\alpha^2\,\beta^2 - \varepsilon^3\,\alpha^2\,\beta^4\right)$$
$$\times \left(1 - 3\varepsilon\,\beta^2 - 5\varepsilon^2\,\alpha^2\,\beta^2 - \varepsilon^3\,\alpha^2\,\beta^4\right)\Big/\left[\alpha^2 - \varepsilon\,\beta^2\left(4\alpha^2 + \beta^2\right)\right.$$
$$\left. - 4\varepsilon^2\,\alpha^2\,\beta^2\left(2\alpha^2 + \beta^2\right) + \varepsilon^4\,\alpha^4\,\beta^4\left(9\alpha^2 + 2\beta^2\right) + \varepsilon^5\,\alpha^4\,\beta^6\left(6\alpha^2 - \beta^2\right)\right] \ .$$

Proof: Assuming the more general form of the conformal mapping

$$z(\zeta, t) = \frac{\alpha(t) + \beta(t)\,\zeta^2}{\zeta\left[1 - \gamma(t)\,\zeta^2\right]} \ ,$$

one obtains

$$z_t(\zeta, t) = \frac{\left[1 - \gamma(t)\,\zeta^2\right]\left[\frac{d\alpha(t)}{dt} + \frac{d\beta(t)}{dt}\,\zeta^2\right] + \left[\alpha(t) + \beta(t)\,\zeta^2\right]\frac{d\gamma(t)}{dt}\,\zeta^2}{\zeta\left[1 - \gamma(t)\,\zeta^2\right]^2} \ ,$$

$$\zeta\,z'(\zeta, t) = -\frac{\alpha(t) - \beta(t)\,\zeta^2 - \left[3\alpha(t) + \beta(t)\,\zeta^2\right]\gamma(t)\,\zeta^2}{\zeta\left[1 - \gamma(t)\,\zeta^2\right]^2} \ ,$$

$$z(1/\zeta, t) = \zeta\,\frac{\beta(t) + \alpha(t)\,\zeta^2}{\zeta^2 - \gamma(t)} \ ,$$

$$(1/\zeta)\,z'(1/\zeta, t) = \zeta\,\frac{\left[\beta(t) + 3\alpha(t)\,\zeta^2\right]\gamma(t) + \left[\beta(t) - \alpha(t)\,\zeta^2\right]\zeta^2}{\left[\zeta^2 - \gamma(t)\right]^2} \ .$$

Furthermore, since

$$W_0[z(\zeta, t), t] = -\frac{\lambda(t)\,\alpha^2(t)}{2\zeta^2}\left[1 + O\left(\zeta^2\right)\right] \ ,$$

$$W_1[z(\zeta, t), t] = -\frac{\lambda(t)\,\varepsilon\,\alpha^3(t)}{2\zeta^3}\left[1 + O\left(\zeta^2\right)\right] \ ,$$

the principal part of $U(\zeta, t)$ is given by

$$\zeta z'(\zeta, t)\,z(1/\zeta, t)\left[\frac{\lambda(t)\,\varepsilon\,\alpha^2(t)\,(1 - \zeta^4)}{\zeta^2} - S(\zeta, t)\right] - \frac{\lambda(t)\,\alpha^2(t)}{\zeta^2} \ ,$$

where

$$S(\zeta, t) = \frac{1}{2\pi}\int_{-\pi}^{\pi}\left|z'\left(e^{i\nu}, t\right)\right|^{-1}\frac{e^{i\nu} + \zeta}{e^{i\nu} - \zeta}\,d\nu \ , \quad |\zeta| < 1 \ .$$

Using the double symmetry of the drop shape, it is straightforward to check that

$$S(\zeta, t) = \hat{S}\left[\zeta^2; \alpha(t), \beta(t), \gamma(t)\right] \ ,$$

where

$$\hat{S}[x; \alpha, \beta, \gamma] = \frac{1}{2\pi}\int_{-\pi}^{\pi}\frac{\left|1 - \gamma\,e^{i\vartheta}\right|^2}{\left|\alpha - \beta\,e^{i\vartheta} - (3\alpha + \beta\,e^{i\vartheta})\,\gamma\,e^{i\vartheta}\right|}\,\frac{e^{i\vartheta} + x}{e^{i\vartheta} - x}\,d\vartheta \ .$$

184

Insofar as

$$\zeta z'(\zeta,t)\, z(1/\zeta,t) = -\frac{\beta(t)+\alpha(t)\,\zeta^2}{[\zeta^2-\gamma(t)]\,[1-\gamma(t)\,\zeta^2]^2}$$
$$\times\left\{\alpha(t)-\beta(t)\,\zeta^2-\left[3\alpha(t)+\beta(t)\,\zeta^2\right]\gamma(t)\,\zeta^2\right\}\,,$$

the function $U(\zeta,t)$ has a double pole at $\zeta=0$ and two simple poles at $\zeta=\pm\sqrt{\gamma(t)}$. Moreover, the function

$$z(1/\zeta,t)\, z_t(\zeta,t) = \frac{\beta(t)+\alpha(t)\,\zeta^2}{[\zeta^2-\gamma(t)]\,[1-\gamma(t)\,\zeta^2]^2}$$
$$\times\left\{\left[1-\gamma(t)\,\zeta^2\right]\left[\frac{d\alpha(t)}{dt}+\frac{d\beta(t)}{dt}\,\zeta^2\right]+\left[\alpha(t)+\beta(t)\,\zeta^2\right]\frac{d\gamma(t)}{dt}\,\zeta^2\right\}$$

has just two simple poles at $\zeta=\pm\sqrt{\gamma(t)}$. Bearing in mind Eq. (15), one obtains

$$\overline{\mathbf{L}\left[z\,(e^{-i\nu},t)\, z_t\,(e^{i\nu},t)+U\,(e^{i\nu},t)\right](\tau)} = 2\tau^2\left[L_0(t)+\frac{L_1(t)+L_2(t)}{[1-\gamma(t)\,\tau^2]^2}\right]\,,$$

where

$$L_0(t)\equiv\lim_{\zeta\to0}\left[U(\zeta,t)\,\zeta^2\right]=\lambda(t)\,\alpha^2(t)\left[\frac{\varepsilon\,\alpha(t)\,\beta(t)}{\gamma(t)}-1\right]\,,$$

$$L_1(t)=\lim_{\zeta^2\to\gamma}\left\{U(\zeta,t)\left[\zeta^2-\gamma(t)\right]\right\}=\frac{\beta(t)+\alpha(t)\,\gamma(t)}{[1-\gamma^2(t)]^2}$$
$$\times\left\{\alpha(t)-\beta(t)\,\gamma(t)-[3\alpha(t)+\beta(t)\,\gamma(t)]\,\gamma^2(t)\right\}$$
$$\times\left\{\hat{S}[\gamma(t);\alpha(t),\beta(t),\gamma(t)]-\lambda(t)\,\varepsilon\,\alpha^2(t)\,\frac{1-\gamma^2(t)}{\gamma(t)}\right\}\,,$$

$$L_2(t)\equiv\lim_{\zeta^2\to\gamma}\left\{z(1/\zeta,t)\, z_t(\zeta,t)\left[\zeta^2-\gamma(t)\right]\right\}=\frac{\beta(t)+\alpha(t)\,\gamma(t)}{[1-\gamma^2(t)]^2}$$
$$\times\left\{\left[1-\gamma^2(t)\right]\left[\frac{d\alpha(t)}{dt}+\frac{d\beta(t)}{dt}\,\gamma(t)\right]+[\alpha(t)+\beta(t)\,\gamma(t)]\frac{d\gamma(t)}{dt}\,\gamma(t)\right\}\,.$$

Using the representation

$$(1/\zeta)\, z'(1/\zeta,t)\, z_t(\zeta,t) = \frac{Z\,(\zeta^2,t)}{[\zeta^2-\gamma(t)]^2}$$

with

$$Z(x,t) = \frac{[\beta(t)+3\alpha(t)\,x]\,\gamma(t)+[\beta(t)-\alpha(t)\,x]\,x}{[1-\gamma(t)\,x]^2}$$
$$\times\left\{[1-\gamma(t)\,x]\left[\frac{d\alpha(t)}{dt}+\frac{d\beta(t)}{dt}\,x\right]+[\alpha(t)+\beta(t)\,x]\frac{d\gamma(t)}{dt}\,x\right\}\,,$$

one derives

$$\Phi_+(\zeta, t) = \frac{Z(\zeta^2, t) - Z(\gamma(t), t) - Z'(\gamma(t), t)[\zeta^2 - \gamma(t)]}{[\zeta^2 - \gamma(t)]^2} ,$$

$$\Phi_-(1/\zeta, t) = -\zeta^2 \frac{Z'(\gamma(t), t)[1 - \gamma(t)\zeta^2] + Z(\gamma(t), t)\zeta^2}{[1 - \gamma(t)\zeta^2]^2} .$$

Thus, Eq. (16) reduces to $F(\zeta^2, t) = 0$, where

$$F(x, t) = \frac{Z(x, t) - Z(\gamma(t), t) - [x - \gamma(t)]Z'(\gamma(t), t)}{[x - \gamma(t)]^2}$$

$$+ x \frac{[1 - \gamma(t)x]Z'(\gamma(t), t) + Z(\gamma(t), t)x}{[1 - \gamma(t)x]^2} - 2x \left[L_0(t) + \frac{L_1(t) + L_2(t)}{[1 - \gamma(t)x]^2}\right] .$$

It is straightforward to check that

$$[1 - \gamma(t)x]^2 F(x, t) = P_0(t) + P_1(t)x + P_2(t)x^2 + P_3(t)x^3 .$$

The condition $P_0(t) = 0$ is equivalent to Eq. (17) which reduces to

$$\frac{\alpha^2(t) - \beta^2(t) - 2\alpha(t)\beta(t)\gamma(t)[1 + \gamma^2(t)] - [3\alpha^2(t) + \beta^2(t)]\gamma^2(t)}{[1 - \gamma^2(t)]^2} = 1 . \tag{21}$$

Putting $P_3(t)$ to zero gives $L_0(t) = 0$ which is equivalent to $\gamma = \varepsilon\,\alpha\,\beta$. In particular, Eq. (21) reduces to a second-order algebraic equation for α^2, whose solution is given by Eq. (19). Finally, $P_2(t) \equiv -\gamma^2(t)P_0(t) = 0$, and hence the last non-trivial equation, after some working, reduces to Eq. (20) with

$$C_\varepsilon(\alpha, \beta) = \frac{\beta\,\hat{S}[\varepsilon\,\alpha\,\beta; \alpha, \beta, \varepsilon\,\alpha\,\beta]}{\alpha\,(1 - \varepsilon^2\,\alpha^2\,\beta^2)} .$$

Q.E.D.

4 Results and discussion

The system of ordinary differential equations with the initial condition $\beta(0) = 0$ ($\alpha(0) = 1$) was solved numerically by the Runge–Kutta algorithm. For definiteness, the capillary number was selected as a piecewise linear function

$$\lambda(t) = 2\lambda_0 \min(t/T, 1 - t/T) , \quad 0 \le t \le T ,$$

with $\lambda_0 = 0.7$ and $T = 1000$. The curve $\{\lambda(t), \delta(t)\}$ for $\varepsilon = 0.001$ is shown in Fig. 4. It is seen that, as $\lambda(t)$ slowly increases with time ($t < T/2$), the unsteady solution is very close to the steady-state solution up to the turning point 1 of Fig. 2, and then it heads to point 2 of the monotonic branch of the response diagram. *Vice versa*, with decreasing $\lambda(t)$, $t > T/2$, the drop deformation is determined by the same branch until it becomes non-monotonic at point 3, and then jumps to point 4. As is expected, some transient effects are seen which become less tangible for greater T.

186

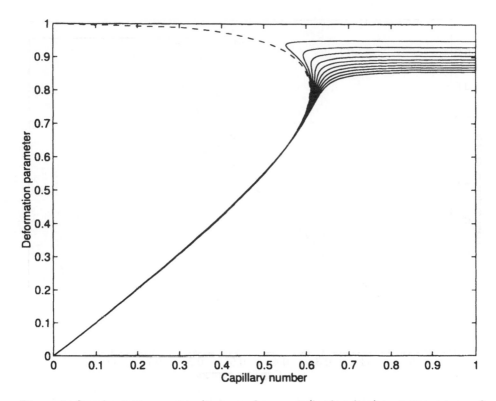

Figure 1: Steady-state response diagrams for $\varepsilon = 0$ (broken line) to 0.01 in steps of 0.001.

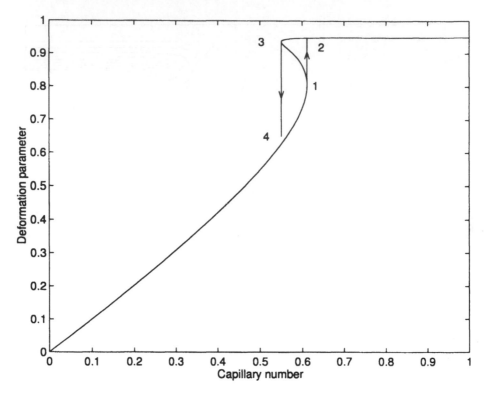

Figure 2: The typical hysteresis loop suggested by the response diagram with $\varepsilon = 0.001$.

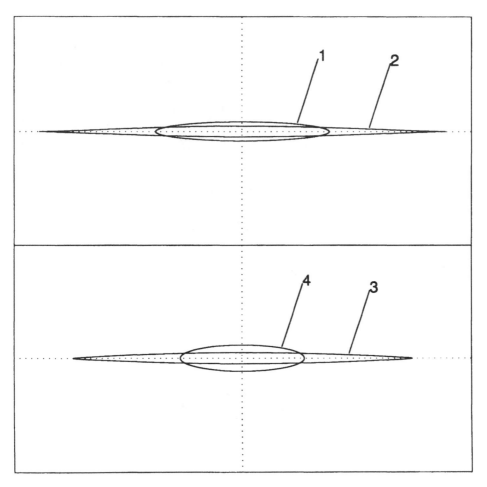

Figure 3: The drop shapes corresponding to the transitions shown in Fig. 2.

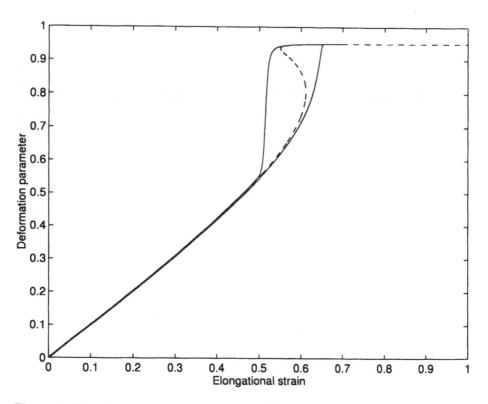

Figure 4: Time-dependent response diagram $\{\lambda(t), \delta(t)\}$ for $\varepsilon = 0.001$; the typical hysteresis loop is seen.

References

[1] ANTANOVSKII, L. K. Formation of a pointed drop in Taylor's four-roller mill. *J. Fluid Mech.* *327* (1996), 325–341.

[2] LANGLOIS, W. E. *Slow Viscous Flow.* Macmillan, New York, 1964.

[3] TAYLOR, G. I. The formation of emulsions in definable fields of flow. *Proc. R. Soc. Lond.* A *146* (1934), 501–523.

Leonid K. Antanovskiĭ
Moldflow International Pty Ltd
259–261 Colchester Rd, Kilsyth
Victoria 3137, Australia

Roger H. J. Grimshaw
Department of Mathematics
Monash University,
Clayton, Victoria 3168, Australia

K. HUTTER

Mathematical Foundation of Ice Sheet and Ice Shelf Dynamics. A Physicist's View

We present an overview of the physical behaviour of ice sheets and ice shelves subject to climate driving. The complicated system is first described in its entirety. Then, we briefly quote the model equations for polythermal ice sheets. These describe free moving boundary problems at various complexities. The shallow ice approximation simplifies these equations. Numerical solutions to these equations have been constructed for many of the largest ice sheets.

1   Introduction

Glaciers and ice sheets are sensitive elements of the climate system. They advance and retreat according to whether the mass and heat balances give rise to a surplus of ice deposition or a devastation due to the dominant melting processes. It is known that Scandinavia, Great Britain and Northern Germany were buried beneath the large Fenno-Scandinavian Ice Sheet, and that Canada and the Northern United States were equally covered by the so-called Laurentide Ice Sheet. Both had so much water of the hydrological cycle locked into their masses that the ocean surface was more than 100 m lower than it is today.

The world's glaciers have similarly retreated during the warming period of the recent past. Some of the climate warming is anthropogenic. It is therefore of considerable interest to know to what extent the man-made greenhouse effect can be made responsible for possible future ocean surface rises.

Mathematical glaciologists have in the past 20 years been concerned with the construction of continuum models for the thermo-mechanical behaviour of ice sheets and ice shelves in their geophysical environment. Both mechanical and thermal processes focus on involving the velocity and thermal fields and the motion of the various surfaces that bound the system. Often phase changes arise along these and thus couple the mechanical and thermal fields in a nontrivial form.

We sketch in §2 the free boundary problem that describes the physical system in its entirety for which it should be solved. In §3 a subsystem with portions being approximated is addressed. This model is then simplified for shallow geometries in §4 where results of a restricted analysis are also presented. The reference list gives an account of our own work on the subject; [2,15,16,17,18,19] review the subject.

2   The free boundary problem that should be solved

Consider Figure 1. It shows a large (hypothetical) ice mass partially resting on the solid earth and partially floating on ocean water. Such a configuration is comprised of an *ice sheet*, defined as an ice mass resting on solid earth – also called *inland ice* – and an *ice shelf* that is primarily fed from the inland ice and is floating on the water. The most important land-based ice sheets are Greenland and Antarctica; the most prominent ice shelves are the Ross and Ronne-Filchner Ice Shelves in Antarctica.

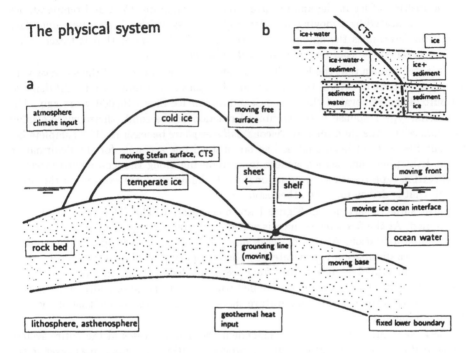

Figure 1 a) Sketch of an ice sheet-ice shelf configuration, partially resting on solid rockbed, partially floating on ocean water. The rockbed also deforms as a combined effect of the solid thermovisco-elastic lithosphere and the viscous fluid asthenosphere. Cold and temperate ice are separated by the cold-temperate transition surface (CTS) which acts as a Stefan-like boundary. The system in focus is delimited by a fixed lower boundary within the lithosphere. The various surfaces: (i) free surface, (ii) ice-ocean interface (iii) base, (iv) front and (v) grounding line are all moving and most are non-material. The system is driven by the climate input from the atmosphere and the geothermal heat from the Earth's interior.

b) Close-up of the ice-rockbed interface in the vicinity of the CTS. On this closer view the interface becomes diffuse and appears as a transition region from ice to ice+sediment to sediment+ice to sediment (cold part) or from ice+water to ice+water+sediment to sediment+water. The extent of this mixture region is just a few meters.

Ice sheets are often *polythermal*, i.e., they consist of *cold ice* regions where the ice temperature is strictly below the melting point and *temperate ice* regions, where the ice temperature is at the melting point. Since in these regions the heat produced by the viscous deformation cannot be stored by raising the temperature of the ice, temperate ice must be a mixture of ice and water, of which the latter is produced by the melting of ice in the amount available by dissipation. Cold and temperate ice are thus separated by a moving *singular surface* across which a Stefan-type boundary condition determines the amount of ice that is melted or water that is frozen. This surface is called the *cold-temperate-transition surface* (CTS).

It is well-known that the Earth's crust deforms under the load of the ice mass with a relaxation time of several thousand years; for instance, Scandinavia is still rising as a result of the removal of the Fenno-Scandinavian Ice Sheet $10,000$ years ago. The relaxation-type response is the combined reaction of the solid (thermovisco-elastic) *lithosphere* and the (non-linearly viscous) *asthenosphere* beneath it. The complete behaviour has still not been analysed so far; instead one accounts for the deformation of the lithosphere and asthenosphere by introducing along the ice-rockbed interface a relaxation-type resistance proportional to the buoyancy force exerted on the lithosphere that is driven into the athenosphere.

Most surfaces sketched in Figure 1 are *non-material*. The CTS has already been mentioned. At the *free surface* mass is added or subtracted by precipitation or melting due to solar irridation. At the *ice-ocean interface* melting or freezing occurs according to how much dissipation is generated by the boundary layer flow in the ocean water; but ice is also added by *accretion of frazil ice* suspended in the arctic water. Ice at the ice-rock interface either adheres to the bed when the ice is cold or slides over the base when it reaches the melting temperature there, also giving rise to melting of basal ice due to sliding friction.

Because it turns out that time-dependent thermal processes in the entire system are slowed down by the heat conduction processes within the upper-most layer in the rockbed, the prescribed geothermal heat flow is postulated to act on a fixed boundary approximately 5 km below the ice-rockbed interface.

Somewhat special is the moving *ice front* at the shelf margin. It is here that the ice sheet loses mass through *calving of icebergs*; so, in addition to a kinematic equation governing its motion, a dynamic equation that quantifies the mass loss through calving must be prescribed.

The *grounding line* defines the location where the ice stops touching the ground and the sheet configuration goes over into the shelf configuration. Its motion is a significant component in coupled dynamics of sheets and shelves, but it is still poorly understood today.

The external world that drives the system acts on it through (i) the climate input from the atmosphere via parameterization of the precipitation function (snowfall minus melting) or a full General Circulation Model of the atmosphere, (ii) a parameterization of the calving rate dependent on the physical processes at the margin of the shelf and (iii) the prescription of the geothermal heat flow at some depth beneath the Earth's surface.

The above description assumes that the interface between the various regions is infinitely thin, i.e., true singular surfaces. Observations indicate that the process at the base of an ice sheet is often more complicated than just the sliding of two surfaces one over the other - the ice sole over the rockbed. A close-up of the basal region in the vicinity of the CTS indicates that the interface is not sharp but diffuse with sediment, ice (and water in the temperate case) coexisting at different concentrations of the constituents, see Figure 1b. Such a soft, diffuse interface is actually a mixture, may deform, and, depending on the water pressure, react as a slurry or hard granular soil. Certain rapid sliding processes require such a formulation, [5,6].

In summary, the ice-sheet, ice-shelf configuration of Figure 1 requires field equations and boundary or transition conditions according to the list below:

Proposals of model equations could be made for all of these, but there is little hope that concrete results could be inferred from them. For these reasons we usually looked at subsystems. One of these will be scrutinized in the next section.

Field equations	Boundary, transition conditions
- Cold ice	- Free surface
- Temperate ice	- Moving base
- Ocean water	- Ice ocean interface
- Sediment interface layer	- Rockbed boundary
- Rockbed/lithosphere	- Cold-temperate-transition surface (CTS)
- Asthenosphere	- Shelf front
	- Grounding line

3   Thermomechanical Stokes flow for polythermal ice sheet

The full problem outlined in the preceding section being out of reach, mathematical glaciologists have focussed attention to restricted subproblems, one of which will subsequently be outlined: a polythermal ice sheet. We present in Figure 2 the formulation of a wholly cold ice sheet and extend this in Figure 3 to a polythermal sheet.

Typical flow velocities U of ice sheets are a few $100 \mathrm{ma}^{-1}$ and distances L travelled by the ice particles are about $10^6 \mathrm{m}$. Thus, Froude numbers $F = U^2/(gL)$ are 10^{-12} and smaller, implying that accelerations are never significant, so that Stokes flow prevails under all circumstances. By definition, cold ice has temperatures below the melting point. Classically, it is an incompressible heat conducting non-linear viscous fluid with equations as stated in the center of Figure 2 and variables as defined in Table 1. These equations are the continuity equation, the force balance (of pressure gradient, stress-deviator divergence and gravity force), the energy equation (for the temperature field) and the constitutive relation for stretching (symgrad $\mathbf{v}$) and stress deviator $\boldsymbol{\sigma}$, in which $A(T)$ is a temperature-dependent rate factor (whose magnitude varies by three powers of 10 when T varies from -50 C to 0 C) and $f(II_\sigma)$, $II_\sigma = \frac{1}{2}\mathrm{tr}(\sigma^2)$ a stress-dependent fluidity, usually a power law $(II_\sigma)^{(n-1)/2}, n = 3$. In the rockbed the heat equation is solved, and is also shown in Figure 2.

Thermomechanical Stokes flow problem

Cold ice

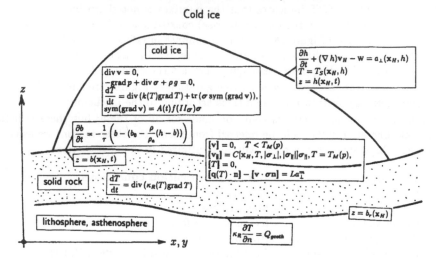

Figure 2. Cold ice sheet resting on solid rock, showing the field equations in the ice and the rock as well as the boundary conditions at the free surface, the ice-rockbed interface and the lower surface of the solid rock. For notation, see also the Table of Notation.

These equations are complemented by boundary conditions at the free surface, the ice-rockbed interface and the lower boundary of the solid rock, all also given in Figure 2. The solid rock lower boundary is prescribed as $z = b(\mathbf{x}_H)$, while the ice-rock bed interface $z = b_r(\mathbf{x}_H, t)$ and the free surface $z = h(\mathbf{x}_H, t)$ are evolving. The corresponding evolution equations are, for the free surface, the kinematic equation $h_t + \dots = a_\perp$, for the ice-rockbed interface, the relaxation equation $b_t = \tau^{-1}[b - (bo - \rho(h - b)/\rho_a)]$ with relaxation time τ. So, the ice-rockbed interface is at rest when the term in braces vanishes; $b = bo$ defines the ice-free ($h = b$) steady interface location, $b = bo - (h - b)/\rho_a$ for a non-trivial ice thickness.

There are further, dynamic conditions that need be fulfilled at the ice-rockbed interface. There is no sliding if $T < T_M$; however, when $T = T_M$, sliding may occur with a viscous law in which the jump of the tangential velocity $[\mathbf{v}_H]$ is proportional to the shear traction $\sigma_\parallel$ with a drag coefficient $C(\mathbf{x}_H, T, |\sigma_\perp|, |\sigma_\parallel|)$ to be explicitly parameterized. Furthermore, the temperature is continuous across it and the latent

196

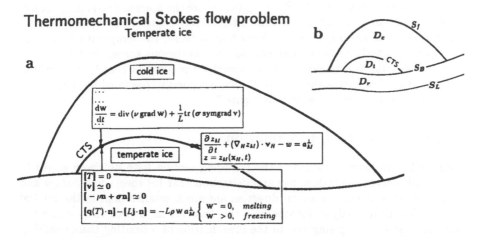

Figure 3 a) Polythermal ice sheet resting on solid rock showing cold and temperate ice regions separating the two by the cold-temperate transition surface (CTS). The figure shows the field equations and boundary conditions that must hold in addition to those of cold ice.

b) Notation used to define the various set regions in the main figure

melting rate is balanced by the jump in heat flow and the power of working due to the sliding of the ice sole over the rockbed. The relevant equations are also shown in Figure 2.

Climate driving is exerted by the accumulation rate, $a_\perp(x_H, t)$ and surface temperature $T_s(x_H, t)$ functions supplied by the climatologist through appropriate parameterizations or GCM-computations of the atmosphere. Further external driving comes from the geothermal heat supplied at the lower boundary of the rockbed, and of course, also by gravity, see [2,15-19].

In *polythermal ice sheets* the field equations for the cold ice and the boundary and transition conditions facing cold ice and at the base are unchanged and are listed in Figure 2. New are only the field equations in the temperate ice zone and the jump conditions along the CTS. The relevant additional equations are stated in Figure 3, [1,13,14,9].

Temperate ice is defined as an incompressible mixture of ice at $T = T_M$ plus water. So, two mass balances for the moisture content ($w = \rho_{water}/\rho_{mixture}$) and the mixture as a whole must be combined with a force and energy balance. The continuity equation and the force balance are the same as for cold ice and the energy equation simply determines the volumetric melting rate $\mathcal{M} = \mathrm{tr}\,(\sigma\,\mathrm{grad}\,v)/L$. So, balance of

197

moisture content is the only field equation stated in Figure 3.

Transition conditions at the CTS $z = z_M(\mathbf{x}_H, t)$ are the kinematic equation involving a surfacial melting rate a_T^m which must alternatively be described by the dynamic conditions involving jump quantities. According to these, the temperature, velocity and tractions (normal and shear stresses) must be continuous across the CTS.

Alternatively, the energy jump condition relates the jump in heat flow minus the jump in moisture flow across the CTS to the surface energy consumed by the melting rate. These equations are also shown in Figure 3.

4 The shallow ice approximation

The full Stokes problems outlined in the preceding section have not been solved to date. Instead, authors introduce the approximation that ice sheets are shallow and variations of fields in the horizontal direction are much smaller than in the vertical direction. A scaling analysis incorporates this fact via an aspect ratio ε and the assumption that $\varepsilon \ll 1$, giving rise to the investigation of a limiting theory valid as $\varepsilon \to 0$, [1,9,18,19]. This limiting theory is governed by the following equations (see Figure 3b for the definitions of the domains):

- Stresses and velocities within $\mathcal{D}_c(t) \cup \mathcal{D}_t(t)$:

$$p = \rho g (h - z),$$

$$\sigma_{xz} = -\rho g (h - z) \frac{\partial h}{\partial x},$$

$$\sigma_{yz} = -\rho g (h - z) \frac{\partial h}{\partial y},$$

$$\sigma = \rho g (h - z) \sqrt{\left(\frac{\partial h}{\partial x}\right)^2 + \left(\frac{\partial h}{\partial y}\right)^2},$$

$$v_x = -\rho g (h - b) C_{(t)} \frac{\partial h}{\partial x} - 2\rho g \frac{\partial h}{\partial x} \int_b^z E A_{(t)}(\cdot) f_{(t)}(\sigma)(h - z') \, dz',$$

$$v_y = -\rho g (h - b) C_{(t)} \frac{\partial h}{\partial y} - 2\rho g \frac{\partial h}{\partial y} \int_b^z E A_{(t)}(\cdot) f_{(t)}(\sigma)(h - z') \, dz',$$

$$v_z = -\int_b^z \left(\frac{\partial v_x}{\partial x} + \frac{\partial v_y}{\partial y}\right) dz' + \frac{\partial b}{\partial t} + v_{x,b} \frac{\partial b}{\partial x} + v_{y,b} \frac{\partial b}{\partial y} - \frac{\partial \dot m_b^w}{\rho}.$$

- Temperature and moisture content within $\mathcal{D}_t(t)$ and $\mathcal{D}_c(t)$, respectively:

$$\frac{\partial T}{\partial t} + v_x \frac{\partial T}{\partial x} + v_y \frac{\partial T}{\partial y} + v_z \frac{\partial T}{\partial z} = \frac{1}{\rho c} \frac{\partial}{\partial z} \left(\kappa \frac{\partial T}{\partial z} \right) + \frac{2}{\rho c} E A(T) f(\sigma) \sigma^2,$$

$$T = T_M = T_0 - \beta(h - z),$$

198

$$\frac{\partial \omega}{\partial t} + v_x \frac{\partial \omega}{\partial x} + v_y \frac{\partial \omega}{\partial y} + v_z \frac{\partial \omega}{\partial z} = \frac{\nu}{\rho} \frac{\partial^2 \omega}{\partial z^2} + \frac{\beta^2}{\rho L} \frac{\partial \kappa}{\partial T}$$

$$+ \frac{c\beta}{L} \left(\frac{\partial h}{\partial t} + v_x \frac{\partial h}{\partial x} + v_y \frac{\partial h}{\partial y} - v_z \right) + \frac{2}{\rho L} E A_t(\omega) f_t(\sigma) \sigma^2 - \frac{1}{\rho} D(\omega).$$

- Evolution equation for the free surface, the CTS and the ice-rockbed interface:

$$\frac{\partial H}{\partial t} = \frac{\partial (h-b)}{\partial t} = -\frac{\partial q_x}{\partial x} - \frac{\partial q_y}{\partial y} + a_\perp^s - a_\perp^m, \quad (q_x, q_y) := \int_b^h (v_x, v_y) \, dz',$$

$$\frac{\partial (z_M - b)}{\partial t} = -\frac{\partial}{\partial x} \int_b^{z_M} v_x \, dz' - \frac{\partial}{\partial y} \int_b^{z_M} v_y \, dz' + a_\perp^M - a_\perp^m,$$

$$\frac{\partial b}{\partial t} = -\frac{1}{\tau} [b - (b_0 - \frac{\rho}{\rho_a} H)].$$

- Temperature in the rockbed within $\mathcal{D}_r(t)$:

$$\frac{\partial T}{\partial t} = \frac{\partial}{\partial z} \left(\kappa_R \frac{\partial T}{\partial z} \right).$$

These equations must be solved subject to the following boundary conditions:

$$T = T_s(x, y, z, t), \quad \text{on} \quad S_f(t),$$

$$T = T_M(x, y, z, t),$$

$$\left[\left[\kappa \frac{\partial T}{\partial z} \right] \right] + L\nu \frac{\partial w}{\partial z} = -\rho L w a_\perp^M, \qquad \left. \begin{array}{c} \\ \\ \end{array} \right\} \quad \text{on} \quad \text{CTM}$$

$$w = 0 \quad \text{(melting)}, \quad w > 0 \quad \text{(freezing)},$$

$$\mathbf{v}_H = -C[\cdot](\sigma_{xx}, \sigma_{yz}),$$

$$-\kappa_R \frac{\partial T_R}{\partial z} + \kappa \frac{\partial T}{\partial z} - \mathbf{v}_H \cdot (\sigma_{xx}, \sigma_{yz}) = L a_\perp^m, \qquad \left. \begin{array}{c} \\ \\ \end{array} \right\} \quad \text{on} \quad S_B(t)$$

$$\kappa_R \frac{\partial T_R}{\partial z} = Q_{geoth}, \quad \text{on} \quad S_L$$

The above free boundary problem has not yet been analysed for a mathematical proof of the existence of solutions; however, numerical solutions were constructed [3,4,7,8,10,11,12,20]. Figure 4 shows the current topography of the Greenland Ice Sheet as observed (left) and as computed with a climate driving through the last

199

250, 000 years (right).

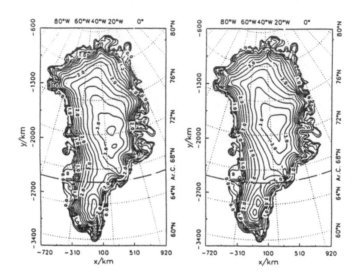

Figure 4. Topography of the Greenland Ice Sheet as observed (left) and as computed for today's climate by R. Greve, [10], [11].

Acknowledgment

The participation in the Conference on **Free Boundary Problems**, held in Heraclion, Crete from June 8 – 13, 1997 and the organization of the **Discussion Session on Mathematical Free Boundary Problems in Glaciology** was made possible by the Max Planck Prize received by us from the A.v. Humboldt Foundation and the Max Planck Society. We thank A. Savvin and D. Baral for TEXperting the manuscript.

References

[1] BLATTER, H. AND HUTTER, K., Polythermal conditions in arctic glaciers, *J. Glaciology*, **37**, 126 (1991), 261 - 269

[2] CALOV, R. AND HUTTER, K., Large scale motion and temperature distributions in land based ice shields - the Greenland Ice Sheet in response to various climatic scenarios,*Archives of Mechanics*, **49** (1997), 919 - 962

[3] CALOV, R. AND HUTTER, K., The thermal regime of the Greenland Ice Sheet to various climate scenarios, *Climate Dynamics*, **12** (1996), 243 - 260

[4] CALOV, R., SAVVIN, A., GREVE, R. AND HUTTER, K., Simulation of the Antarctiv with a 3d polythermal ice sheet model - a support of the EPICA project, *Annals of Glaciology*, **27** (1998), 201 - 206.

[5] DELL' ISOLA, F. AND HUTTER, K., Continuum mechanical modelling of the dissipative processes in the sediment-water layer below glaciers, *Comptes rendue de l'Acad. Francaise* t.325, Ser. IIb (1997), 449 - 456.

[6] DELL' ISOLA, F. AND HUTTER, K., What are the dominant thermomechanical processes in the basal sediment layer of large ice sheets? *Proc. Royal Soc. London* A **454** (1998), 1169 -1196

[7] GREVE, R. AND HUTTER, K., Polythermal three-dimensional modelling of the Greenland Ice Sheet with varied geothermal heat flux. *Annals of Glaciology*, **21** (1995), 8 - 12

[8] GREVE, R. AND MACAYEAL, D. R., Dynamic/thermodynamic simulations of Laurentide ice sheet instability, *Annals of Glaciology*, **23** (1996), 328 - 335

[9] GREVE, R., A continuum-mechanical formulation for shallow polythermal ice sheets, *Phil. Trans. Royal. Soc. London*, A **355** (1997), 921 - 974

[10] GREVE, R., Application of a polythermal three-dimensional ice sheet model to the Greenland Ice Sheet: response to steady-state and transient climate scenarios, *J. Climate*, **10** (1997), 901 - 918

[11] GREVE, R., WEIS, M. AND HUTTER, K., Palaeoclimatic evolution and present conditions of the Greenland Ice Sheet in the vicinity of Summit: An approach by large-scale modelling, *Palaeoclimates*, **2** (2-3) (1998), 133 - 161

[12] HANSEN, I., GREVE, R. AND HUTTER, K., *Application of a polythermal ice model to the Antarctic Ice Sheet. Steady state solution and response to Milankovi'c cycles.* Proc. Fifth International Symposium on Thermal Engineering and Science for Cold Regions, Ottawa 19 - 22 May, 1996, 89 - 96 (eds.: Y. Lee & W. Hallet)

[13] HUTTER, K., A mathematical model of polythermal glacier and ice sheets, *Geophys. Astrophys. Fluid Dynamics*, **21** (1982): 201 - 224

[14] HUTTER, K., BLATTER, H. AND FUNK, M., A model of polythermal glaciers, *J. Geophys. Res.* **93**, 10B (1988): 12205 - 12214

201

[15] HUTTER, K., Dynamics of glaciers and large ice masses, *Annual Reviews of Fluid Mechanics*, **14** (1982): 183 - 230

[16] HUTTER, K., Glacier flow, *American Scientist*, **70** (1982): 26 - 34

[17] HUTTER, K., Mathematical foundation of flow of glaciers and large ice masses, in: *Mathematical Models and Methods in Mechanics*, Banach Center, Publications, Volume **15**, PWN-Polish Scientific Publishers, Warsaw (1985): 277 - 322

[18] HUTTER, K., *Theoretical Glaciology*, Reidel Book Comp., Dordrecht-Boston, 1983

[19] HUTTER, K., Thermo-mechanically coupled ice sheet response: Cold, polythermal, temperate, *Journal of Glaciology*, **39**, 131, (1993), 65 - 86

[20] WEIS, M., HUTTER, K. AND CALOV, R., 250,000 years in history of the Greenland Ice Sheet, *Annals of Glaciology*, **23** (1996), 359 - 363

Table 1. Notation

$A_t \cong A(T), A(\omega)$	Rate factor, depending on temperature or moisture content
$a_\perp^s(\mathbf{x}_H, t)$	Accumulation rate function at the ice surface
$a_\perp^m$	Melting rate at the basal surface of the ice sheet
$a_\perp^M$	Melting rate at the ice-bedrock interface
$b(\mathbf{x}_H, t)$	Ice-bedrock interface function defining its topography
$b_0(\mathbf{x}_H)$	Ice-bedrock interface function for steady basal topography without ice load
$b_r(\mathbf{x}_H)$	Geodetic coordinate of the lower boundary of the solid rockbed
$C[.]$	Viscous drag coefficient for sliding along the ice-rockbed interface
CTS	Cold-temperature transition surface
$f(II_\sigma)$	Creep response function, fluidity depending on the second stress deviator invariant $II_\sigma = \frac{1}{2}\mathrm{tr}(\sigma^2)$
$g, \mathbf{g}$	Gravity constant, vector
$h(\mathbf{x}_H)$	Geodetic coordinate of the free surface of the ice
$\mathbf{j} = -\rho\nu\,\mathrm{grad}\,\omega$	Moisture flux vector
L	Latent heat of fusion
$\mathcal{M}$	Volumetric melting rate of mass in temperate ice
$\mathbf{n}$	Unit normal vector defined on surfaces
p	Pressure, a free variable
Q_{geoth}	Geothermal heat flow $(40 - 100) \times 10^{-3}$ Watts
$\mathbf{q}$	Heat flux vector
T, T_M	Temperature, melting temparature
$T_s(\mathbf{x}_H, t)$	Surface temperature function
t	Time
$\mathbf{v}$	Three-dimensional velocity vector
$\mathbf{v}_H$	Projection of the velocity vector $\mathbf{v}$ into the horizontal plane

$\mathbf{v}_{\|}$	Projection of the velocity vector to the ice-rockbed-interface surface
w	Vertical component of the velocity vector
$\mathbf{w}$	Moisture content
$\mathbf{x}$	Three-dimensional position vector
$\mathbf{x}_H$	Projection of $\mathbf{x}$ into the horizontal plane
x, y, z	Cartesian coordinates
$z_M(\mathbf{x}_H, t)$	Geodetic coordinate of the CTM
κ, κ_R	Thermal diffusivity of the ice, rockbed
ν	Moisture diffusivity
ρ, ρ_a	Density of ice, of the asthenosphere
$\boldsymbol{\sigma}$	Stress deviator
$\boldsymbol{\sigma}_{\|}$	Projection of $\boldsymbol{\sigma} \cdot \mathbf{n}$ on the surface with unit normal $\mathbf{n}$, shear traction in this surface
$\boldsymbol{\sigma}_\perp = (\mathbf{n} \cdot \boldsymbol{\sigma}\mathbf{n})\mathbf{n}$ $= \boldsymbol{\sigma} - \boldsymbol{\sigma}_{\|}$	Stress traction parallel to $\mathbf{n}$
σ_{xz}, σ_{yz}	Horizontal shear stresses

Kolumban Hutter
Institute of Mechanics
Darmstadt University
Hochschulstr. 1, D-64289 Darmstadt, Germany

F. DELL'ISOLA AND K. HUTTER

A Free Moving Boundary Problem for the Till Layer Below Large Ice Sheets

Abstract

We formulate a free moving boundary problem for the till (i.e., soil + water) layer that may form below glaciers or large ice sheets and is thought to be responsible for their catastrophic advance when the water content makes such layers slippery against shear deformations. We indicate how the FMBP is formulated, specialize it to steady plane flow and deduce an ordinary differential equation which describes the distribution of the solid's volume fraction across the layer. This differential equation is second order and gives rise to a singular perturbation solution procedure. This problem can be analysed under the assumption that the fluid viscosity is a monotonic function of the solid's volume fraction. However, in this paper we prove that by choosing a constant fluid viscosity and vanishing thermodynamic pressure the emerging solid volume fraction turns out to be physically meaningless.

1 Introduction

Glaciers and ice sheets which rest on the ground may respond critically to the local conditions of the materials of which this ground is composed. Typically one differentiates between *hard* and *soft* beds. By the latter one means that the ice rests on a thin layer (till layer) of a mixture of sediment and water which separates the ice from the rock bed. The horizontal motion of ice, which may slide along the upper interface, generates a (shear) deformation of this till layer while the ice overburden pressure as well as the melt water at the ice-till-layer interface in combination with the drainage of water and abrasion of till at the till-layer-rock-bed interface give rise to a vertical flow of water and sediment.

The dissipative mechanisms are the frictional heat due to the sliding of the till over the rock bed and of the ice over the upper interface, plus the dissipation provided by the stress power and the work done by the sediment-water-interaction force within the layer. The heat generated by them and the geothermal heat supplied from below are conducted and convected to the top interface to provide the heat necessary to melt ice that may become available once transformed into water to be pressed into the till layer to form the lubricand for its shear deformations.

A binary mixture model for the viscous behaviour of true density preserving constituents, granules and water, has been deduced from first principles of thermodynamics by Svendsen and Hutter (1995), and the general FMBP for the soil-water mixture bounded by the top and bottom interfaces as described above has been formulated by dell'Isola and Hutter (1997, 1998a). Here we simply quote the relevant equations for the plane flow problem in which the thermal processes are ignored.

It is obvious from this description that the mechanical problem *per se* is already significant and that the water motion from above must play a significant role in the description of the distribution of the solid's volume fraction. The water that is pressed into the layer must loosen the granular matrix and will therefore provide the mechanism which reduces the resistance of the layer to shear deformations. It shall be demonstrated in this paper that plane flow is described by a FMBP. By specializing this problem to steady conditions, a two-point boundary value problem can be deduced for the solid's volume fraction with fixed boundaries. Dell'Isola and Hutter (1998b) have shown that this boundary value problem is stiff, and they have analysed it for nonvanishing thermodynamic pressure β_s when the fluid viscosity is constant. In this paper we prove that with vanishing thermodynamic pressure and constant fluid viscosity the emerging analytical profiles of solid volume fraction are physically meaningless.

2   Governing equations

Let $Oxyz$ be a Cartesian coordinate system with origin O, horizontal axes (x, y) and vertical axis z. Consider a layer of a saturated binary mixture of granules and a fluid bounded by $z = f_b(x, y, t)$ and $z = h(x, y, t)$ and let $f_b < h$. Assume the material within this layer to be subject to plane deformations such that $\partial(\cdot)/\partial y = 0$ for any field variable, and suppose that all fields obey the condition $\partial(\cdot)/\partial x = 0$ except the saturation pressure $p(x, t)$. Consider the following list of variables and parameters:

$$
\begin{aligned}
&\nu && \text{solid volume fraction,} \\
&u_s, u_f && \text{horizontal velocity components of the solid and} \\
& && \text{fluid, respectively,} \\
&w_s, w_f && \text{vertical velocity components of the solid} \\
& && \text{and fluid, respectively,} \\
&p && \text{saturation pressure,} \\
&\beta_s(\nu) && \text{themodynamic pressure,} \\
&\hat{\rho}_f, \hat{\rho}_s && \text{true fluid and sediment mass densities,} \\
&\mu_f, \mu_s && \text{(apparent) fluid and solid viscosities,} \\
&\tilde{\alpha} =: \hat{\rho}_f g / K, \\
&a := \tfrac{\hat{\rho}_f}{\hat{\rho}_s} && \xi_s := \tfrac{\rho_s}{\rho} = \tfrac{\nu\hat{\rho}_s}{\nu\hat{\rho}_s + (1-\nu)\hat{\rho}_f} = \tfrac{\nu}{\nu + (1-\nu)a}, \\
&g && \text{gravity constant,} \\
&K && \text{soil permeability.}
\end{aligned}
\tag{1}
$$

Dell'Isola and Hutter (1998b) explain that plane deformation of the sediment water

205

mixture is described by the following set of partial differential equations:

$$\frac{\partial \nu}{\partial t} + (\nu w_s)' = 0,$$
$$-\frac{\partial \nu}{\partial t} + ((1 - \nu) w_f)' = 0,$$
$$(-\nu \beta_s - \nu p + \mu_s w_s')' + (p + (1 - \xi_s) \beta_s) \nu' +$$
$$-\nu (1 - \nu) \tilde{\alpha} (w_s - w_f) = 0,$$
$$(- (1 - \nu) p + \tfrac{4}{3} \mu_f w_f')' - (p + (1 - \xi_s) \beta_s) \nu' +$$
$$+\nu (1 - \nu) \tilde{\alpha} (w_s - w_f) = 0,$$
$$-\nu \frac{\partial p}{\partial x} + (\mu_s u_s')' - \nu (1 - \nu) \tilde{\alpha} (u_s - u_f) = 0,$$
$$- (1 - \nu) \frac{\partial p}{\partial x} + (\mu_f u_f')' + \nu (1 - \nu) \tilde{\alpha} (u_s - u_f) = 0,$$
$$\text{in } z \in (f_b, h), t > 0 \tag{2}$$

in which the prime denotes differentiation with respect to z, and β_s and μ_f are prescribed functions of ν,

$$\begin{cases} \beta_s = \beta_0 + \check{\beta}_s(\nu), & \check{\beta}_s(0) = 0, \\ \mu_f = \mu_f(\nu), \end{cases} \tag{3}$$

to be specified below. Physics suggests μ_f to monotonically increase with ν. Moreover, β_s must satisfy the inequality

$$\frac{d\beta_s}{d\nu} - \beta_s \frac{1}{\nu + b} \geq 0, \quad b = \frac{a}{1 - a}, \tag{4}$$

Note that equations $(2)_{1-4}$ are uncoupled from $(2)_{5,6}$; a similar decoupling also applies when abrasion is ignored as seen in dell'Isola and Hutter (1998b). We thus ignore $(2)_{5,6}$, only deal with $(2)_{1-4}$ and quote the following boundary conditions from the aforementioned paper:

- At the top interface $z = h(t)$

$$w_f (1 - \nu) = -V_{f0},$$
$$\frac{\partial h}{\partial t} = w_s,$$
$$-\alpha \nu^l p_i = -\nu (\beta_s + p) + \mu_s w_s', \tag{5}$$
$$- (1 - \alpha \nu^l) p_i = - (1 - \nu) p + \tfrac{4}{3} \mu_f w_f'.$$

in which V_{f0} is the absolute value of the volume flow per unit area of water from the top into the layer and $\alpha \in (0, 1)$, while $l \simeq 2/3$.

- At the bottom surface $z = f_b(t)$

$$\frac{\partial f_b}{\partial t} = -w_b = -\frac{M_s^b}{\rho_r},$$
$$(1 - \nu) (w_f - w_b) = \frac{M_f^b}{\hat{\rho}_f} = -\frac{m_f^b}{\hat{\rho}_f} \sigma_f^4, \tag{6}$$
$$\nu (w_s - w_b) = -\frac{M_s^b}{\hat{\rho}_s}.$$

with

$$\sigma_f := -(1 - \nu)\,p + \frac{4}{3}\mu_f w_f',\tag{7}$$

in which $\mathcal{M}_s^b/\rho_r$ is the abrasion rate, henceforth set to be zero, while $\mathcal{M}_f^b/\hat{\rho}_f$ is the drainage rate of the water into the rock bed which we assume to be proportional to the fourth power of σ_f, the fluid pressure normal to the bottom surface.

Moreover, in dell'Isola and Hutter (1998b) it is shown that the above equations, complemented by initial conditions for ν, f_b and h, are likely a well-posed initial value FMBP for the fields p, ν, w_s and w_f, provided V_{f0}, the incumbent pressure and the abrasion rate function are *a priori* known.

3   Steady-state conditions with vanishing abrasion.

Let abrasion be zero and V_{f0} as well as p_0 be time independent. Then, as show in dell'Isola and Hutter (1998b), the above FMBP transforms into the following two-point boundary-value problem for the solids volume fraction ν expressed in the new variable

$$y := \frac{1}{1 - \nu}, \quad y \in [1, \infty]\tag{8}$$

$$
\begin{aligned}
-\left(\tilde{\mu}_f \delta_f\right) y'' + \left(h(y) - \tilde{\mu}_f' \delta_f\right) y' + \eta_f &= 0, \quad \tilde{z} \in (0, 1)\\
-\delta_f \tilde{\mu}_f y' + \tfrac{\nu-1}{y^2}\tilde{\beta}_s + \tfrac{1}{y}\tfrac{p_i}{[\beta_1]} - \mathcal{F} &= 0, \quad \text{at } \tilde{z} = 0,\\
-\delta_f \tilde{\mu}_f y' + \tfrac{\nu-1}{y^2}\tilde{\beta}_s + \left(\tfrac{\nu-1}{y} - \alpha\left(\tfrac{\nu-1}{y}\right)'\right)\tfrac{p_i}{[\beta_1]} &= 0, \quad \text{at } \tilde{z} = 1,
\end{aligned}\tag{9}
$$

in which

$$
\left\{
\begin{aligned}
\eta_f &:= \tilde{\alpha}\frac{[V_{f0}][H]}{[\beta_1]} = O(10^{-7} - 10^{-4}),\\
h(y(\nu)) &= (1 - \nu)^3\left(\tilde{\beta}_s\frac{1}{\nu+b} + \frac{d}{d\nu}\tilde{\beta}_s\right),\\
\delta_f &:= \tfrac{4}{3}\frac{[\mu_f][V_{f0}]}{[H][\beta_1]} = O\left(10^{-20} - 10^{-17}\right),\\
\mathcal{F} &:= \left(\frac{V_{f0}\hat{\rho}_f}{m_f^b}\right)^{1/4}\frac{1}{[\beta_1]} \simeq O\left(10^{-1} - 10^0\right),
\end{aligned}
\right.\tag{10}
$$

and $[\cdot]-$ quantities are scales for the variables in brackets, $[H]$ being the constant thickness of the layer. Furthermore, $z = [H]\,\tilde{z}$ and primes now denote differentiations with respect to $\tilde{z} : (\cdot)' = d(\cdot)/d\tilde{z}$. Note that the differential equation $(9)_1$ has the following properties:

when $\tilde{\beta}_s \equiv 0$, i.e., when the thermodynamic pressure is neglected, then (9) reduces to the differential equation

$$(\tilde{\mu}_f(y)y')' = \frac{3}{4}\tilde{\alpha}\frac{[H]^2}{[\mu_f]} =: \frac{1}{\varepsilon} \quad (\simeq 10^{15}) \tag{11}$$

the integration of which, subject to the boundary conditions

$$y(0) = y_0, \quad y(1) = y_h, \text{ with } y_h < y_0 \tag{12}$$

yields

$$M(y; y_h) = \frac{1}{2\varepsilon}\left(\tilde{z}^2 - \tilde{z}\right) + M(y_0; y_h)\left(1 - \tilde{z}\right) =: \tilde{M}(\tilde{z}) \tag{13}$$

with

$$M(y; y_h) := \int_{y_h}^{y} \tilde{\mu}_f(\bar{y})d\bar{y}. \tag{14}$$

Plotting $\tilde{M}(\tilde{z})$ in the interval $0 \le \tilde{z} \le 1$ gives the parabola with values zero at $\tilde{z} = 1$ and $M(y_0; y_h)$ at $\tilde{z} = 0$, and the minimum $\tilde{M}(\tilde{z}^*) = -\frac{1}{2\varepsilon}\left(\frac{1}{2} - \varepsilon M(y_0; y_h)\right)^2$ at $\tilde{z}^* = \left[\frac{1}{2} + \varepsilon M(y_0; y_h)\right]$. Whether or not $\tilde{z}^*$ lies in $[0, 1]$ depends upon the functional dependence of $\tilde{\mu}_f$ on y. For constant $\tilde{\mu}_f = 1$, $\tilde{z}^* = \left[\frac{1}{2} + \varepsilon(y_0 - y_h)\right]$ and (13) becomes

$$y(\tilde{z}) = \frac{1}{2\varepsilon}\left(\tilde{z}^2 - \tilde{z}\right) - (y_0 - y_h)\tilde{z} + y_0. \tag{15}$$

The function $y(\tilde{z})$ has values y_0 and y_h at $\tilde{z} = 0$ and at $\tilde{z} = 1$, respectively, and decreases into the interval $\tilde{z} \in [0, 1]$ to reach the negative minimum $y_{\min} = y_h - \frac{1}{2\varepsilon}\left(1 - \varepsilon(y_0 - y_h)\right)^2$ at $\tilde{z} = \tilde{z}^*$. Correspondingly, the function

$$\nu(\tilde{z}) = 1 - \frac{1}{y(\tilde{z})} \tag{16}$$

shows in $\tilde{z} \in [0, 1]$ three branches separated by two singularities when $y(\tilde{z}_s) = 0$. The two roots $\tilde{z}_s^1$, $\tilde{z}_s^2$ are in the neighbourhood of $\tilde{z} = 0$ and $\tilde{z} = 1$, respectively, and define two boundary layers with thicknesses of order ε. The properties of $\nu(\tilde{z})$ are

- For $\tilde{z} \in [0, \tilde{z}_s^1]$, $\nu(\tilde{z})$ starts with a positive value at $\tilde{z} = 0$ and approaches $\mp\infty$ as $\tilde{z} \to \tilde{z}_s^{1\mp}$.

- For $\tilde{z} \in [\tilde{z}_s^2, 1]$, $\nu(\tilde{z})$ is positive at $\tilde{z} = 1$ and decreases with decreasing $\tilde{z}$, approaching $\mp\infty$ as $\tilde{z} \to \tilde{z}_s^{2\pm}$.

- For $\tilde{z} \in [\tilde{z}_s^1, \tilde{z}_s^2]$, $\nu(\tilde{z})$ is larger than unity and singular, i.e., $\nu(\tilde{z})$ tends to $+\infty$ for $\tilde{z} \to \tilde{z}_s^{1+}$ and $\tilde{z} \to \tilde{z}_s^{2-}$, respectively.

208

Such behaviour is physically unacceptable not only because $\nu(\tilde{z})$ is larger than 1 in the middle of the layer but equally so because it also attains twice an infinite modulus within the layer.

This result proves that $\tilde{\mu}_f$ cannot be constant, or the thermodynamic pressure must be assumed to be nonvanishing, or both. This is demonstrated elsewhere.

4 Acknowledgments

We thank Kurt Frischmuth for intriguing discussions on viscosity. K.H. acknowledges financial support by the A.v. Humboldt foundation and the Max Planck Society through its Max Planck Prize.

References

[1] DELL'ISOLA F. AND HUTTER K. 1997 Continuum mechanical modelling of the dissipative processes in the sediment-water layer below glaciers *C. R. de l'Acad. Sci. Paris*, t. 325, Ser IIb (1997), 449 - 456.

[2] DELL'ISOLA F. AND HUTTER K. 1998a What are the dominant thermomechanical processes in the basal sediment layer of large ice sheets? *Proc. R. Soc. London*, **A 454** (1998), 1169 - 1196.

[3] DELL'ISOLA F. AND HUTTER K. 1998b A qualitative analysis of the dynamics of a sheared and pressurized layer of saturated soil *Proc. R. Soc. London* in press.

[4] SVENDSEN B. AND HUTTER K. 1995 On the thermodynamics of a mixture of isotropic materials with constraints *Int. J. Engng. Sci.* **33, 14,** p. 2021-2054.

Francesco dell'Isola
Dipartimento di Ingegneria Strutturale e Geotecnica
Universita' di Roma LA SAPIENZA
Via Eudossiana 18, I-00184 Roma

Kolumban Hutter
Institut für Mechanik,
Technische Universität Darmstadt,
D-64289 Darmstadt, Deutschland

D. D. Joseph
New Ideas about Flow-Induced Cavitation of Liquids

The problem of the inception of cavitation is formulated in terms of a comparison of the breaking strength or cavitation threshold at each point of a liquid sample with the principal stresses there. A criterion of maximum tension is proposed which unifies the theory of cavitation, the theory of maximum tensile strength of liquid filaments and the theory of fracture of amorphous solids. It is argued that the liquid ruptures in tension at nucleation sites; the cavity then fills with gas and the liquid flows. Liquids at atmospheric pressure which cannot withstand tension will cavitate when and where tensile stresses due to motion exceed one atmosphere. A cavity will open in the direction of the maximum tensile stress which is 45 ffi from the plane of shearing in pure shear of a Newtonian fluid. An analysis of capillary collapse based on viscous potential flow leads to the total collapse of a capillary filament in a finite time; before this the filament enters into tension and presumably would break under tension. For water the critical radius is about 1.5 microns.

Conventional Cavitation

A fluid will cavitate when the local pressure falls below the cavitation pressure

- The cavitation pressure is the vapor pressure in a pure liquid

- Natural liquids have nucleation sites defined by impurities and may cavitate at higher pressures

What is Pressure?

- In an incompressible Newtonian fluid "pressure" is the mean normal stress. A fluid cannot average its stresses, even though you can. The fluid knows its state of stress at a point.

- In non-Newtonian fluids the pressure is an unknown flow variable, usually not even the mean normal stress, and the definition of it is determined by the constitutive equation. This "pressure" has no intrinsic significance. The fluid doesn't recognize such a "pressure" and knows its state of stress.

Nonconventional Cavitation Based on Principal Stresses

Look at the state of stress at each point in the fluid in principal axis coordinates. Identify the largest of the stresses. Suppose a static fluid cavitates at zero pressure. It will cavitate in flow wherever the maximum tensile stress is positive.

If it cavitates statically when the pressure falls below the vapor pressure, it will cavitate in flow even when the maximum tensile stress is only slightly negative.

Suppose you do an experiment in your lab where the ambient pressure is

one atmosphere = 10^5 Pa.

Then if you get tensile stresses due to flow larger than this, the fluid will cavitate.

Stress, Principal Axes, Deviator

- Stress in two dimensions

$$\begin{bmatrix} T_{11} & T_{12} \\ T_{12} & T_{22} \end{bmatrix}$$

"Pressure" cannot be recognized in a liquid; it sees a state of stress.

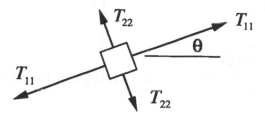

Figure 1. The direction of maximum tension. A cavitation must open in the direction θ; then it can rotate away.

- Principal coordinates (Figure 1)

- Mean normal stress and deviator

$$T = -p\mathbf{1} + \mathbf{S} \qquad \mathbf{S} \text{ is the extra stress}$$

$$p = -\frac{T_{11} + T_{22}}{2} \qquad \text{the fluid cannot average stresses}$$

$$S = \begin{bmatrix} S_{11} & 0 \\ 0 & S_{22} \end{bmatrix}, \qquad S_{11} + S_{22} = 0$$

The extra stress is good because it has positive and negative components.

Cavitation Criteria

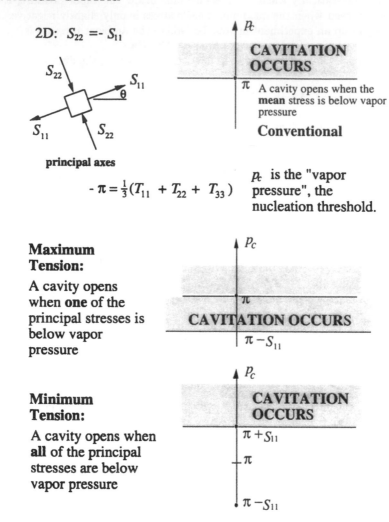

2D: $S_{22} = -S_{11}$

principal axes

$$-\pi = \tfrac{1}{3}(T_{11} + T_{22} + T_{33})$$

p_c

CAVITATION OCCURS

π A cavity opens when the **mean** stress is below vapor pressure

Conventional

p_c is the "vapor pressure", the nucleation threshold.

Maximum Tension:

A cavity opens when **one** of the principal stresses is below vapor pressure

P_c

π

CAVITATION OCCURS

$\pi - S_{11}$

Minimum Tension:

A cavity opens when **all** of the principal stresses are below vapor pressure

P_c

CAVITATION OCCURS

$\pi + S_{11}$

π

$\pi - S_{11}$

Figure 2. Three criteria for cavitation could be proposed, but the one based on maximum tension is the only one consistent with fracture of solids and solid-like liquids.

Maximum Tension

*All books on cavitation have sections on "tensile strength of liquids." A stress tensor is **never** introduced.*

212

"If a cavity is to be created in a homogeneous liquid, the liquid must be ruptured, and the stress required to do this is not measured by the vapor pressure but is the tensile strength of the liquid at that temperature." (Knapp et al. 1970)

FIRST **THE FLUID RUPTURES**
THEN **VAPOR FILLS THE CAVITY**

Cavitation in Shear

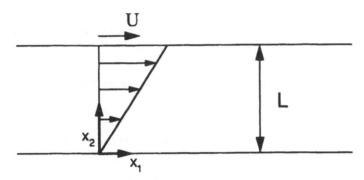

Figure 3. Simple shear between walls.

The stress in this flow is given by

$$\begin{bmatrix} T_{11} & T_{12} & 0 \\ T_{12} & T_{22} & 0 \\ 0 & 0 & T_{33} \end{bmatrix} = -\pi \begin{bmatrix} 1 & 0 & 0 \\ 0 & 1 & 0 \\ 0 & 0 & 1 \end{bmatrix} + \eta \begin{bmatrix} 0 & \frac{U}{L} & 0 \\ \frac{U}{L} & 0 & 0 \\ 0 & 0 & 0 \end{bmatrix}$$

where $\pi = \frac{1}{3}(T_{11} + t_{22} + T_{33})$ is determined by the "pressurization" of the apparatus. The angle which diagonalizes T is given by

$$\theta = 45°$$

(In the break-up of viscous drop experiments in plane shear flow done by G.T. Taylor [1934], the drops first extend at 45° from the direction of shearing.) In principal coordinates, we have

$$\begin{bmatrix} T_{11} + \pi & 0 & 0 \\ & T_{22} + \pi & 0 \\ 0 & 0 & T_{33} + \pi \end{bmatrix} = \eta \frac{U}{L} \begin{bmatrix} 1 & 0 & 0 \\ 0 & -1 & 0 \\ 0 & 0 & 0 \end{bmatrix}$$

where

$$T_{11} + \pi = S_{11} = \eta \frac{U}{L}$$

213

This tension is of the order of one atmosphere of pressure if

$$\eta \frac{U}{L} = 10^6 \frac{\text{dynes}}{\text{cm}^2} = 10^5 \, Pa$$

If $\eta = 1000$ poise, $U = 10$ cm/sec and $L = 10^{-1}$cm, we may achive such a stress. *A shear stress of this magnitude is enough to put the liquid into tension.*

The production of cavitation in pure shear appears to have been realized recently (1997)

Abstract, "Fracture" phenomena in shearing flows of viscous liquids, L.A. Archer, D. Ternet and R. Larson:

In start-up of the steady shearing flow of two viscous unentangled liquids, namely, low-molecular-weight polystyrene and α-D-glucose.

The shear stress catastrophically collapses if the shear rate is raised above a value corresponding to a critical initial shear stress of around 0.1 – 0.3 Mpa. The time-dependence of the shear stress during this process is similar for the two liquids, but visualization of samples *in situ* and after quenching reveals significant differences. For α-**D-glucose, the stress collapse evidently results from debonding of the sample from the rheometer tool, while in polystyrene, bubbles open up within the sample; as occurs in cavitation.** Some similarities are pointed out between these phenomena and that of "lubrication failure" reported in the tribology literature.

We have adhesive and cohesive fracture, 0.1–0.3 Mpa = 1–3 atm. This is enough to put the sample into tension 45° from the direction of shearing.

Breaking Strength of Polymer Strands

The strand breaks at the thinnest cross-section of the strand when the tensile stress

$$\sigma = \frac{FV}{A_0} \approx 10^6 \, \text{Pa} = 10 \text{ atmospheres}$$

214

for many kinds of polymeric liquids. They say that the breaking stress is a material constant.

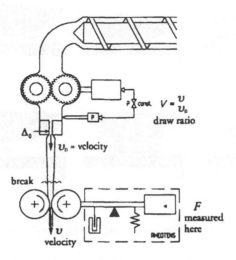

Figure 4. Drawdown apparatus (Wagner, Schulze and Gottfert [1996]).

Breaking Time & Flow Time Vacuum Cavities

Experiments of Israelachvili and coworkers (Chen & Israelachvili [1991], Kuhl et al. [1994]) on ultrathin (nanometer) films show that cavities open in tension at a threshold value of the extensional stress $2\eta\overset{\circ}{S}$

$$\left(\begin{array}{c} \text{which I estimate as} \\ 3.6 \times 10^5 \text{ Pa} < 2\eta\overset{\circ}{S} < 3.6 \times 10^6 \text{Pa} \end{array} \right)$$

and that the formation of cavities is analogous to the fracture of solids except *after* fracture, vapor flows into the cavity " ... When a cavity initially forms and grows explosively, it is essentially a VACUUM CAVITY since dissolved solute molecules or gases have not had time to enter the rapidly growing cavity."

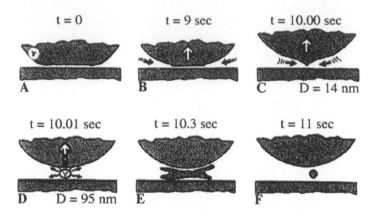

Figure 5. Surfaces separating at high speed, $v = v_c$

Kuhl et al. 1994 describe the experiment shown in Figure 5 as

"If the speed of separation is increased, the surfaces become increasingly more pointed just before they rapidly move apart. Then, above some critical speed ν_c (here about 100 μm/s) a completely new separation mechanism takes over, as shown in Figure 5. Instead of separating smoothly, the liquid 'fractures' or 'cracks' open like a solid. It is known that when subjected to very high shear rates, liquids begin to behave mechanically like solids, for example, fracturing like a brittle solid. In our experiments, the point and time at which this 'fracture' occurred just as the surfaces were about to separate from their most highly pointed configuration (Fig. 5C) - for had the separation velocity been any smaller than ν_c they would have separated smoothly without fracturing. We consider that in the present case, the 'fracturing' or 'cracking' of the liquid between the surfaces must be considered synonymous with the "nucleation" or "inception" of a vapor cavity."

Capillary Collapse and Rupture

It might be thought that flow stress induced cavitation is restricted to rather viscous liquids, where high stress levels can be achieved. However, such high stresses can be reached even in water, as the following analysis of Lundgren & Joseph [1997] shows:

"The breakup of a liquid capillary filament is analyzed as a viscous potential flow near a stagnation point on the centerline of the filament towards which the surface collapses under the action of surface tension forces. We find that the neck is of parabolic shape and its radius collapses to zero in a finite time. During the collapse the tensile stress due to viscosity increases in value until at a certain finite radius, which is about 1.5 microns for water in air, the stress in the throat passes into tension, presumably inducing cavitation there."

Potential flows satisfy the Navier-Stokes equations, though they slip at solid boundaries, the viscosity of the fluid never has to be and should not be put to zero (see Joseph and Liao [1994], Liao and Joseph [1994] for a complete discussion). The flow at the stagnation point of a collapsing capillary can be treated as a viscous potential flow.

$$[u_z, u_r] = a(t)[z, -\frac{r}{2}] \text{ Stagnation flow}$$

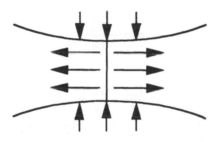

$$\varphi = \frac{1}{2}az^2 - \frac{1}{4}ar^2 \qquad \text{Potential flow}$$

Bernoulli Equations:

$$\frac{\partial t}{\partial t} + \frac{1}{2}(u_r^2 + u_z^2) + \frac{p}{\rho} = \frac{p_0}{\rho}$$

$$\frac{p - p_0}{\rho} = -\left(\frac{1}{2}\dot{a} + \frac{1}{2}a^2\right)z^2 + \left(\frac{1}{4}\dot{a} - \frac{1}{8}a^2\right)r^2$$

$$T_{zz} = -p + 2\mu\frac{\partial z}{\partial z} = -p + 2\mu a,$$
$$T_{rr} = -p + 2\mu\frac{\partial r}{\partial r} = -p - \mu a \qquad \text{Stresses}$$

217

On $R(z,t) = r_0(t) + R_2(t)z^2 + O(z^4)$

$$-T_{nn} - p_a = \sigma\kappa$$

$$T_{nn} = n_r^2 T_{rr} + n_z^2 T_{rr} + n_z^2 T_{zz}$$

$$\kappa = -\frac{\frac{\partial_z^2}{\partial_z^2}}{\left(1+\left(\frac{\partial_z}{\partial_z}\right)^2\right)^{3/2}} + \frac{1}{R\left(1+\left(\frac{\partial_z}{\partial_z}\right)^2\right)^{1/2}}$$

$$= \frac{1}{R_0} - 2R_z + O(Z^2)$$

Normal stress balance

It's zero at $z = 0$
because $\partial u_z/\partial r = 0$ | shear stress

$$u_r = \frac{\partial t}{\partial t} + u_z \frac{\partial z}{\partial z}$$

$$-\tfrac{1}{2}aR = \frac{\partial t}{\partial t} + az\frac{\partial Z}{\partial Z}$$

$$-\tfrac{1}{2}aR_0 = \overset{\circ}{R_0} \quad \text{Solve for } a$$

$$-\tfrac{5}{2}aR_2 = \overset{\circ}{R_2} \quad R_2 = CR_0^5$$

Kinematic Condition

The parabola flattens

as it collapses

$$a = -\frac{2\overset{\circ}{R_0}}{R_0}$$

To lowest order in z^2

$$\frac{p_a - \sigma\kappa}{\rho} = \frac{T_{nn}}{\rho} = \frac{T_{rr}}{\rho} = -\frac{p}{a} - va$$

$$\frac{p_a}{\rho} + \frac{\sigma}{\rho}\left(\frac{1}{R_0} - 2R_2\right) = -\frac{p_0}{\rho} - \left(\frac{1}{4}\overset{\circ}{a} - \frac{1}{8}a^2\right)R_0^2 - va$$

$$R_2 = CR_0^5 \qquad a = 2\overset{\circ}{R}/R_0$$

"Rayleight Plesset" type of equation:

$$\frac{p_0 - p_a}{\rho} - \frac{1}{2}R_0\ddot{R}_0 - \frac{2v\dot{R}_0}{R_0} = \frac{\sigma}{\rho}\left(\frac{1}{R_0} - 2CR_0^5\right).$$

$$\overset{\circ}{R_0} = \frac{1}{2}\frac{\sigma}{\rho v} \qquad \begin{array}{l}\text{Small } R_0\text{, balances}\\ \text{viscosity against inertia}\end{array}$$

to leading
order in
$t_* - t$

$R_0 = \frac{\sigma}{2\rho v}(t_* - t)$ R_0 collapses to
$a = \frac{2}{t_* - t}$ zero in finite time
$\frac{p}{\rho} = \frac{p_0}{\rho} - \frac{3z^2}{(t_* - t)^2}$

AXIAL STRESS (to leading order)

$$\frac{T_{zz}}{\rho} = -\frac{2p_0 - p_a}{\rho} + \frac{4v}{t_* - t} = \frac{2p_0 - p_a}{\rho} + \frac{2\sigma}{\rho R_0(t)}$$

T_{zz} turns for small R_0

$$R_{ocr} = \frac{2\sigma}{2p_0 - p_a}$$

Estimating $p_0 = p_a = 10^6$ dynes/cm^2
$\sigma = 75$ dyne/cm

$R_0 = 1.5$ microns

At the collapse condition
the Reynolds number is
about 55 based on $\overset{\circ}{R_0}$

The capillary thread cavitates before it collapses to zero

References

[1] D.D. JOSEPH, Cavitation and the state of stress in a flowing liquid, under review, see: *http://www.aem.umn.edu/people/faculty/joseph/papers/download.html*; (1997).

[2] T.S. LUNDGREN AND D.D. JOSEPH, Capillary collapse and rupture, under review, see: *http://www.aem.umn.edu/people/faculty/joseph/papers/downcap.html*; (1997).

[3] D.D. JOSEPH AND T. LIAO, Potential flows of viscous and viscoelastic fluids, *J. Fluid Mech.*, 265, 1–23, (March, 1994).

[4] T. LIAO AND D.D. JOSEPH, Viscous and viscoelastic potential flow, "Trends and Perspectives in Applied Mathematics," Applied Mathematical Sciences, 100, Springer-Verlag, New York (1994).

[5] Y. CHEN AND J. ISRAELACHVILI, New mechanism of cavitation damage, *Science*, 252, 1157–1160, (1991).

[6] R. KNAPP, J.W. DAILY, F. HAMMIT, Cavitation, McGraw Hill, N.Y., (1970).

[7] T. KUHL, M. RUTHS, Y.L. CHEN, J. ISRAELACHVILI, Direct visualization of cavitation and damage in ultrathin liquid films, *The Journal of Heart Valve Disease*, 3, (suppl. I) 117127, (1994).

[8] G.I. TAYLOR, The formation of emulsions in definable fields of flow, *Proc. R. Soc. London Ser. A*, 146: 501–23, 1934.

[9] M.H. WAGNER, V. SCHULTZE AND A. GÖTTFERT, Rheotens – Mastercurves and drawability of polymer melts, *Polymer Eng. Sci*, 36(7), 925–935, (April 1996).

Acknowledgment. This work was supported by NSF/CTS 9523579 and ARO grant DA/DAAH04.

Daniel D. Joseph
Aerospace Eng. and Mechanics
UMTC
University of Minnesota
Minneapolis, MN 55455, USA

J.S Lowengrub, J. Goodman, H. Lee, E.K. Longmire, M.J. Shelley and
L. Truskinovsky
Topological Transitions in Liquid/Liquid Interfaces

1 Introduction

A set of fundamental yet ill-understood phenomena in fluid dynamics involves changes
in the topology of interfaces between partially miscible or nominally 'immiscible' fluids. Such changes occur, for example, when continuous jets pinch off into droplets,
when sheared interfaces atomize, and when droplets of one fluid reconnect with one
another. These topological transitions occur in many practical applications involving
transport, mixing, and separation of petroleum, chemical, and food products as well
as contaminated waste streams.

The dynamics of topological transitions are difficult to understand and model for
several reasons. For one, the fluids in which these transitions occur are complex. A
second problem associated with topological transitions is caused by the short time
scales over which they occur. In practical flows, the transition time scales are much
shorter than the local flow time scales making the transitions difficult to characterize
experimentally or compute numerically. A third problem associated with transitions is
purely numerical: how does one handle the change in interface topology in a physically
justified way? In this paper, we will address the last problem in the context of
incompressible fluid flows.

Many researchers (see [17, 20], for example) have tried using ad hoc methods to
change the topology of interfaces. While this approach, often referred to as "contour
surgery," allows topological transitions to be overcome, it is difficult to justify the
reconnection conditions based on physical principles. In a few special cases, involving
fluid-gas interfaces, it is possible to develop physically based reconnection conditions
by using special similarity solutions of the Navier-Stokes equations (see [13, 5, 4]). For
flows involving liquid/liquid interfaces, however, the dynamics are more complicated
and no such similarity solutions have been constructed.

In an attempt to derive a physically based theory of sharp liquid/liquid interfaces near topological transitions, several researchers (see [2, 3, 19], for example) have
proposed representing the interface as the level set of a higher-order function. Thus
different level sets (or fixed values of the function) could exhibit different topologies.
In this approach, the interface is effectively given a finite thickness by smoothing the
flow discontinuities (density, viscosity) over a narrow region. Although this procedure
generally yields a smooth evolution through topological changes, one can demonstrate
that the results can depend essentially on the type of smoothing chosen [16]. Within
the context of this "level set" method, it is not clear *which* types of smoothing are
physically justified.

221

Three of the authors (JG, JL, and MS) have proposed an alternative: introduce an explicit order parameter (e.g., concentration) and allow limited mixing across the interfacial zone. In this approach, the sharp interface is replaced by a smooth, narrow transition layer (in the order parameter) and the resulting system consists of the Navier-Stokes (NS) equations coupled to either a Ginzburg-Landau (nonconserved, GL) or Cahn-Hilliard (conserved, CH) equation for the order parameter. Gradients in the parameter produce reactive stresses in the fluid which mimic surface tension. The above authors gave a physical derivation of the equations in a special case (models of type E and H in the nomenclature of Hohenberg & Halperin [9]). Subsequently, Lowengrub & Truskinovsky in [16] gave a more general and systematic derivation of the equations, with the conserved mass concentration as the order parameter, and gave an analysis of the equations in some special cases including a simple model of a topology transition.

Here, the mass concentration is the physically relevant order parameter and the limited mixing is due to chemical diffusion between the different fluids. In physical chemistry, it is well-known that limited molecular mixing occurs between macroscopically "immiscible" fluids. This limited mixing provides a *physical mechanism* to smooth the flow discontinuities and to yield smooth evolutions through topological changes. Thus, the NSCH system can be viewed as a partial miscibility regularization (PMR) of the sharp interface model.

The NSCH model fits naturally into the general framework of the so-called phase field models which have been widely used in the study of free boundary problems. We refer the reader to the recent article by Anderson, McFadden & Wheeler [1] for a review of recent work using phase-field models in fluid mechanics.

In this paper, we present preliminary numerical results using the PMR model in two interfacial flow regimes: (1) viscously dominated flows in Hele-Shaw cells (unstably stratified fluid layers), and (2) 2-d inertially dominated flows (liquid/liquid jets). In both cases, topological transitions are smoothly captured by the PMR model. Vorticity is produced at the pinchoff point and the interfaces 'snap' back after the transition. In the Hele-Shaw case, the vorticity remains bound to the interface through the transition. This suggests that it may be possible to use the PMR to formulate 'topological jump conditions' to reconnect sharp interfaces within the context of a boundary integral simulation, for example. This is currently under investigation. In the inertially dominated case, simulations of inviscid liquid/liquid jets show that some vorticity *separates* from the interface after pinchoff and there is a nontrivial flow inside the newly created drops. In both cases before pinchoff, the results show good agreement with boundary integral simulations.

2   Equations of Motion

We begin our presentation of the PMR equations by first reviewing the classical theory of surface tension. Let two immiscible, incompressible fluids be separated by a sharp interface Γ and let $\Gamma = \{x|\phi(x,t) = 0\}$ where $\phi > 0$ denotes the region with fluid 1 and $\phi < 0$ denotes the region with fluid 2. Then, one can introduce the characteristic function χ of the fluid 1 region by

$$\chi(\phi) = H(\phi) = \begin{cases} 1 & \text{if } \phi > 0, \\ 0 & \text{otherwise} \end{cases} \tag{1}$$

where H is the Heaviside function. One can then define the fluid quantities in terms of χ. For example, let the density $\rho = \rho_1 \chi(\phi) + \rho_2(1 - \chi(\phi))$ where ρ_1 and ρ_2 are the constant densities of fluids 1 and 2, respectively. The other material parameters are defined analogously. The mass and momentum balance equations are given by

$$\nabla \cdot \mathbf{u} \;=\; 0 \quad \text{and} \quad \dot{\phi} = 0, \tag{2}$$
$$\rho \dot{\mathbf{u}} \;=\; \nabla \cdot \mathbf{P}, \quad \text{and} \quad \mathbf{P} = -p\mathbf{I} - \sigma\,(\mathbf{n} \otimes \mathbf{n} - \mathbf{I})\,\delta_\Gamma + 2\nu\mathbf{D} \tag{3}$$

where $\dot{} = \partial_t + \mathbf{u} \cdot \nabla$ is the advective time derivative, p is the pressure, σ is the surface tension, $\mathbf{n} = \nabla\phi/|\nabla\phi|$ is the normal vector to Γ, δ_Γ is the surface delta function, $\nu = \nu(\phi)$ is the viscosity and $\mathbf{D} = (\nabla\mathbf{u} + \nabla\mathbf{u}^T)/2$ is the rate of strain tensor. As is well-known, the surface delta function can be related to the 1-d delta function δ by $\delta_\Gamma = \delta(\phi)|\nabla\phi|$. This formulation (2)-(3) guarantees that the classical boundary conditions

$$[\mathbf{u} \cdot \mathbf{n}]_1^2 \;\equiv\; \mathbf{u} \cdot \mathbf{n}|_2 - \mathbf{u} \cdot \mathbf{n}|_1 = 0, \quad V = \mathbf{u} \cdot \mathbf{n}|_\Gamma \tag{4}$$
$$[\mathbf{P} \cdot \mathbf{n}]_1^2 \;=\; 2\sigma\kappa \tag{5}$$

hold across Γ. In the above, V is the normal velocity of Γ and κ is its mean curvature.

The solutions to Eqs. (2)-(3) generically develop singularities in both 2 and 3 dimensions. These singularities typically develop due to topological changes in the flow such as the collision of material interfaces Γ. When material surfaces collide, or self-intersect, velocity gradients necessarily diverge [11] and the curvature tends to blow up as well. For 2-d, see [10, 11]. In 3-d, it is the classical Rayleigh instability that drives the singularity formation.

To bridge the transition and continue the flow beyond the singularity, one has several choices. One can deal directly with the equations (2)-(3) and try to obtain and match similarity solutions as was done in [4] for liquid/gas interfaces or one can regularize the equations. Here, we follow the latter approach.

The simplest regularization of the sharp interface Eqs. (2)-(3) is to smooth $\chi \to \chi_\epsilon$ and $\delta_\Gamma \to \delta_\Gamma^\epsilon$. For example, one can set

$$\delta_\Gamma^\epsilon = \epsilon|\nabla\phi|\,(\chi_\epsilon')^2 \quad \text{with} \quad \sigma = \lim_{\epsilon \to 0} \int_{-\infty}^{+\infty} \epsilon\,[\chi_\epsilon'(\phi)]^2 \; d\phi. \tag{6}$$

Then, one can repose Eq. (3) using the smoothed stress tensor $\mathbf{P}^\epsilon$. This method is basically the level set method described in [2, 3, 19, 18]. While this method yields smooth evolutions through topology changes, in our view, there are several potential drawbacks. First, it is the numerical diffusion, which is fluid-independent, that actually controls the reconnection process and yields the smooth evolution through the topology change. Second, the solutions can depend essentially on the type of smoothing (see [16] for an explicit example for spherical drops). Thus, while we can imagine χ_ϵ as an artificial concentration field, there is no physical chemistry in this model.

2.1 Navier-Stokes-Cahn-Hilliard Equations

One can try to account for the physical chemistry by introducing a mass concentration field $c = M_1/M$ which is conserved, consistently coupled to the fluid equations and evolves according to a diffusion equation; this is the PMR. A derivation of the equations is given in [16] and we only present the nondimensional result here:

$$\rho_0 \dot{c} = \frac{1}{\text{Pe}} \Delta \mu, \quad \nabla \cdot \mathbf{u} = \frac{\alpha}{\text{Pe}} \Delta \mu \tag{7}$$

$$\tag{8}$$

$$\rho_0 \dot{\mathbf{u}} = -\frac{1}{\text{M}} [\nabla p + \mathbf{C} \nabla \cdot (\rho_0 \nabla c \otimes \nabla c)] + \frac{1}{\text{Re}} \nabla \cdot [2\eta(c)\mathbf{D}] + \frac{1}{\text{Fr}} \rho_0 \mathbf{g} \tag{9}$$

where $\rho_0 = \rho_0(c)$ is the simple mixture density [12] defined by $1/\rho_0(c) = c/\rho_1 + (1 - c)/\rho_2$, μ is the chemical potential

$$\mu = \frac{df_0}{dc}(c) - \frac{\rho_0'}{\rho_0^2} p - \frac{\mathbf{C}}{\rho_0} \nabla \cdot (\rho_0 \nabla c) \tag{10}$$

and $\mathbf{C}$ is the Cahn number which is a measure of the interface thickness and $f_0(c)$ is a non-convex, non-negative function. Pe is the diffusional Peclet number, $\alpha = 1/\rho_1 - 1/\rho_2$ is a constant, M is a generalized Mach number, Re is a Reynolds number, Fr is a Froude number and $\mathbf{g}$ is unit vector pointing in the direction of the gravitational field. We refer the reader to [16] for explicit definitions of these nondimensional quantities.

We remark in the density-matched case $\rho_1 = \rho_2$, Eqs. (7)-(10) reduce to Model H given in [9]. The equation for c is a $4th$ order diffusion equation of Cahn-Hilliard type. Eq. (9) is a generalization of the Navier-Stokes equation in which gradients in c produce reactive stresses in the fluid which mimic surface tension. We refer to this as the NSCH system. It is interesting to note that if $\rho_1 \neq \rho_2$, then diffusion creates density variation so that $\nabla \cdot \mathbf{u} \neq 0$ which introduces compressibility effects even when the original fluids are incompressible. We refer to this case as *quasi-incompressible*. The pressure also explicitly appears in the chemical potential. Some consequences of this are discussed in [16]. Finally, Eqs. (7)-(10) have an associated non-increasing energy functional which is the sum of the kinetic, potential and chemical energies.

In [16], the sharp interface limit of (7)-(10) was discussed using matched asymptotic expansions. We present the result here. Let γ be a small parameter that measures the thickness of the interface and let $\mathbf{C} = \gamma^2$, $\mathbf{M} = \gamma$ and $\text{Pe} = 1/\gamma$. Then, if the mean curvature $\kappa << 1/\gamma$, the system (7)-(10) converges to the sharp interface system (2)-(3) with ϕ replaced by c, $\Gamma = \{\mathbf{x}|\ c(\mathbf{x}, t) = 1/2\}$ and the surface tension given by

$$\sigma = \int_0^1 \rho_0(c)\sqrt{2f_0(c)}dc \tag{11}$$

In this paper, we always use $f_0(c) = Ac^2(1-c)^2/4$ and use A to match the sharp interface surface tension.

2.2   Hele-Shaw-Cahn-Hilliard Equations

We next present the PMR model appropriate for quasi-incompressible binary fluids in a Hele-Shaw cell. A Hele-Shaw cell consists of two parallel plates separated by a narrow gap. The flow takes place in the gap and it is assumed that the fluids are highly viscous so that inertial effects are negligible. In addition, because the gap between the plates is narrow, the flow is only weakly three-dimensional. Our primary motivation for discussing and eventually studying this flow is that it is much simpler mathematically and physically than the general NSCH system and therefore provides an excellent case for testing the effects of parametric variations.

The sharp interface formulation is as follows. In each fluid domain, the velocity is given by Darcy's law:

$$\mathbf{u}_i = -\frac{1}{12\eta_i}\left[\nabla p - \rho_i \mathbf{g}\right], \quad \text{in } \Omega_i \tag{12}$$

and the fluids are incompressible $\nabla \cdot \mathbf{u}_i = 0$ for $i = 1, 2$. The boundary conditions across the interface $\Gamma = \partial\Omega_1 \cap \partial\Omega_2$ are exactly as in (4) and (5).

In recent work, Lee & Lowengrub [14] derived the PMR appropriate for flow in a Hele-Shaw cell. In this case, the equations are referred to as the Hele-Shaw-Cahn-Hilliard (HSCH) system and are given by

$$\rho_0 \dot{c} = \gamma \Delta\mu, \quad \nabla \cdot \mathbf{u} = \alpha\gamma\Delta\mu \tag{13}$$

$$\mathbf{u} = -\frac{1}{12\nu(c)}\left[\nabla p + \gamma\nabla \cdot (\rho_0 \nabla c \otimes \nabla c) - \rho_0 \mathbf{g}\right] \tag{14}$$

$$\mu = \frac{df_0}{dc}(c) + \alpha\gamma p - \frac{1}{\rho_0}\nabla \cdot (\gamma^2 \rho_0 \nabla c) \tag{15}$$

where $\mathbf{u}$ and c are the gap-averaged velocity and concentration fields, respectively (2-dimensional), ∇ is the two-dimensional gradient and $\mathbf{g}$ is gravity. Note that the gap width has been scaled out of the formulation and that we have explicitly used the sharp interface scaling as described in the previous section.

There are three important parameters in the HSCH model. There is the Bond number $\mathbf{B} = g(\rho_1 - \rho_2)/\sigma$, the Atwood number for viscosity $\mathbf{A}_\nu = (\nu_1 - \nu_2)/(\nu_1 + \nu_2)$ and the interface thickness γ.

This system is much simpler than the NSCH equations. There is no dynamical equation for the velocity **u**; the velocity is determined through a generalized Darcy's law (14). Once c is determined, then **u** is found by solving Eqs. (13)b and (14). Nevertheless, there are still many common features with the NSCH model. In particular, the compressibility effects due to chemical diffusion are still present. As in the NSCH case, it can be shown [14] that as $\gamma \to 0$, the system (13)-(15) converges to the sharp interface equations with surface tension σ given by Eq. (12).

3 Results

In this section, we present preliminary numerical results for the HSCH and NSCH models. We begin by considering thin, unstably stratified layers in a Hele-Shaw cell.

3.1 HSCH Simulations

Let us suppose that $\rho_1 - \rho_2 << 1$ but $g(\rho_1 - \rho_2) = O(1)$. Then, we can assume ρ_0 is constant in the equations everywhere except in the gravitational term where we take ρ_0 to be a linear function of c. This is a Boussinesq approximation and there are no compressibility effects in this limiting case. In the simulations we present here, we use this approximation of the HSCH equations and we further set $A_\nu = 0$ so that the fluids are viscosity matched. More complicated scenarios are discussed in [14]. By considering this case first, we can isolate the effects of diffusion from those of compressibility.

For our numerical methods, we use periodic boundary conditions and pseudo-spectral spatial discretizations of Eqs. (13)-(15). We use a non-stiff time-stepping algorithm to solve the concentration equation in Fourier space. In this approach, the equation is reformulated by using an integrating factor associated with the $4th$ order term in Fourier space. The reformulated equations are discretized using a $3rd$ order Adams-Bashforth method. See [14].

Now, consider an unstably stratified layer which consists of a layer of light fluid surrounded by a heavy fluid. There are two interfaces; the upper interface is unstable while the lower interface is stable. We take **B** = 25 for the upper interface and **B** = -25 for the lower. A boundary integral simulation of the evolution is given in Figure 1. Periodic boundary conditions are used and only 1 periodic box is shown. The simulation suggests that at a time slightly beyond $t = 7.7$, the layer pinches off at two points leaving the flow consisting of two large bubbles with a single narrow bubble between. Flows of this type have been studied extensively both theoretically and numerically in [8, 7] and we refer the reader there for additional background.

We now repeat the flow using the HSCH model with $\rho_0 = 1 - 0.1c$. In addition, we use the reference frame of the sharp interface model. That is, a constant adverse pressure gradient (to the gravitational field) is introduced whose strength is equal to the mean density (in a single periodic box). With this pressure gradient, flat interfaces are motionless. The resulting simulation is shown in Figure 2. Three x-periods are plotted. In the upper graphs, concentration is plotted and in the lower graphs, the vorticity is plotted. In this simulation, we have used $\gamma = 0.05$, $N = 256$ and $\Delta t = 10^{-3}$. There are approximately 8 computational points across the interface. The evolution smoothly passes through the topological change and the layer breaks up into droplets. Interestingly, there is a secondary break-up as the flat droplet splits into three small 'round' drops. Oppositely signed vorticity is produced at the pinching point and acts to 'snap' back the drop tips.

That vorticity is produced by the transition can be seen by considering the maximum vorticity as a function of time. This is shown in Figure 3. The four graphs correspond to the four interface thicknesses $\gamma = 0.05$, $\gamma = 0.06$, $\gamma = 0.08$ and $\gamma = 0.10$. The second peak in the $\gamma = 0.05$ curve is due to the secondary pinchoff. The computations with larger γ do not exhibit this secondary pinchoff and so only one peak is observed. In those computations, the diffusion smears those details enough so that only a single bubble is observed [14]. Note that the primary pinchoff time is a stable function of γ.

Finally, in Figure 4, results from the sharp and HSCH models are compared. Good agreement is observed between the sharp interface (solid line) and the $c = 1/2$ contour line (dashed line), although the layer in the HSCH model pinches off slightly earlier.

3.2 NSCH Simulations

In this section, we present a single simulation of the break-up of a 2-d density-matched, inviscid jet using the NSCH model. We also compare the result to a sharp interface simulation.

As in the Hele-Shaw case, we use periodic boundary conditions and pseudo-spectral spatial discretization. The time-stepping for both the concentration and velocity equation uses $2nd$ order Adams-Bashforth. In the concentration equation, the integrating factor in Fourier space is used. In the velocity equation, high-order Fourier filtering is used to maintain stability [15].

We begin by comparing the results from a sharp interface simulation with those from the NSCH model. The surface tension $\sigma = 1/2$ and the jump in tangential velocity equal 1 across the upper interface and -1 across the lower interface. Thus, the flow inside the jet is from left to right. A time sequence of a single periodic box is shown in Figure 5. Again, the solid curve is the sharp interface and the dashed curve is the $c = 1/2$ contour line for the NSCH simulation. The NSCH simulation uses $\rho_0 = 1$, $\gamma = 0.06$, $N = 256$; there are approximately 12 points across the interfacial zone.

Good agreement is seen between the two simulations, although the NSCH jet pinches at a slightly later time than the sharp interface jet. In the picture on the lower right, a close-up of the pinching region is shown and an additional NSCH simulation using $\gamma = 0.04$ is included (dot-dashed curve) which suggests convergence to the sharp interface result.

The NSCH simulations continue smoothly through the pinching and the long-term evolution of this NSCH jet is shown in Figure 6. Again, three x-periods are shown; the upper graphs are concentration and the lower graphs are vorticity. As in the Hele-Shaw case, oppositely signed vorticity is produced at the pinching point, but now some vorticity separates from the concentration interface creating a complicated flow-field both inside and outside the newly created drops which oscillate in time as they travel from left to right through the flow.

4 Conclusions and Future Work

In this paper, we have presented two new systems of equations to describe the motion of binary fluids in 2-d inertially dominated flows and viscously dominated flows in a Hele-Shaw cell. We have solved these equations numerically and demonstrated that topology transitions are smoothly captured. Before pinchoff, there is good agreement with sharp interface models.

In the future, we will consider axisymmetric and 3-d flows, incorporate local grid refinement and adaptivity and consider compressibility effects. We will compare our results to those obtained using other models, such as the level set method, and to those from actual experiments of liquid/liquid jets performed in E. Longmire's laboratory where quantified measurements will be made of velocity, vorticity and concentration. Preliminary experiments have already been performed and the results are encouraging [6].

5 Acknowledgments

J.S.L. and H.L. were supported in part by the National Science Foundation (NSF) Grants DMS-940310, DMS-9706931, the Sloan Foundation and the Minnesota Supercomputer Institute. J.G. acknowledges the support of the NSF and the Office of Naval Research. E.K.L. acknowledges the support of the NSF Grant CTS-9457014. M.J.S. acknowledges the support of the Department of Energy Grant DE-FG02-88ER25053, NSF Grants DMS-9396403 (PYI) and DMS-9404554 and the Exxon Educational Foundation. L.T. acknowledges the support of NSF Grant DMS-9501433.

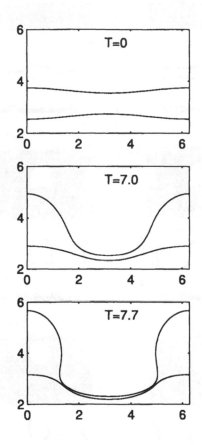

Figure 1: Boundary integral simulation with $\mathbf{B} = \pm 25$, $N = 1024$, $\Delta t = 5 \times 10^{-4}$.

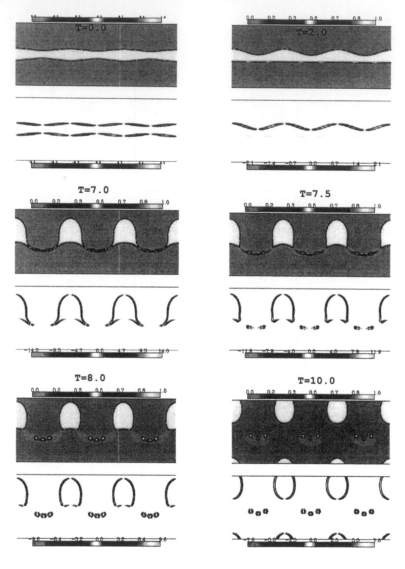

Figure 2: HSCH simulation with $\mathbf{B} = \pm 25$, $\rho_0 = 1 - 0.1c$, $N = 256$ and $\Delta t = 10^{-3}$. Three x-periods are shown.

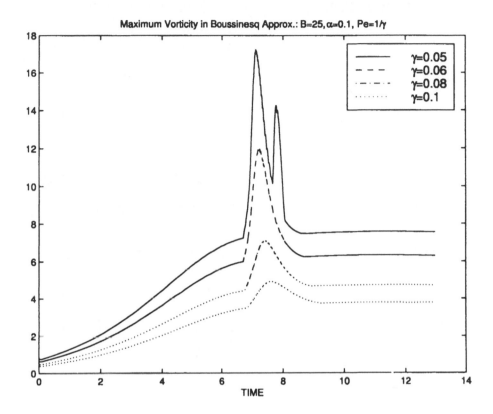

Figure 3: Maximum vorticity vs. time for HSCH simulation.

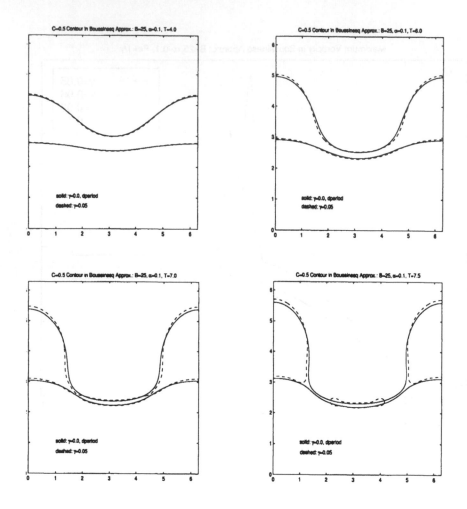

Figure 4: Comparison of sharp and HSCH simulations.

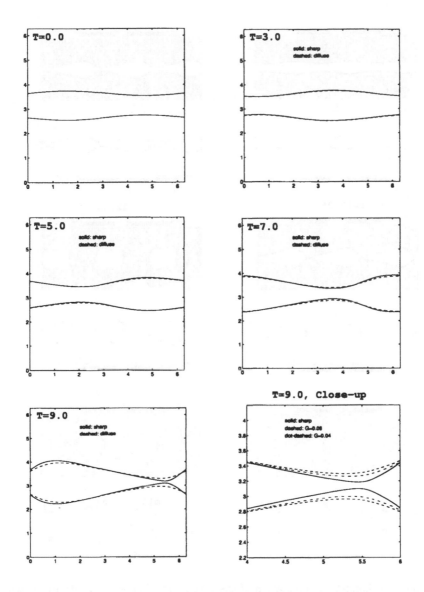

Figure 5: Comparison of sharp and NSCH simulations. The sharp interface method uses $N = 1024$, $\Delta t = 10^{-3}$, $\sigma = 1/2$. The NSCH model uses $\rho_0 = 1$, $\gamma = 0.06$ and $N = 256$.

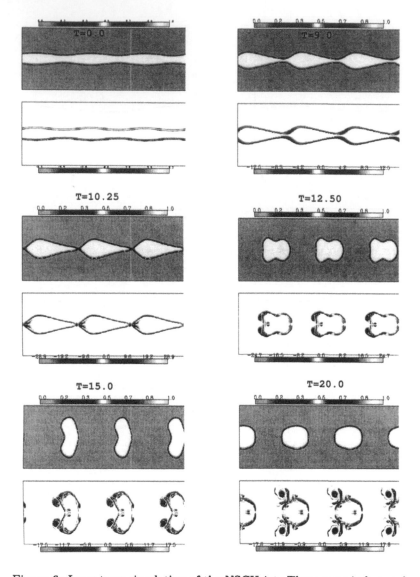

Figure 6: Long-term simulation of the NSCH jet. Three x-periods are shown.

References

[1] D.M. ANDERSON, G.B. MCFADDEN, AND A.A. WHEELER. Diffuse interface methods in fluid mechanics. Annual Review of Fluid Mechanics, 30:1, 1998.

[2] J.U. BRACKBILL, D.B. KOTHE, AND C. ZEMACH. A continuum method for modeling surface tension. J. Comp. Phys., 100:335, 1994.

[3] Y.C. CHANG, T.Y. HOU, B. MERRIMAN, AND S. OSHER. A level set formulation of eulerian interface capturing methods for incompressible fluid flows. J. Comp. Phys. In press.

[4] J. EGGERS. Theory of drop formation. Phys. Fluids, 7:941, 1995.

[5] J. EGGERS AND T.F. DUPONT. Drop formation in a one dimensional approximation of the Navier-Stokes equations. J. Fluid Mech., 262:205, 1994.

[6] D. GEFROH, E.K. LONGMIRE, D. WEBSTER AND J.S. LOWENGRUB. Dynamics of pinchoff in liquid jets flowing into immiscible liquid ambients. In Proceedings of Third International Conference on Multiphase Flow. 1998.

[7] R. GOLDSTEIN, A. PESCI, AND M. SHELLEY. Instabilities and instabilities in Hele-Shaw flow. To appear Phys. Fluids A.

[8] R. GOLDSTEIN, A. PESCI, AND M. SHELLEY. Attracting manifold for a viscous topology transition. Phys. Rev. Lett., 75:3665, 1995.

[9] P.C. HOHENBERG AND B.I. HALPERIN. Theory of dynamic critical phenomena. Rev. Mod. Phys., 49:435, 1977.

[10] T.Y. HOU, J.S. LOWENGRUB, AND M.J. SHELLEY. Removing the stiffness from interfacial flows with surface tension. J. Comp. Phys., 114:312, 1994.

[11] T.Y. HOU, J.S. LOWENGRUB, AND M.J. SHELLEY. The long-time evolution of vortex sheets with surface tension. Phys. Fluids A, 9:1933, 1997.

[12] D.D. JOSEPH. Fluid dynamics of two miscible liquids with diffusion and gradient stresses. Eur. J. Mech. B/Fluids, 9:565, 1990.

[13] J. B. KELLER AND M. MIKSIS. Surface tension driven flows. SIAM J. App. Math, 43:268, 1983.

[14] H. LEE AND J.S. LOWENGRUB. Topological transitions in Hele-Shaw cells. In preparation.

[15] J.S. LOWENGRUB, M.J. SHELLEY, J. GOODMAN, AND L. TRUSKINOVSKY. Topological transitions in CahnHilliard fluid-fluid jets. In preparation.

[16] J.S. LOWENGRUB AND L. TRUSKINOVSKY. Cahn-Hilliard fluids and topological transitions. To appear Proc. R. Soc. London A.

[17] N. MANSOUR AND T. LUNDGREN. Satellite formation in capillary jet breakup. Phys. Fluids A, 2:1141, 1990.

[18] M. SUSSMAN AND P. SMEREKA. Axisymmetric free boundary problems. J. Fluid Mech., 341:269, 1997.

[19] M. SUSSMAN, P. SMEREKA, AND S. OSHER. A level set approach for computing solutions to incompressible two-phase flow. J. Comp. Phys., 114:146, 1994.

[20] S.O. UNVERDI AND G. TRYGGVASON. A front tracking method for viscous, incompressible multifluid flows. J. Comp. Phys., 100:25, 1992.

J.S Lowengrub
Department of Mathematics
University of Minnesota
Minneapolis, MN 55455, USA

E.K. Longmire
Department of Aerospace and Mechanics
University of Minnesota
Minneapolis, MN 55455, USA

J. Goodman
Courant Institute
New York University
New York, NY 10012, USA

M.J. Shelley
Courant Institute
New York University
New York, NY 10012, USA

H. Lee
Department of Mathematics
University of Minnesota
Minneapolis, MN 55455, USA

L. Truskinovsky
Department of Aerospace and Mechanics
University of Minnesota
Minneapolis, MN 55455, USA

J. F. RODRIGUES AND J. M. URBANO
On the Mathematical Analysis of a Valley Glacier Model

In this note we solve a free boundary problem concerning the unidirectional motion of a glacier in a valley, subject to a mass flux condition.

We consider the ice to be an incompressible and isotropic material, with a constant density ρ. As is usual in theoretical glaciology, we assume that the flow of the glacier is so slow that we can restrict the analysis to the stationary situation. The glacier will be taken to be isothermal, which means that we neglect heat transfer.

In this setting, the basic equations of conservation of mass and momentum are (see [4])

$$\nabla \cdot \mathbf{u} = 0 \tag{1}$$

$$\mathbf{u} \cdot \nabla \mathbf{u} = -\nabla p + \nabla \cdot \tau + \rho \mathbf{g} \tag{2}$$

where $\mathbf{u} = (u_i)$ is the velocity field, p is the hydrostatic pressure, $\mathbf{g}$ the gravity vector and $\tau = [\tau_{ij}]$ the deviatoric part of the stress tensor $\sigma = [\sigma_{ij}]$:

$$\sigma_{ij} = -p\delta_{ij} + \tau_{ij} \ .$$

We take a constitutive relation between stress and strain rate of the form

$$\tau_{ij} = \eta \dot{\epsilon}_{ij} \tag{3}$$

where $\dot{\epsilon}_{ij} = \frac{1}{2}(\frac{\partial u_i}{\partial x_j} + \frac{\partial u_j}{\partial x_i})$ and η is the effective viscosity, a function only of the second invariant of one of the tensors, according to the generalised Von Mises criterion.

The flow law that we consider is the nonlinear law of Glen [4]:

$$\dot{\epsilon}_{ij} = \frac{1}{2}\tau^{n-1}\tau_{ij} \ . \tag{4}$$

Here, τ is the second invariant of stress, given by $2\tau^2 = \tau_{ij}\tau_{ij}$. The physical constant n is obtained from experimental data and the value $n = 3$ is now commonly accepted. The limit cases correspond to a viscous Newtonian fluid ($n = 1$) and an ideal plastic material ($n = \infty$). In our mathematical analysis, we will take $1 < n < \infty$.

Using the deformation law (4), we can express the viscosity in the form $\eta = 2\tau^{1-n}$ and the relation (3) becomes

$$\tau_{ij} = 2\tau^{1-n}\dot{\epsilon}_{ij} \ . \tag{5}$$

237

A major simplification occurs with the shallow ice approximation. The idea is that the depth of the glacier is very small compared to its length. After an appropriate rescaling of the variables, this corresponds to considering a laminar unidirectional flow, in which the velocity is parallel to the axis of the valley (see, for instance, [7], [4] and [3]).

Our model considers a simple geometric setting. We take the x axis, $O\vec{e_1}$, in the direction of the valley axis, the y axis across stream and the z axis upward. The glacier bed will be described by a Lipschitz function $h : \mathbf{R} \to \mathbf{R}$ such that $h(y) \to +\infty$ when $|y| \to \infty$. We also assume, for simplicity, that $\min_{y \in \mathbf{R}} h(y) = 0$. The graph of h will be called H. Denoting with α the mean valley slope, we have $\mathbf{g} = (g \sin \alpha, 0, -g \cos \alpha)$.

The new feature of the problem considered here is the fact that the height of the top surface of the glacier, taken to be flat, is not known. A cross-section is thus a set of the form

$$\Omega_\xi = \{(y,z) \in \mathbf{R}^2 \ : \ h(y) < z < \xi\} \ ,$$

where ξ is unknown. The number ξ is to be determined so that a prescribed mass flux condition is satisfied, namely, [3]

$$\int_{\Omega_\xi} \mathbf{u} \cdot \vec{e_1} = \int_{\Omega_\xi} u \, dy dz = Q \ ,$$

where $Q > 0$ is given.

We then have $\mathbf{u} = (u, 0, 0)$. From (1), we see that $\partial u / \partial x = 0$ and u is a function of y and z alone. In particular, $\mathbf{u} \cdot \nabla \mathbf{u} = 0$. From (5) we then have

$$\tau_{11} = \tau_{22} = \tau_{33} = \tau_{23} = 0 \ ,$$

$$\tau_{12} = \tau^{1-n} \frac{\partial u}{\partial y} \quad \text{and} \quad \tau_{13} = \tau^{1-n} \frac{\partial u}{\partial z}$$

and so, with $\nabla = (\frac{\partial}{\partial y}, \frac{\partial}{\partial z})$,

$$\tau = |\nabla u|^{\frac{1}{n}} \ .$$

Now the equations of motion (2) read

$$\frac{\partial p}{\partial x} = \nabla \cdot (|\nabla u|^{\frac{1-n}{n}} \nabla u) + \rho g \sin \alpha \ , \quad \frac{\partial p}{\partial y} = 0 \ , \quad \frac{\partial p}{\partial z} = -\rho g \cos \alpha \ .$$

From the second and third equations, we find that $p(x,z) = (-\rho g \cos \alpha)z + \beta(x)$, and replacing on the first one, we obtain

$$\beta'(x) = \nabla \cdot (|\nabla u|^{\frac{1-n}{n}} \nabla u) + \rho g \sin \alpha \ .$$

Since the right-hand side of this equation does not depend on x, from the boundary conditions for the pressure we see that $\beta'(x) = -c$, where c is a positive constant called drop in pressure per unit length. We have then determined the pressure up

238

to this constant and, dividing by $c + \rho g \sin \alpha$, we are left with the singular elliptic equation

$$\Delta_p u_\xi \equiv \nabla \cdot (|\nabla u|^{p-2} \nabla u) = -1 \quad \text{in} \quad \Omega_\xi \ ,$$

where $p = 1 + \frac{1}{n}$, $1 < p < 2$.

The boundary conditions for u are of mixed type. Along the glacier bed we take an homogeneous Dirichlet condition for no slip at the base, which is adequate to deal with a cold glacier but inadequate for a temperate one:

$$u = 0 \quad \text{on} \quad \Gamma_H^\zeta = H \cap \overline{\Omega_\zeta} \ .$$

On the contact surface of the glacier with the atmosphere we consider an homogeneous Neumann condition:

$$\frac{\partial u}{\partial z} = 0 \quad \text{on} \quad \Gamma_\Pi^\zeta = \Pi_\zeta \cap \overline{\Omega_\zeta} \ ,$$

where $\Pi_\zeta = \{(y, z) \in \mathbf{R}^2 \ : \ z = \zeta\}$.

A similar problem in a fixed (*a priori* known) domain has been considered by Pélissier and Reynaud [7] (see also [6]), where, instead of the no slip condition at the base, an imposed Neumann condition is taken.

The mathematical problem is then the following:

Problem (P): Given $Q > 0$, find a positive real number ξ and a function u_ξ such that

$$
\begin{cases}
(1) \begin{cases} \Delta_p u_\xi = -1 & \text{in} \quad \Omega_\xi = \{(y, z) \ : \ h(y) < z < \xi\} \\[4pt] u_\xi = 0 & \text{on} \quad \Gamma_H^\xi \\[4pt] \frac{\partial u_\xi}{\partial z} = 0 & \text{on} \quad \Gamma_\Pi^\xi \end{cases} \\[30pt]
(2) \ \displaystyle\int_{\Omega_\xi} u_\xi = Q \ .
\end{cases}
$$

Since classical solutions are not to be expected in general, we now consider the problem in a weak form. The appropriate space of test functions is

$$W_H^{1,p}(\Omega_\xi) = \{v \in W^{1,p}(\Omega_\xi) \ : \ v = 0 \quad \text{on} \quad \Gamma_H^\xi\}, \quad 1 < p < 2 \ ,$$

where a Poincaré type inequality is valid; the usual $W^{1,p}$-norm of a function is then equivalent to the L^p norm of its gradient and this is the one that will be used.

A standard formal integration by parts, taking the boundary conditions into account, leads to the following:

239

Definition. *A weak solution to (P) is a pair $(\xi, u_\xi) \in \mathbf{R}^+ \times W_H^{1,p}(\Omega_\xi)$ such that*

$$\int_{\Omega_\xi} |\nabla u_\xi|^{p-2} \nabla u_\xi \cdot \nabla v = \int_{\Omega_\xi} v \; , \; \forall v \in W_H^{1,p}(\Omega_\xi) \tag{6}$$

and

$$\int_{\Omega_\xi} u_\xi = Q \; . \tag{7}$$

Remark 1 *For each fixed ζ, we will call (P_ζ) the problem consisting of solving (6). The well-known properties of the p-Laplacian, namely, strict monotonicity, ensure that each (P_ζ) has a unique solution. Moreover, (6) is the Euler-Lagrange equation for the minimization problem*

$$\min_{v \in W_H^{1,p}(\Omega_\zeta)} \left[\frac{1}{p} \int_{\Omega_\zeta} |\nabla v|^p - \int_{\Omega_\zeta} v \right] \; , \tag{8}$$

so the solution u_ζ is the corresponding minimizer (see [5], for example).

We also have regularity results for the solution. In fact, in [1] it is shown that $u_\zeta \in C_{loc}^{1,\alpha}(\Omega_\zeta)$, that is, the local Hölder continuity of its gradient is the best regularity that is to be expected.

As is easily seen, we can reduce the problem to the (nontrivial) study of the real function Φ defined by

$$\mathbf{R}^+ \ni \zeta \longmapsto (\Omega_\zeta, u_\zeta) \longmapsto \int_{\Omega_\zeta} u_\zeta \in \mathbf{R} \; ,$$

where u_ζ is the solution of (P_ζ), and to the question: Does the range of Φ contain the given $Q > 0$?

We can consider the problem in the nonphysical one-dimensional case just to get some insight from this simpler situation. We have $\Omega_\zeta = (0, \zeta)$ and the problem (P_ζ) consists of solving the ordinary boundary value problem

$$(|u_\zeta'|^{p-2} u_\zeta')' = -1 \text{ in } (0, \zeta) \; , \quad u_\zeta(0) = 0 \; , \quad u_\zeta'(\zeta) = 0 \; .$$

We find

$$u_\zeta(z) = \frac{1-p}{p} \left[(-z + \zeta)^{\frac{p}{p-1}} - \zeta^{\frac{p}{p-1}} \right] \quad \text{and} \quad \Phi(\zeta) = \int_0^\zeta u_\zeta = \frac{p-1}{2p-1} \zeta^{\frac{2p-1}{p-1}} \; ,$$

240

so, being Φ bijective, given $Q > 0$, we just have to choose $\xi = \left(\frac{2p-1}{p-1}Q\right)^{\frac{p-1}{2p-1}}$, and the solution of the problem is

$$u(z) = \frac{1-p}{p}\left[\left\{-z + (\frac{2p-1}{p-1}Q)^{\frac{p-1}{2p-1}}\right\}^{\frac{p}{p-1}} - (\frac{2p-1}{p-1}Q)^{\frac{p}{2p-1}}\right] \text{ in } \left(0, \left(\frac{2p-1}{p-1}Q\right)^{\frac{p-1}{2p-1}}\right).$$

The analysis of the general case cannot be done explicitly for a general graph h. Therefore, we fix ζ and choose $v = -u_\zeta^-$ in (6) to find

$$\int_{\Omega_\zeta \cap \{u_\zeta < 0\}} |\nabla u_\zeta|^p = -\int_{\Omega_\zeta} |\nabla u_\zeta|^{p-2}\nabla u_\zeta \cdot \nabla u_\zeta^- = -\int_{\Omega_\zeta} u_\zeta^- \leq 0 .$$

This implies $\int_{\Omega_\zeta} |\nabla u_\zeta^-|^p = 0$; hence $u_\zeta^- = 0$ and $u_\zeta \geq 0$. So $\Phi(\zeta) \geq 0$, $\forall \zeta \in \mathbf{R}^+$ and it would be zero only if $u_\zeta = 0$ a.e. in Ω_ζ, which is not possible. So the range of Φ is contained in $\mathbf{R}^+$, thus choosing $Q > 0$ is necessary. It turns out that it is also sufficient by the following:

Theorem. *The function Φ is a continuous, strictly increasing function, whose range is the set $(0, +\infty)$. Thus, for every $Q > 0$, there exists a unique weak solution to* (P).

Proof. Observe that, for each $\zeta \in (0, \infty)$, the unique solution of (P_ζ) satisfies

$$\int_{\Omega_\zeta} |\nabla u_\zeta|^p = \int_{\Omega_\zeta} u_\zeta . \tag{9}$$

It is obvious that the second assertion of the theorem follows from the first one.

Monotonicity: For each $\hat\zeta > \zeta$, put $\hat u \equiv u_{\hat\zeta}$ and $u \equiv u_\zeta$. Let $\alpha = \hat\zeta - \zeta > 0$ and define the translation of the set Ω_ζ

$$\omega_\alpha = T_\alpha(\Omega_\zeta) = \{(y, z) \in \mathbf{R}^2 : h(y) + \alpha < z < \zeta + \alpha\} .$$

Now let u^α be the function defined for $(y, z) \in \omega_\alpha$, by $u^\alpha(y, z) = u(y, z - \alpha)$. It is clear that this function satisfies

(P$_\alpha$) $\quad u^\alpha \in W^{1,p}_{H+\alpha}(\omega_\alpha) : \int_{\omega_\alpha} |\nabla u^\alpha|^{p-2}\nabla u^\alpha \cdot \nabla v = \int_{\omega_\alpha} v , \forall v \in W^{1,p}_{H+\alpha}(\omega_\alpha) .$

We show that

$$\int_{\Omega_{\hat\zeta}} \hat u > \int_{\omega_\alpha} \hat u \geq \int_{\omega_\alpha} u^\alpha = \int_{\Omega_\zeta} u . \tag{10}$$

The equality is an obvious consequence of translation and the strict inequality follows from the fact that $\hat{u} \geq 0$ is nonvanishing in the open set $\Omega_{\hat{\zeta}} \setminus \omega_\alpha$, since it solves $-\Delta_p u = 1$ there. To conclude, it suffices to show that $\hat{u} \geq u^\alpha$ a.e. in ω_α. Take the function

$$
v = \begin{cases} (\hat{u} - u^\alpha)^- & \text{in} \quad \omega_\alpha \\[2mm] 0 & \text{in} \quad \Omega_{\hat{\zeta}} \setminus \omega_\alpha \end{cases},
$$

which can be used as a test function both in (P_α) and $(P_{\hat{\zeta}})$. Subtracting and using a well-known inequality for $1 < p < 2$, we get

$$
C_p \frac{\|\nabla(\hat{u} - u^\alpha)^-\|_{L^p(\omega_\alpha)}^2}{\|\nabla \hat{u}\|_{L^p(\omega_\alpha)}^{2-p} + \|\nabla u^\alpha\|_{L^p(\omega_\alpha)}^{2-p}} \leq \int_{\omega_\alpha} (|\nabla u^\alpha|^{p-2}\nabla u^\alpha - |\nabla \hat{u}|^{p-2}\nabla \hat{u}) \cdot \nabla(\hat{u} - u^\alpha)^- = 0.
$$

Hence, we can conclude that

$$
\|\nabla(\hat{u} - u^\alpha)^-\|_{L^p(\omega_\alpha)} = 0 \ , \quad \text{i.e.} \quad \hat{u} \geq u^\alpha \ \text{a.e. in } \omega_\alpha.
$$

Continuity: We start with left continuity, i.e., $\Phi(\zeta - \eta) \to \Phi(\zeta)$ when $\eta \to 0^+$. Since we can use u_ζ as a test function in $(P_{\zeta-\eta})$, i.e. in (6) for $u_{\zeta-\eta}$, we so obtain

$$
\int_{\Omega_{\zeta-\eta}} u_\zeta = \int_{\Omega_{\zeta-\eta}} |\nabla u_{\zeta-\eta}|^{p-2}\nabla u_{\zeta-\eta} \cdot \nabla u_\zeta \leq \left(\int_{\Omega_{\zeta-\eta}} |\nabla u_{\zeta-\eta}|^p\right)^{\frac{1}{p'}} \left(\int_{\Omega_{\zeta-\eta}} |\nabla u_\zeta|^p\right)^{\frac{1}{p}},
$$

using Hölder's inequality. Using (9), this can be written in the form

$$
\Phi(\zeta - \eta) \geq \frac{\left(\int_{\Omega_{\zeta-\eta}} u_\zeta\right)^{p'}}{\left(\int_{\Omega_{\zeta-\eta}} |\nabla u_\zeta|^p\right)^{\frac{p'}{p}}} \ .
$$

Taking the limit when $\eta \to 0^+$, this implies

$$
\lim_{\eta \to 0^+} \Phi(\zeta - \eta) \geq \Phi(\zeta) \ .
$$

Now we may conclude simply by observing that, due to the monotonicity, $\Phi(\zeta - \eta) \leq \Phi(\zeta), \forall \eta > 0$.

Concerning right continuity, we show that, for fixed ζ, $\Phi(\zeta + \eta) \to \Phi(\zeta)$ when $\eta \to 0^+$. By using reasoning similiar to the one that led to (10), we obtain

$$
\int_{\Omega_{\zeta+\eta} \setminus \Omega_\zeta} u_{\zeta+\eta} \longrightarrow 0 \ .
$$

Then from

$$\Phi(\zeta + \eta) = \int_{\Omega_{\zeta+\eta} \setminus \Omega_\zeta} u_{\zeta+\eta} + \int_{\Omega_\zeta} u_{\zeta+\eta} \, ,$$

we need only show the second term converges to $\Phi(\zeta)$. Observe that $\|\nabla u_{\zeta+\eta}\|_{L^p(\Omega_\zeta)} \leq C$, independently of $\eta < 1$, so we can extract a subsequence such that $u_{\zeta+\eta} \rightharpoonup u_*$ in $W^{1,p}(\Omega_\zeta)$-weak, for some $u_* \in W^{1,p}(\Omega_\zeta)$. If we show that u_* is a solution to the limit problem (P_ζ), we get as a consequence, due to the uniqueness, that $u_* = u_\zeta$. Since the imbedding $W^{1,p}(\Omega_\zeta) \hookrightarrow L^1(\Omega_\zeta)$ is compact, this gives $\int_{\Omega_\zeta} u_{\zeta+\eta} \to \int_{\Omega_\zeta} u_\zeta = \Phi(\zeta)$. So take an arbitrary $v \in W_H^{1,p}(\Omega_\zeta)$ and extend it to a function in $W_H^{1,p}(\Omega_{\zeta+\eta})$, first by reflection with respect to Π_ζ, and then by zero, neglecting the portion of the domain above $\Pi_{\zeta+\eta}$. Call this extension $\tilde{v}$ and use it as a test function in $(P_{\zeta+\eta})$ to obtain

$$0 \leq \int_{\Omega_{\zeta+\eta}} \left(|\nabla u_{\zeta+\eta}|^{p-2} \nabla u_{\zeta+\eta} - |\nabla \tilde{v}|^{p-2} \nabla \tilde{v} \right) \cdot (\nabla u_{\zeta+\eta} - \nabla \tilde{v}) =$$

$$= \int_{\Omega_{\zeta+\eta} \setminus \Omega_\zeta} (u_{\zeta+\eta} - \tilde{v}) - \int_{\Omega_{\zeta+\eta} \setminus \Omega_\zeta} |\nabla \tilde{v}|^{p-2} \nabla \tilde{v} \cdot (\nabla u_{\zeta+\eta} - \nabla \tilde{v})$$

$$+ \int_{\Omega_\zeta} (u_{\zeta+\eta} - v) - \int_{\Omega_\zeta} |\nabla v|^{p-2} \nabla v \cdot (\nabla u_{\zeta+\eta} - \nabla v) \ \ .$$

Finally, take the limit as $\eta \to 0^+$ and, since the integrals over $\Omega_{\zeta+\eta} \setminus \Omega_\zeta$ vanish, use Minty's Lemma (see [5]) to conclude that

$$\int_{\Omega_\zeta} |\nabla u_*|^{p-2} \nabla u_* \cdot \nabla w = \int_{\Omega_\zeta} w \, , \quad \forall w \in W_H^{1,p}(\Omega_\zeta) \, ,$$

thus proving that u_* is a solution to (P_ζ).

Range: First, we have $\Phi(\zeta) \to 0$ when $\zeta \to 0$. In fact, using the inequalities of Hölder and Poincaré,

$$\Phi(\zeta) = \|\nabla u_\zeta\|_{L^p(\Omega_\zeta)}^p = \int_{\Omega_\zeta} u_\zeta \leq |\Omega_\zeta|^{\frac{1}{p'}} \|u_\zeta\|_{L^p(\Omega_\zeta)} \leq C \, |\Omega_\zeta|^{\frac{1}{p'} + \frac{1}{2}} \|\nabla u_\zeta\|_{L^p(\Omega_\zeta)} \, ,$$

where C depends only on p. We ultimately obtain

$$0 \leq \Phi(\zeta) \leq C \, |\Omega_\zeta|^{1 + \frac{p}{2(p-1)}} \, ,$$

and let ζ converge to zero to conclude.

Next, we prove that $\Phi(\zeta) \to \infty$ when $\zeta \to \infty$. In fact, due to Remark 1, we have, for any $v \in W_H^{1,p}(\Omega_\zeta)$,

$$\Phi(\zeta) \geq \left[\int_{\Omega_\zeta} v - \frac{1}{p} \int_{\Omega_\zeta} |\nabla v|^p \right] \, , \forall v \in W_H^{1,p}(\Omega_\zeta) \, . \tag{11}$$

243

We now make an appropriate choice of a sequence v_ζ, such that the second member of (11) will tend to infinity with ζ, thus concluding that the same happens with Φ. Without loss of generality, we may choose $\rho > 0$ and $\sigma > 0$ such that, for all $\zeta > \sigma + 1$, the rectangle $\mathcal{R}_\zeta = (-\rho, \rho) \times (\sigma, \zeta)$ is contained in Ω_ζ. Define $v_\zeta \in W_H^{1,p}(\Omega_\zeta)$, as $v_\zeta(y, z) = f(y)g_\zeta(z)$ for $(y, z) \in \mathcal{R}_\zeta$ and $v_\zeta(y, z) = 0$ in $\Omega_\zeta \setminus \mathcal{R}_\zeta$, where $f : (-\rho, \rho) \to \mathbf{R}$ and $g_\zeta : (\sigma, \zeta) \to \mathbf{R}$ are the functions

$$
f(y) = \left\{ \begin{array}{ccc} 2 + \frac{2}{\rho}y & \text{if} & -\rho < y < -\frac{\rho}{2} \\ 1 & \text{if} & -\frac{\rho}{2} \leq y \leq \frac{\rho}{2} \\ 2 - \frac{2}{\rho}y & \text{if} & \frac{\rho}{2} < y < \rho \end{array} \right.
\quad \text{and} \quad
g_\zeta(z) = \left\{ \begin{array}{ccc} z - \sigma & \text{if} & z \in (\sigma, \sigma + 1] \\ \\ 1 & \text{if} & z \in (\sigma + 1, \zeta) \end{array} \right.
$$

Using the theorem of Fubini, we find from (11)

$$
\Phi(\zeta) \geq (\zeta - \sigma) \left[\frac{3}{2}\rho - \frac{2^p}{p}\rho^{1-p} \right] + C ,
$$

with C a constant independent of ζ. Now the second member in this inequality goes to infinity with ζ provided $\frac{3}{2}\rho - \frac{2^p}{p}\rho^{1-p} > 0$. This holds true if we choose $\rho > (2^{p+1}/3p)^{1/p}$, which is always possible due to our assumptions on h and choosing σ also big enough.

Remark 2 *This argument extends to the case $p \geq 2$, as well as to a more general class of elliptic problems (see [8]).*

Acknowledgments. This work was partially supported by FCT(JNICT), Praxis XXI and Praxis/2/2.1/MAT/125/94 projects. J. M. Urbano is also indebted to CMAF for the warm hospitality.

References

[1] DiBENEDETTO, E., $C^{1+\alpha}$ Local regularity of weak solutions of degenerate elliptic equations, *Nonlinear Analysis*, T.M.A., Vol. 7, No. 8 (1983), 827-850.

[2] FOWLER, A.C. AND LARSON, D.A. On the flow of polythermal glaciers I. Model and preliminary analysis, *Proc. R. Soc. London*, A 363 (1978), 217-242.

[3] FOWLER, A.C. Glaciers and ice sheets, in *"The Mathematical Models for Climatology and Environment"* (Ed. Jesús Ildefonso Díaz), NATO ASI Series I, Vol. 48, Springer-Verlag, Berlin, 1997.

[4] HUTTER, K. *Theoretical Glaciology*, D. Reidel, Dordrecht, 1983.

[5] LIONS, J.L. *Quelques Méthodes de Résolution des Problèmes aux Limites Non Linéaires*, Dunod, Gauthier-Villars, Paris, 1969.

[6] PÉLISSIER, M.C. Sur quelques problèmes non linéaires en glaciologie, *Public. Math. d'Orsay*, 110 / 75-24, Univ. Paris-Sud, 1975.

[7] PÉLISSIER, M.C. AND REYNAUD, M.L. Etude d'un modèle mathématique d'écoulement de glacier, *C.R. Acad. Sc. Paris*, t.279 (1974), 531-534.

[8] RODRIGUES, J.F. AND URBANO, J.M. Degenerate elliptic problems in a class of free domains, *J. Math. Pures Appl.* (1999)(in press).

José Francisco Rodrigues
C.M.A.F. / Universidade de Lisboa
Av. Prof. Gama Pinto, 2
1649-003 Lisboa Codex, Portugal

José Miguel Urbano
C.M.U.C. / Universidade de Coimbra
Apartado 3008
3000 Coimbra, Portugal

A. WAGNER

On the Bernoulli Free Boundary Problem with Surface Tension

Introduction. We consider a bounded domain $\Omega \in \mathbb{R}^n$, partially filled with water. The water occupies a fixed part of the domain at an initial time, while the rest is filled with a gas of constant pressure. We assume surface tension at the water-gas interface. Given also an initial velocity distribution we ask how the fluid and the shape of the gas bubble evolve in time.

This problem was considered by Chen and Friedman in [1]. They investigated the quasistationary case, i.e., the situation where the fluid velocity does not depend on time. They proved the existence of a unique classical solution local in time and space. In [2] and [3] the author considered the case without surface tension. Using the notion of a "generalized geometric evolution" introduced by Giga and Takahashi in [4] we were able to prove an existence result which was global in time. However, we could not prove uniqueness, since there was no control on the $n - 1$-dimensional Hausdorff measure of the water-gas interface. Physically speaking, without surface tension, we could not exclude a fattening effect of the interface during some time interval.

In this article we include surface tension and we obtain a unique solution, which is global in time.

1. As in [3], we assume that the fluid is ideal, incompressible and irrotational. We describe the velocity by a harmonic potential u. p will always denote the pressure. Due to surface tension, the pressure will jump at the interface (due to Laplace's law):

$$p = p_0 + \sigma H. \tag{1}$$

H denotes the mean curvature of the interface and σ is the coefficient of surface tension, which we assume to be constant.

The interface is an unknown itself. We assume that it moves with the fluid, i.e., if its velocity is denoted by V, we assume

$$V = \nu \cdot \nabla u \tag{2}$$

on the interface. ν is the normal vector on the interface. To describe the interface we will use the description of level sets. We introduce an auxillary function $\Psi(x, t)$ having opposite sign in the domain occupied by the fluid resp. the gas. The interface is then the zero level set of Ψ. Since we require this for all time, Ψ has to satisfy the differential equation

$$\partial_t \Psi + \nabla u \cdot \nabla \Psi = 0 \tag{3}$$

on the interface $\{\Psi \equiv 0\}$ with $\Psi(x,0) = \Psi_0(x)$ in Ω. Comparing this with (2) and noting that $\nu = \frac{\nabla\Psi}{|\nabla\Psi|}$ we see that $V = \frac{\partial_t\Psi}{|\nabla\Psi|}$.

With this notation H is a function of Ψ. It takes the form

$$H(\Psi) = \frac{\Delta\Psi}{|\nabla\Psi|} - \frac{\nabla\Psi \cdot D^2\Psi \cdot \nabla\Psi}{|\nabla\Psi|^3}$$

on $\Psi = 0$. Thus H is a second order operator, which degenerates if $|\nabla\Psi| = 0$.

The underlying equations which couple the fluid velocity with the pressure inside the fluid are the Euler equations. They imply Bernoulli's law which states that

$$\partial_t u + \frac{1}{2}|\nabla u|^2 + p = b(t) \tag{4}$$

for any function $b(t)$, i.e., the left-hand side of the equation only depends on time. Thus knowing the potential u (4) gives us the pressure up to a function that depends only on time. We will always assume that b is a continuous function in t.

We denote the domain occupied by the fluid at time t by B_t and the unknown gas bubble at time t by A_t. Thus $\Gamma_t = \overline{\Omega} \setminus (A_t \cup B_t)$ denotes the interface.

Taking into account (1) we now may formulate a nonlinear boundary condition on Γ_t, namely,

$$\partial_t u + \frac{1}{2}|\nabla u|^2 + p_0 + \sigma H - b(t) = 0.$$

Thus we consider the system

$$\Delta u = 0 \quad \text{in} \bigcup_{0<t<T} B_t \times \{t\} \tag{5}$$

$$\partial_t u + \frac{1}{2}|\nabla u|^2 + p_0 + \sigma H - b(t) = 0 \quad \text{in} \bigcup_{0<t<T} \Gamma_t \times \{t\} \tag{6}$$

$$\partial_t \Psi + \nabla u \cdot \nabla \Psi = 0 \quad \text{in} \bigcup_{0<t<T} \Gamma_t \times \{t\} \tag{7}$$

$$\Psi(x,0) = \Psi_0(x) \quad \text{in } \Omega \tag{8}$$

$$u(x,0) = u_0(x) \quad \text{in } B_0. \tag{9}$$

We choose Ψ_0 to be a continuous function such that

$$\Psi_0(x) \begin{cases} > 0 & : \quad x \in B_0 \\ = 0 & : \quad x \in \Gamma_0 \\ < 0 & : \quad x \in A_0 \end{cases}$$

for initially prescribed sets A_0, B_0 and Γ_0. u_0 is assumed to be harmonic in B_0.

We simplify our model by choosing $\overline{\Omega}$ to be the unit square $[0,1]^n$, thereby identifying opposite faces. Thus we consider $\overline{\Omega}$ as a n-torus which is a compact manifold without boundary. This avoids the difficulty which arises if the interface Γ_t touches the boundary of Ω. To keep the notation short we will write $Q_T = \Omega \times (0, T)$.

Remark: The main problem of our model (5)-(9) is the high regularity needed to understand (6) in a classical sense on one hand, and the rather low regularity for the free boundary provided by equation (7). The key idea is to understand (5)-(9) in a suitable weak sense. The weak formulation of our model will be given by a set of equations which hold in all Q_T and for which we prove the global existence in time. If all unknowns are sufficiently smooth, a weak solution is also a solution of (5)-(9) by restriction.

2. In this paragraph we recall some existence results for degenerate parabolic equations as they were proved in [5]. The model problem we have in mind is a modified equation for prescribed mean curvature flow

$$\partial \Psi - \sigma H(\Psi) + \nabla \cdot \Psi = f \quad \text{in} \quad Q_T \tag{10}$$

where F is a prescribed inhomogenity defined in Q_T, depending on x and t. Furthermore we will assume that f satisfies a Hölder condition in space and time.
We also assume some regularity for u, namely, we choose u to be in the space

$$X^q \equiv L^\infty(0, T; W^{2,q}(\Omega)) \cap W^{1,\infty}(0, T; L^q(\Omega)) \quad \forall n < q < \infty.$$

In particular, by imbedding (see [7] Lemma 3.3, Chapter II) we have that $\tilde{u}$ is bounded in $C^{0,\alpha}(0, T; C^{1,\alpha}(\overline{\Omega}))$ for $0 \leq \alpha < 1$.
In [5] the authors introduce the notion of viscosity solutions for a very general class of parabolic equations. In particular, they formulate conditions under which a comparison principle for sub- and supersolutions can be found. It is easy to check that (10) falls into their class and satisfies all the conditions. Hence, by Perron's method we may assume the unique existence of a viscosity solution. For a survey article on this theory see [6].
We now give a precise definition of what we mean by a solution to (10). For that we follow [5]. In the first step we introduce a regularized mean curvature $\tilde{H}$ which allows us to deal with the case $|\nabla \Psi| = 0$. In a slight abuse of notation we will use the notation $H(\Psi) = H(\nabla \Psi, D^2 \Psi) = H(p, X)$.

Definition 1:

$$\tilde{H}_*(q, X) = \liminf_{\epsilon \to 0} \{ H(r, Y) : r \neq 0, |r - q| \leq \epsilon, |X - Y| \leq \epsilon \}$$

is called the lower semicontinuous envelope of H and $\tilde{H}^* = -(-H)_*$ is called the upper semicontinuous envelope of H.

248

Since H is continuous on the space of all n-vectors q with $|q| \neq 0$ and on the space of all $n \times n$ matrices, the lower (and upper) semicontinuous envelope of H gives an extension of H which includes the singular case $|q| = 0$.

Definition 2: A function $\Psi : Q_T \to \mathbf{R}$ is called viscosity sub- (super-) solution of (10), if $\Psi^* < \infty$ (resp. $\Psi_* > -\infty$) on Q_T and

$$s - \sigma \tilde{H}_*(q, X) + \nabla u \cdot q - f \leq 0 \quad \forall (s, q, X) \in \mathcal{P}^{2,+}_{Q_T} \Psi^*(t, x), \quad (t, x) \in Q_T,$$
$$(resp. s - \sigma \tilde{H}^*(q, X) + \nabla u \cdot q - f \geq 0 \quad \forall (s, q, X) \in \mathcal{P}^{2,-}_{Q_T} \Psi_*(t, x), \quad (t, x) \in Q_T).$$

Here $\mathcal{P}^{2,+}_{Q_T} \Psi^*(t, x)$ is the set of $(s, q, X) \in \mathbf{R} \times \mathbf{R}^n \times S^n$ such that

$$\Psi(\tau, y) \leq \Psi(t, x) + s(\tau - t) + q \cdot (y - x) + \frac{1}{2} X(y - x) \cdot (y - x) + o(|\tau - t| + |y - x|^2)$$

as $(\tau, y) \to (t, x)$ in Q_T.
Here S^n denotes the space of all symmetric $n \times n$ matrices.

We call Ψ a solution of (10) if it is a sub- and a supersolution.

Based on the comparison result in [5] the following theorem holds.

Theorem 1: Under the above assumption, there exists a unique solution Ψ to (10).

With the help of Ψ we may now define the geometric evolution of initially given sets B_0 and A_0.

Definition 3: The evolution of B_0 (resp. A_0) is given by

$$B_t := \{(x, t) \in Q_T : \Psi(x, t) > 0\}$$
$$A_t := \{(x, t) \in Q_T : \Psi(x, t) < 0\}$$

and these sets are uniquely defined.

3. We will now define our notion of a weak solution to (5)-(9). The following observation motivates our notion. If we assume in equation (7) u to be defined in all Q_T, equation (7) is defined in Q_T and thus we obtain Ψ as a function in Q_T. Consequently, if Ψ is smooth enough, $H(\Psi)$ is also defined in Q_T and hence we may (at least formally) consider equation (6) as an equation in Q_T.

Definition 4: (u, Ψ) is a weak solution to (5)-(9) if $u \in X^q$ and Ψ continuous

249

satisfy

$$-\sigma \Delta H = |D^2 u|^2 \quad \text{in } Q_T \tag{11}$$

$$\partial_t u - n\Delta u + \frac{1}{2} \mid \nabla u \mid^2 + p_0 + \sigma H = b(t) \quad \text{in } Q_T \tag{12}$$

$$\partial_t \Psi - \sigma H(\Psi) + \nabla u \cdot \nabla \Psi = \partial_t u + \frac{1}{2} \mid \nabla u \mid^2 + p_0 - b(t) \quad \text{in } Q_T \tag{13}$$

$$\Psi(x,0) = \Psi_0(x) \quad \text{in } \Omega \tag{14}$$

$$u(x,0) = u_0(x) \quad \text{in } B_0 \tag{15}$$

for n sufficiently large.

The following observations illustrate the idea behind this notion. The first equation is a Poisson equation on a compact manifold without boundary. Thus if $u \in X^q$ we have $H \in L^\infty(0,T;W^{2,q}(\Omega))$. Furthermore, the estimate

$$\| H \|_{L^\infty(0,T;W^{2,q}(\Omega))} \le c \| u \|_{X^q} \, .$$

If we rewrite

$$\partial_t u - n\Delta u + \frac{1}{2} \mid \nabla u \mid^2 = b(t) - p_0 - \sigma H,$$

we may use parabolic theory to deduce the unique existence of a solution $u \in X^q$ satisfying the estimate

$$\| u \|_{X^q} \le c(n) \| H \|_{L^\infty(0,T;W^{2,q}(\Omega))} \tag{16}$$

with $c(n) \to 0$ as $n \to \infty$, see [7].

Combining the two results we may conclude with the help of Banach's fixed point theorem (and n sufficiently large) the existence of a unique (u, H) satisfying (11) - (12).

If (11)-(12) hold, we observe that by applying the Laplace operator to (12) and using (11) we are left with

$$\partial_t \Delta u - n\Delta^2 u = 0 \quad \text{in } Q_T.$$

Thus if the initial data are harmonic so will be $u(x,t)$ for all $t > 0$.

Finally, we turn to (13). This is a special case of equation (10) in **2.** where we have set

$$f = \partial_t u + \frac{1}{2} \mid \nabla u \mid^2 + p_0 - b(t).$$

Since $u \in X^q$ f is continuous in x and t. Thus all results of **2.** hold.

250

We are now in the following situation. For a given $v \in X^q$ we defined (by (10)) a geometric evolution of initially given sets A_0 and B_0. This evolution gives us a unique solution to equation (11)-(12). The composition of these two maps is a continuous compact map $S : C^0(0, T; C^1(\overline{\Omega})) \to \{v \in X^q : \| v \|_{X^q} \leq c\}$.

By (16) this map is also a contraction, thus we may conclude the unique existence of a solution.

Theorem 2: Assume that initially two disjoint sets A_0 and B_0 are given in Ω. Furthermore, there is a harmonic function $u_0 \geq 0$ defined in B_0 and a continuous function $b(t)$. Then there exists a unique potential $u(x, t)$ and two uniquely defined evolutions A_t and B_t such that u is a weak solution to (5)- (9).

References

[1] CHEN, X.; FRIEDMAN, A.: A bubble in ideal fluid with gravity; *J. Differ. Equations* **81**, No.1, p.136-166, (1989).

[2] WAGNER, A.: The Nonstationary Bernoulli Problem; *Arch. Math.* **72** (1999), p.1-8.

[3] WAGNER, A.: On the Bernoulli Free Boundary Problem; will appear in *Differential and Integral Equations*.

[4] GIGA, Y.; TAKAHASHI, S.: On global weak solutions of the nonstationary two phase Stokes flow; *SIAM J. Math. Anal.* **25**, No.3, p.876-893, (1994).

[5] GIGA, Y.; GOTO, S.; ISHII, H.; SATO, M.-H.: Comparison Principle and Convexity Preserving Properties for Singular Degenerate Parabolic Equations on Unbounded Domains: *Indiana University Mathematics Journal* **40**, No.2, p.443-470, (1991).

[6] CRANDALL, M.G.; ISHII, H.; LIONS, P.: User's Guide to Viscosity Solutions of Second Order Partial Differential Equations; *Bulletin of the American Mathematical Society*, **27**, No.1, p.1-67, (1992).

[7] LADYZENSKAYA, O.A.; SOLONNIKOV, V.A.; URAL'TZEVA, N.N.: Linear and Quasilinear Equations of Parabolic Type; *Transl. Math. Mono.* Vol.23 AMS, Providence, RI (1986).

Alfred Wagner
Institut für Mathematik
Universität Köln
Weyertal 86-90
50931 Köln
Germany

Part 4.

Phase Change in Material Science

V. Alexiades

Alloy Solidification with Convection in the Melt

Abstract

Most models of alloy solidification are severely limited by the assumption of constant density, thus excluding all convective effects. We present a thermodynamically consistent model for binary alloy solidification that incorporates energy, species and momentum conservation, constitutional supercooling, as well as temperature, concentration, and pressure dependence of thermophysical parameters. The crucial aspect is the development of an Equation of State capturing the thermochemistry of the phases. A numerical algorithm will also be outlined.

1 Introduction

We outline the main features of a macroscopic model for solidification of a binary alloy with convective and diffusive heat and mass transfer.

Given a binary melt $A_{1-x}B_x$ and initial and boundary conditions, the goal is to describe macroscopically the evolution of the phases as the melt undergoes solidification by modeling the heat and mass transfer in the melt and solid.

The basic physical assumptions underlying the model are
• rather slow cooling (not quenched), so that *local* thermodynamic equilibrium prevails (the lever rule applies to the phase diagram of the A-B binary);
• thermophysical properties may depend on temperature, composition, and pressure;
• negligible nucleation difficulties (freezing starts at the liquidus);
• negligible surface tension effects.

The mathematical model of the solidification process is an extension of Alexiades [1], [2] to include convective effects, further generalizing the well-known "enthalpy formulation" ([3]).

The main features of the model include
• coupled heat and mass transfer (conduction, diffusion and convection), with possible cross effects (Soret, Dufour);
• solidification with constitutional supercooling;
• thermochemistry of the phases incorporated via an actual Equation of State;
• macroscopic description in terms of local variables: C = mass fraction of component B, u = internal energy, T = temperature, P = pressure, $\vec{v}$ = velocity, Λ = liquid fraction;
• conservation laws valid everywhere in the weak (integral) sense; phases are distinguished only by values of the liquid fraction ($\Lambda = 1$ in melt, $\Lambda = 0$ in solid,

$0 \leqslant \Lambda \leqslant 1$ in two-phase, constitutionally supercooled regions);
- front "capturing" (not tracking).

2 Conservation Laws

Let C_i = mass fraction of component $i = A, B$, $\rho_i = \rho C_i$ = partial density of i, $\rho = \rho_A + \rho_B$ = total density, u = internal energy (per gram), $\vec{v}$ = velocity, P = pressure, $\vec{J}_i$ = mass flux of species i, $\vec{Q}$ = energy flux, τ = stress tensor of fluid.

The conservation laws for species, energy, and momentum may be expressed as follows, understood in a weak (distributional) sense, throughout the material:

$$\partial_t(\rho_A) + \nabla \cdot (\rho_A \vec{v} + \vec{J}_A) = 0 , \quad \partial_t(\rho_B) + \nabla \cdot (\rho_B \vec{v} + \vec{J}_B) = 0, \tag{1}$$

$$\partial_t(\rho u) + \nabla \cdot (\rho u \vec{v} + \vec{Q}) = -P \nabla \vec{v}, \tag{2}$$

$$\partial_t(\rho \vec{v}) + \nabla \cdot (\rho \vec{v} \vec{v} - \tau) = -\nabla P + \rho \vec{g}, \tag{3}$$

$$\partial_t(\rho) + \nabla(\rho \vec{v}) = 0. \tag{4}$$

The fluxes are specified by the constitutive relations:

Fick's Law, including pressure-diffusion and thermal-diffusion (Soret) effects:

$$\vec{J}_i = -\rho \left(D \nabla C_i + \delta_i^P \nabla P + \delta_i^T \nabla T \right), \quad i = A, B, \qquad \text{with } \vec{J}_A + \vec{J}_B = 0, \tag{5}$$

where D = interdiffusion coefficient, and δ^P, δ^T are the pressure-diffusion and Soret coefficients, T = temperature.

Fourier's Law for conduction, with interdiffusion, and Dufour effect:

$$\vec{Q} = -k \nabla T + \bar{h}_A \vec{J}_A + \bar{h}_B \vec{J}_B + Q^D, \tag{6}$$

with k = thermal conductivity (tensor), $\bar{h}_i :=$ partial (specific) enthalpy of i, and Q^D = Dufour energy flux (diffusion-thermo effect).

Note that the Soret and Dufour cross effects are usually negligible, but the other flux terms are principal couplings and cannot be neglected *a priori*.

Finally, an Equation of State (see below) relating the energy to the fields C_i, T, P is needed to close the system of equations.

The thermophysical parameters depend on the local state, characterized by the triplet (C_i, T, P), and this dependence is generally different in different phases, with possible jump discontinuities across interfaces. Thus the system is highly nonlinear.

In addition, the system is undergoing a phase transition, as dictated by the phase diagram of the A-B binary, the simplest cases of which, and under fixed pressure, are shown in Figure 1. The liquidus and solidus curves seen in the figure are constant-pressure level curves of surfaces in 3-dimensional (C, T, P) space. The phase diagram describes the phase of each (C, T, P) state at thermodynamic equilibrium, and the possible coexistence of phases. States lying between the solidus and liquidus are "constitutionally supercooled", thermodynamically metastable, and are actually mixtures of solid and liquid, referred to as "mushy". The phase diagram encapsulates the thermochemistry of the binary system, and both are encoded into the Equation of State, of the form:

$$
h = \begin{cases} h^L & \text{in liquid} \\ h^M & \text{in mushy} \\ h^S & \text{in solid} \end{cases} \tag{7}
$$

presented in the next section. Here h is the enthalpy from which the internal energy can be found via

$$
u = h - P/\rho. \tag{8}
$$

Remark on Fickian diffusion:

There are, of course, many details glossed over in the brief description above. In particular, one must be careful with the definition of the Fickian diffusion term "$-D\,\nabla C$" in 2-phase (mushy) regions.

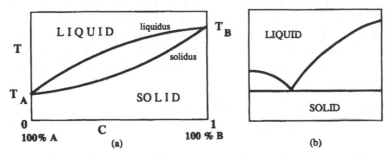

Figure 1. Simple phase diagrams: (a) A and B soluble in all proportions; (b) eutectic.

To make it precise, define the fields $(C := C_B)$:

$$
C^L = \begin{cases} C & \text{in liquid} \\ \text{Liquidus composition} & \text{in mushy} \\ 0 & \text{in solid,} \end{cases}
$$

257

$$C^S = \begin{cases} 0 & \text{in liquid} \\ \text{Solidus composition} & \text{in mushy} \\ C & \text{in solid,} \end{cases}$$

and

$$\Lambda = (C - C^L)/(C^S - C^L) = \text{liquid fraction by weight.}$$

Then

$$\text{diffusion flux} = \Lambda(-D^L \nabla C^L) + [1 - \Lambda](-D^S \nabla C^S)$$

with D^L and D^S the interdiffusion coefficients in liquid and solid, respectively.

Thus, diffusion is driven by gradients of the Liquidus and Solidus concentrations and **not** by gradients of the mean concentration C, which is

$$C = \Lambda C^L + [1 - \Lambda]C^S.$$

In isothermal mushy regions, C^L and C^S will be uniform, so their gradients will be zero, and therefore there will be no Fickian diffusion in such regions, even if C is not uniform (due to different Λ's).

3 Equation of State

The state is determined by (C, T, P) and the phase diagram. Let $T = T^L(C,P)$, $T = T^S(C,P)$ be the liquidus and solidus temperatures at composition C, pressure P, and let $C = C^L(T,P)$, $C = C^S(T,P)$ be the liquidus and solidus compositions at temperature T, pressure P. Choose a *reference state* (C_0, T_0, P_0), *say* solid at $(C = 0$, $T = T_A$, $P = P_{atm})$, and integrate along appropriate C, T, and P paths the basic Gibbs relation:

$$\begin{aligned} du(C,T,P) &= (u_C)dC + (u_T)dT + (u_P)dP \\ &= [\bar{u}_B - \bar{u}_A]dC + [c_p - \alpha_T P/\rho]dT - [\alpha_T T/\rho - \alpha_P P/\rho]dP, \end{aligned} \qquad (9)$$

with $\bar{u}_i$ the partial u's, c_p the specific heat, α_T the thermal expansion coefficient, α_P the compressibility $(= 0$ here), of the appropriate phase. The enthalpy, $h = u - P/\rho$, in each phase can be written as follows:

Liquid: $T \geqslant T^L(C,P)$:

$$h^L(C,T,P) = H^L(C,P) + \int_{T^L(C,P)}^{T} c_p^L(C,\tau,P)d\tau \qquad (10)$$

258

with $H^L(C, P)$ = enthalpy of liquidus at the point $(C,\ T^L(C, P^L), P)$ given by

$$
\begin{aligned}
H^L(C, P) = h_0 + \Delta h_0^{fus} + \frac{P - P_0}{\rho_0}[1 - T_0 \alpha_T^L] \\
+ \int_{C_0}^{C} \bar{h}^L(\xi, T^L(C, P), P) d\xi \\
+ \int_{T_0}^{T^L(C,P)} c_p^L(C_0, \tau, P) d\tau
\end{aligned}
\tag{11}
$$

h_0 being the enthalpy of formation and Δh_0^{fus} the heat of fusion at the reference state.
Solid: $\quad T \leqslant T^S(C, P)$:

$$
h^S(C, T, P) = H^S(C, P) - \int_{T}^{T^S(C,P)} c_p^S(C, \tau, P) d\tau
\tag{12}
$$

with $H^S(C, P)$ = enthalpy of solidus at the point $(C, T^S(C, P^S), P)$ given by

$$
\begin{aligned}
H^S(C, P) = h_0 + \frac{P - P_0}{\rho_0}[1 - T_0 \alpha_T^S] \\
+ \int_{C_0}^{C} \bar{h}^S(\xi, T^S(C, P), P) d\xi \\
+ \int_{T_0}^{T^S(C,P)} c_p^S(C, \tau, P) d\tau
\end{aligned}
\tag{13}
$$

Mushy: $\quad T^S(C, P) \leqslant T \leqslant T^L(C, P)$:

$$
h^M(C, T, P) = \Lambda\, h^L(C^L(T, P), T, P) + [1 - \Lambda] h^S(C^S(T, P), T, P)
\tag{14}
$$

with $\Lambda(C, T, P)$ the liquid fraction from the "lever rule":

$$
\Lambda(C, T, P) = \frac{C - C^L(T, P)}{C^S(T, P) - C^L(T, P)}.
\tag{15}
$$

Relations (10), (12), (14) define the Equation of State (7), namely, the enthalpy h as a function of the triplet (C, T, P). Thus, knowing (C, T, P) we can determine the phase and the enthalpy (hence also the energy u). However, the conservation laws update the quantities C, h, P, ρ, but *not* the temperature T, which has to be found from the Equation of State. The particular way we have expressed the EoS enables us to use the enthalpy as the phase indicator, and then determine the temperature. This can be accomplished as follows.

4   Enthaply as Phase Indicator

Given C, h, P, we compute the quantities $H^L(C, P)$ and $H^S(C, P)$ from (11) and (13), and we compare h with them.

If $h \geq H^L(C, P)$, then the phase is **liquid**: set $\Lambda(C, T, P) = 0$ and find T ($\geq T^L(C, P)$) by solving

$$\int_{T^L(C,P)}^T c_p^L(C, \tau, P) d\tau = h - H^L(C, P) \quad \geq \quad 0.$$

If $h \leq H^S(C, P)$, then **solid**: set $\Lambda(C, T, P) = 1$ and find T ($\leq T^S(C, P)$) by solving

$$-\int_T^{T^S(C,P)} c_p^S(C, \tau, P) d\tau = h - H^S(C, P) \quad \leq \quad 0.$$

If $H^S(C, P) \leq h \leq H^L(C, P)$, then **mushy**: set $\Lambda = \frac{h - H^S(C,P)}{H^L(C,P) - H^S(C,P)}$ and find T by solving

$$\Lambda h^L(C^L(T, P), T, P) + [1 - \Lambda] h^S(C^S(T, P), T, P) = h.$$

5   Outline of Updating Scheme

An algorithm for updating the values of $C, T, P, \Lambda, \rho, \vec{v}, u, h$ from a time t to time $t + \Delta t$ would go as follows:

Step 1. The balance laws update the quantities ρ_A , ρ_B , (ρu) , $(\rho \vec{v})$, P (for each control volume) to new time.

Step 2. From these we deduce:

$$\rho = \rho_A + \rho_B, \qquad C_A = \frac{\rho_A}{\rho}, \qquad C_B = \frac{\rho_B}{\rho},$$

$$u = \frac{(\rho u)}{\rho}, \qquad h = u + P/\rho.$$

Step 3. Determine the new phase of each control volume from the Equation of State and find T and (weight) liquid fraction Λ, as described in Section 4.

Step 4. The consistency of the updated values may be checked as follows: Find the volume fraction of liquid:

$$\beta = \frac{\Lambda \rho^L}{\Lambda \rho^L + [1 - \Lambda] \rho^S}$$

and the "fluid" (non-void) fraction (of each control volume):

$$f = \frac{(\rho u)}{\beta \rho^L u^L(T, P, C^L) + [1 - \beta] \rho^S u^S(T, P, C^S)};$$

check the consistency of the value of density:

$$\rho^* := f\left[\beta \rho^L + (1 - \beta)\rho^S\right] + (1 - f)\rho^{gas} = \rho ?$$

This may be used as criterion for convergence of an iterative numerical scheme.

References

[1] V. ALEXIADES, "Casting of HgCdTe, Part II: Conduction - Diffusion Model and its Numerical Simulation", Oak Ridge National Laboratory Report, ORNL/TM-11753, 1991.

[2] V. ALEXIADES AND G.K. JACOBS, "Solidification modeling of in situ vitrification melts", pp.11-16 in *Proceedings of Dynamic Systems and Applications*, Vol. 1, editors G.S. Ladde and M. Sambandham, Dynamic Publishers, 1994.

[3] V. ALEXIADES AND A.D. SOLOMON, *Mathematical Modeling of Melting and Freezing Processes*, Hemisphere Publ. Co., Washington DC, 1993.

Vasilios Alexiades
Mathematics Department
University of Tennessee
Knoxville TN 37996
and
Math. Sciences Section
Oak Ridge Nat'l Lab.
Oak Ridge TN 37831
USA

J. W. DOLD

Implications of Non-Monotonic Curvature-Dependence in the Propagation Speed of an Interface

Abstract

When an interface or free-boundary propagates normal to itself at a speed that depends non-monotonically on curvature alone, an anti-diffusive form of instability appears in the shape of the interface wherever the speed-curvature relationship has negative slope. If this slope is restricted to windows of curvature, then a remarkable degree
of regularisation appears. An interface can be subjected to the anti-diffusive instability for an indefinite time but remain well-posed with respect to initial conditions for
all time. This result is connected with the theorem: a diffusion equation $u_t = D\nabla^2 u$,
having time-dependent diffusion coefficient $D(t)$, is ill-posed with respect to initial
conditions when the integral of the diffusion coefficient becomes negative, which is
not necessarily the same as the time when the diffusion coefficient itself becomes negative. On the other hand, the interface is structurally ill-posed when subject to the
anti-diffusive instability, as described in [1]. With a non-monotonic speed curvature
relationship, discontinuities in curvature can usually be expected to arise, rendering
inappropriate some otherwise elegant results based on assuming continuity of curvature. These discontinuities separate curvature into ranges which appear to have much
in common with separation into phases in material systems, although the separation
of an interface into different "phases" of curvature involves important additional features. Some further discussion of the nature and consequence of non-monotonicity in
the speed-curvature relationship is given in [1].

1   Introduction

An interface moving in two dimensions can be represented by a vector-valued function
of arclength s and time t in the form $r = R(t, s)$. An example of the kind of interface
that may be considered is sketched in Figure 1a. The unit tangent vector can be
represented by $\hat{s} = R_s$ and we can take $\hat{n}$ to represent the unit normal in the direction
of positive propagation of the interface. Choosing to represent the curvature κ of the
interface as being positive when it is concave as seen from the side of the interface
into which $\hat{n}$ points, the curvature can be identified as $\kappa = \hat{s}_s \cdot \hat{n} = -\hat{n}_s \cdot \hat{s}$. If the
speed of propagation is represented by the function $V(\kappa)$, then the interface moves
such that

$$R_t \cdot \hat{n} = V(\kappa) \tag{1}$$

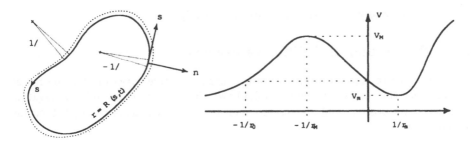

Figure 1. Sketches of: a) a propagating interface, shown left, illustrating the sign convention in the definition of curvature κ and a sample movement of the interface from the solid line to the dashed line over some time interval; and b) a possible speed-curvature relationship, shown right.

beginning with a suitable initial shape $R(0, s) = R_0(s)$. This simple law has a number of surprisingly rich features, especially if the speed-curvature law $V(\kappa)$ is non-monotonic.

Models for the movement of free-boundaries in the form of (1), with non-monotonic dependence on curvature, arise in the theoretical description of flames in some mixtures, especially near stoichiometry [2]. They can also arise in other circumstances, and may model the meandering of rivers. A discussion of the context in which this model may prove valuable is provided by Dold and Crighton [1], where an extended analysis of the results presented in this article can also be found. The implications of non-monotonicity in the dependence of normal propagation speed on curvature are surprising, and may be far-reaching. This article and the more general description given in [1] provide an outline of a range of aspects that are accessible through simple analytical investigation. Further understanding is currently being sought through a combination of numerical and analytical studies, along with generalisations of the basic model (1) through including higher-order dependencies of propagation-speed on such things as derivatives of curvature. The discussion here is based solely on the simplest model (1).

To illustrate the main features resulting from non-monotonicity of $V(\kappa)$, this article will mostly consider a speed-curvature law of the form illustrated in Figure 1b. We take $V > 0$ for all values of κ, one local maximum $V_M = V(-1/r_M)$ and one local minimum $V_m = V(1/r_m)$, at a negative and a positive curvature, respectively, V decreasing as $\kappa \to -\infty$, and V increasing without bound (or at least above the local maximum value) as $\kappa \to +\infty$. It is also worth identifying the negative curvature $-1/r_0$ at which the speed is the same as it is for a flat interface, $V(-1/r_0) = V(0)$. Many other forms could be considered in exactly the same way as is done below. For the overall shape of the law which we adopt, it is not necessary to provide a more exact formulation, because the results obtained are general enough to be extended

263

readily to wide classes of speed-curvature relationship.

2 Diffusive and anti-diffusive aspects

An outwardly propagating circular interface has a negative curvature that decreases in magnitude (increases positively in value) over time, while an inwardly propagating interface has a positive and increasing curvature. If the radius is $r = a(t)$, then the two cases satisfy

$$\frac{da}{dt} = V(-1/a) \qquad \Longrightarrow \qquad t = \int_{a_0}^{a} \frac{da}{V(-1/a)} \qquad (2)$$

and

$$\frac{da}{dt} = -V(1/a) \qquad \Longrightarrow \qquad t = \int_{a}^{a_0} \frac{da}{V(1/a)} \qquad (3)$$

respectively, where a_0 is the initial radius. The first formulation (2) includes the latter if, without loss of generality, we simply consider inward propagation to correspond to a negative radius and so we need not consider inward propagation separately. In polar coordinates, if the interface lies at $r = a(t) + u(\theta, t)$, then the interface satisfies the equation

$$\frac{da/dt + u_t}{(1 + u_\theta^2/r^2)^{1/2}} = V\left(\frac{ru_{\theta\theta} - 2u_\theta^2 - r^2}{r^3(1 + u_\theta^2/r^2)^{3/2}}\right) = V\left(-\frac{1}{a} + \frac{u + u_{\theta\theta}}{a^2} + O\left(u^2\right)\right) \qquad (4)$$

The final form for $V(\kappa)$ in this equation includes the linearised expression for curvature when u is small.

Linearising, to represent asymptotically small deviations in u from a perfectly circular outwardly propagating interface, the model (4) becomes

$$u_t = \frac{V'(-1/a)}{a^2}(u_{\theta\theta} + u) \qquad \text{with} \qquad \frac{da}{dt} = V(-1/a) \qquad (5)$$

This linearised equation for u is, essentially, a diffusion equation having a non-constant "diffusion coefficient", $\kappa^2 V'(\kappa)$ to leading order, whatever the sign of κ. The coefficient is negative wherever the function $V(\kappa)$ has a negative gradient. Although negative diffusivity is normally associated with ill-posedness, we can note that $V'(\kappa)$ is not universally negative for the speed-curvature law we are considering. Being negative only in a finite window of curvature space, namely, $-1/r_M < \kappa < 1/r_m$, has some significant consequences.

These are easily demonstrated by introducing a transformation to a pseudo-time variable $\tau(t)$, defined as the time-integral of the diffusion coefficient. The equation (5) can then be rewritten as

$$u_\tau = u_{\theta\theta} + u \qquad \text{with} \qquad \tau = \int_0^t V'(-1/a)\frac{dt}{a^2} \qquad (6)$$

264

which is a diffusion equation having a constant positive coefficient of precisely unity. Such an equation is well-posed with respect to initial conditions at the time $t = \tau = 0$, for all positive values of τ. It remains only to evaluate τ, which can be done exactly in terms of $a(t)$, since

$$\tau = \int_{a_0}^{a} \frac{V'(-a^{-1})}{V(-a^{-1})} \frac{da}{a^2} = \int_{-1/a_0}^{-1/a} \frac{V'(\kappa)}{V(\kappa)} d\kappa = \ln \frac{V(-1/a)}{V(-1/a_0)} \tag{7}$$

Clearly, τ is positive as long as the propagation speed at radius a exceeds the initial speed at radius a_0. If the initial radius is small enough, less than r_0 for the speed-curvature law sketched in Figure 1b, then this is true for all time; and so the problem is then well-posed for all time in spite of the "diffusion coefficient" being negative for an unlimited time!

Inherent in this result is a little-known **theorem** of diffusion or "heat" equations [1] (which the author has not seen quoted elsewhere in any form). That is

The linear diffusion equation $u_t = D\nabla^2 u$, with time-dependent coefficient $D(t)$, becomes ill-posed with respect to initial conditions when the integral of the diffusion coefficient with respect to time becomes negative (which is not necessarily when the diffusion coefficient itself becomes negative).

In fact, if $D(t)$ starts off being positive, then by the time that D changes sign, the solution $u(t, \theta)$ will have undergone a great deal of smoothing, which may be expressed in terms of the spectrum of coefficients in a Fourier expansion of the interface, as described in [1]. The solution can then survive an equivalent degree of negative diffusivity before becoming ultra-sensitive to initial conditions, as it is for any negative value of τ. As equation (7) shows, this provides a remarkable and very simple regularising effect for propagating interfaces.

This result must be tempered by noting that the problem is structurally ill-posed whenever $V'(\kappa)$ is negative. That is, small changes in the equation, for example, through the addition of any effects of "noise", can lead to dramatically large variations in the solution in arbitrarily short times; solutions do not vary continuously with changes in the structure of the model (1) when $V'(\kappa) < 0$. This aspect is explored further in [1].

Under such conditions, a more regularised model would require the addition of higher order stabilising terms. An example, for which the Kuramoto-Sivashinsky equation [3] is a weakly nonlinear equivalent when $V(\kappa) = -\kappa$, is

$$\boldsymbol{R_t} \cdot \hat{n} = V(\kappa) - \delta^2 \nabla^2 \kappa \tag{8}$$

As discussed in [1] this equation is actually a variant of the Cahn-Hilliard equation, having an additional nonlinearity due to normal propagation. The Cahn-Hilliard equation models phase separation [4], suggesting that the equation (8), or its limit as $\delta \to 0$, may also display features of phase separation in the shape of the interface.

265

This aspect is explored further in Section 4. In this article we will continue the discussion using only the straightforward model (1), presumably corresponding to the limit $\delta \to 0$.

3 Total variation

In considering larger amplitude, fully nonlinear, behaviour it is useful to examine the evolution of the total variation $C(t)$, defined as the integral of the absolute value of curvature over the entire interface. That is,

$$C = \int_0^L |\kappa|\, ds \tag{9}$$

where $L(t)$ is the total arclength of the interface. For any simple closed curve in which κ does not change sign (therefore making it uniformly negative) the value of C is exactly 2π. If κ does change sign, being positive in any interval, then C must be larger than 2π. The evolution of C is an indication of how the interface might be distorting or smoothing itself out as it propagates.

Sethian [5, 6] derived a remarkable result concerning the evolution of total variation. Taking the interface to be sufficiently differentiable throughout, and using the relations, $\hat{s}_s = \kappa\,\hat{n}$, $\hat{n}_s = -\kappa\,\hat{s}$ and $\hat{n}_t \cdot \hat{n} = \hat{s}_t \cdot \hat{s} = 0$, the model (1) can be differentiated twice to give

$$\kappa_t = V_{ss} + \kappa^2 V + \boldsymbol{R}_t \cdot \hat{s}\,\kappa_s = (V_s + \kappa\,\boldsymbol{R}_t \cdot \hat{s})_s \tag{10}$$

Splitting the integral for C into integrals over intervals where κ is either positive or negative, it takes the form

$$C = \sum_n C_n \qquad \text{with} \qquad C_n = \frac{\kappa}{|\kappa|} \int_{s_n}^{s_{n+1}} \kappa\, ds \tag{11}$$

where κ has one sign only, represented by $\kappa/|\kappa|$, or is zero between successive values of s_n, and n ranges over all of the countably many such intervals that make up the interface. Then, taking curvature to be at least continuous so that κ is zero where it changes sign

$$\frac{dC_n}{dt} = \frac{\kappa}{|\kappa|} \int_{s_n}^{s_{n+1}} \kappa_t\, ds = \frac{\kappa}{|\kappa|} \left[V_s + \kappa\,\boldsymbol{R}_t \cdot \hat{s} \right]_{s_n}^{s_{n+1}} = -V'(0)\left| \kappa_s(s_{n+1}, t) - \kappa_s(s_n, t) \right| \tag{12}$$

The final part of this formula arises because of the sign restriction that κ_s must have when entering or leaving any one interval. To obtain dC/dt, these derivatives must be added and account taken of the appearance or disappearance of intervals through κ changing sign locally. The latter possibility was not considered in Sethian's article, but if κ is continuously differentiable, then κ_s is zero when, and where, an interval appears or disappears and so this aspect makes no contribution to dC/dt.

266

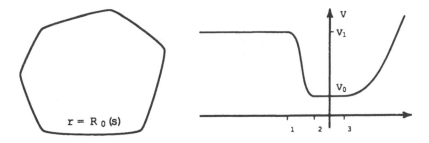

Figure 2. Sketches of: a) a propagating interface, shown left; and b) a possible speed-curvature relationship, shown right. Under these conditions, the interface will undergo changes in the sign of curvature.

It follows that, subject to certain assumptions, any increase or decrease of $C(t)$ is determined entirely by the gradient of the speed-curvature relationship at zero curvature, namely, $V'(0)$. If $V'(0) > 0$, then $dC/dt \leq 0$; if $V'(0) < 0$, then $dC/dt \geq 0$; and if $V'(0) = 0$, then C is a constant. If this remarkable conclusion should hold in general, then global features of interfaces would be greatly simplified. Indeed, the interface could possess anti-diffusive properties, having $V'(\kappa) < 0$, at all curvatures except at $\kappa = 0$, without total variation ever increasing. If κ were uniformly negative to start with, it would never be able to change sign. In view of the strong nature of the anti-diffusive instabilities, such a restriction would be truly remarkable if it could be sustained.

On the other hand, foremost among the assumptions leading to Sethian's conclusion is the continuity of curvature. Other assumptions might possibly be dispensable in weak formulations, but it is clear that the arguments contained in the derivation (12) are easily violated if κ changes sign discontinuously with arclength.

In fact, there are propagation functions $V(\kappa)$ and interfaces for which Sethian's prediction is violated. This can be demonstrated using an interface of the type sketched in Figure 2a and the propagation law sketched in Figure 2b. This law has a zero gradient at $\kappa = 0$ so C should stay constant. However, for negative curvatures $\kappa < \kappa_1 < 0$ it has a constant propagation speed V_1. Around $\kappa = 0$, in the interval $\kappa_2 < \kappa < \kappa_3$ which straddles the origin, the speed is $V_0 < V_1$. For larger values of κ we can (for example) take V to be increasing, but this makes no difference. The initial shape in Figure 2a can be chosen to be almost a polygon, having regions of high negative curvature ($\kappa < \kappa_1$) and regions of nearly zero curvature ($\kappa_2 < \kappa \leq 0$) joined by any smoothly varying negative curvature. Clearly, $C = 2\pi$ initially. However, the regions of near-zero curvature propagate at speeds lower than the regions of large-negative curvature. The curve can easily be chosen such that this discrepancy in propagation speeds brings about an inversion in curvature in arbitrarily brief times, causing C to increase above 2π.

267

This amounts to a proof that the evolution cannot maintain the degree of smoothness in curvature required for Sethian's results to remain valid. Moreover, because any smooth curve can be approximated arbitrarily closely by another smooth shape that approaches a high-order polygon, it follows that arbitrarily small, although still smooth disturbances to any initial shape can cause Sethian's result to be violated arbitrarily quickly. The actual evolution of total variation in any one example must therefore depend on the nature of the speed-curvature law over much more of its range than simply the origin.

It is clear that Sethian's results would be maintained for strictly monotonic speed-curvature laws, by which $V'(\kappa)$ remains either strictly positive or strictly negative for all accessible curvatures κ. In other cases, the example above shows that his conclusions may be violated. In doing so, of course, the assumptions underlying Sethian's deduction must be violated and in particular, it is likely that curvature becomes discontinuous.

4   Discontinuous solutions; propagation without change of shape

The appearance of discontinuous solutions can be demonstrated by considering interfaces that propagate without change of shape. If an interface propagates in the y-direction at a speed c and follows the path $y = ct + v(x)$, then the shape of the interface is determined by

$$\frac{c}{(1+v'^2)^{1/2}} = V\left(\frac{v''}{(1+v'^2)^{3/2}}\right) \tag{13}$$

Any solution can be characterised (modulo translations in x and y) by the initial conditions $v(0) = v'(0) = 0$ and the equation (13) makes it clear both that solutions would then be symmetric about $x = 0$ and that $V(v'') = c$ at $x = 0$. Since v'^2 increases away from $v' = 0$, the second derivative v'' must vary such that V decreases away from $x = 0$.

In general, for the speed-curvature law of Figure 1b, with $V_m < c < V_M$, there are three possible solutions determined by the three intersections of the curve with the line $V = c$, as sketched in Figure 3a. Labelling these I, II and III from left to right the qualitative features of each of these solutions can be described as follows:

Type I: Type I solutions have negative curvature which increases negatively away from their maximum point (where $v' = 0$). They must be bounded above by a semi-circle of curvature equal to $v''(0) < 0$ and so they have bounded support approaching infinite negative curvature at their end-points. Along the lines of the analysis in Section 2, these solutions should be stable to small-scale disturbances, since $V'(\kappa) > 0$ for all curvatures found along the curve.

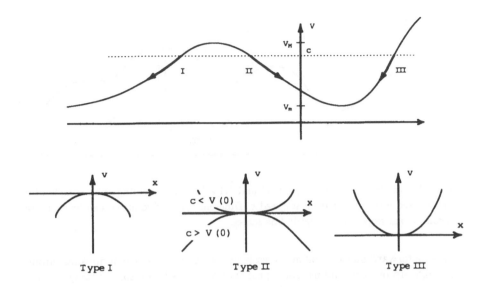

Figure 3. Sketches of: a) intersections between the propagation law of Figure 1 with the line $V = c$, leading to three types of steadily propagating solutions (shown uppermost); and b) each of the different qualitative shapes of steady solutions $v(x)$ that can arise.

Type II: Type II solutions have three alternative forms depending on whether $c > V(0)$, $c < V(0)$ or $c = V(0)$. In the latter case they are simply constant, $v \equiv 0$. These curves are all unstable to small-scale disturbances, since $V'(\kappa) < 0$. For $c > V(0)$, curvature is negative and increases toward zero away from a maximum value, so that the solutions have unbounded support. For $c < V(0)$, curvature is positive and decreasing, but because the curvature is bounded below by the value $v''(0) > 0$ (and the solution is therefore bounded below by a semi-circle of that curvature), the solutions have bounded support with curvature approaching $1/r_m$ at their end-points.

Type III: Type III solutions have positive curvature, decreasing away from the min-imum of $v(x)$ where $v' = 0$. However, because the curvature is bounded below by the value $1/r_m$, these solutions also have bounded support and approach a curvature of $1/r_m$ at their end-points. These solutions should be stable to small-scale disturbances, since $V'(\kappa) > 0$.

Each type of solution is sketched qualitatively in Figure 3b.

We can note that if $V(\kappa)$ tends to zero as $\kappa \to -\infty$, and to infinity as $\kappa \to \infty$, the system cannot dynamically sustain infinite curvatures (that is, discontinuities in direction, or cusps). An excessively sharp positive curvature diminishes through

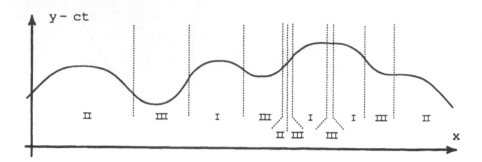

Figure 4. An amalgamation of Type I, II and III solutions providing one example of an interface of fixed form propagating at speed c. Different types of solutions can be connected with a continuous slope on arbitrarily fine scales.

propagating forward quickly and an excessively negative curvature diminishes through being caught up by surrounding faster propagation. This being so, the tangent vector $\hat{s}$ must be continuous even though the curvature might be discontinuous.

In the context of solutions propagating with constant form, this means that quite complicated overall solutions can be constructed out of suitable translations of Type I, II and III solutions for a given value of c. Any suitable portion of one type of solution needs only to be joined together with a suitable portion of the next at a point where both solutions share the same slope. This portion can, in turn, be joined to another in the same manner. At points where different types of solution meet, each has the same normal propagation speed and normal direction vector $\hat{n}$, so that the overall solution is therefore consistent. It can be noted that there is no lower limit to the size of the interval occupied by each successive solution, so solutions can be joined together in an arbitrarily fine structure.

Other forms of behaviour become possible if $V(\kappa)$ remains finite and below V_M in the limit of infinite positive curvature. Cusps can then be sustained above a certain speed c. Likewise if $V(\kappa)$ were to increase toward c in a suitable manner as $\kappa \to -\infty$, sharp leading cusps could be generated. Indeed, if the variation of the function $V(\kappa)$ were to be modified to generate more or fewer intersections with $V = c$, or if the locations of the maxima or minima were to be changed, then other types of solutions would arise in the same kind of way. There are clearly many alternative scenarios which could all be examined in a similar manner.

A sketch of one possible array of Type I, II and III solutions is shown in Figure 4. A composite solution of this form can be regarded as having separated into different possible "phases" where each type of solution represents a different phase. Within constraints set by the nature of the propagation law $V(\kappa)$ that controls the system, such a separation would permit the interface to propagate at speeds that are different from the propagation speed $V(0)$ of a flat interface. Interestingly, because phases can

270

be joined together on arbitrarily fine scales, it is possible to design composite solutions that are arbitrarily close to a flat interface but which propagate at a quite different speed.

5 Concluding remarks

It is clear that non-monotonicity in the speed-curvature relationship is responsible for a wide range of phenomena, including the fact that anti-diffusive instability for $V'(\kappa) < 0$ may then be restricted only to a window of curvatures, giving rise to a surprising degree of regularisation. Questions of ill-posedness emerge as anti-diffusive features appear. We have seen that non-monotonicity can in some circumstances provide a *complete regularisation* of the linearised problem, although the system is also very sensitive to the presence of noise or more deterministic forms of forcing as described in [1]. On the other hand, it has been demonstrated that discontinuities in curvature are likely to arise naturally, rendering irrelevant results that are based on assuming continuity of curvature. Multiple, steadily propagating forms of solutions are found that can co-exist on different parts of the same interface, satisfying suitable conditions of consistency where the solutions meet. This resembles a form of "phase separation" in curvature space, on scales that can be arbitrarily fine, an association that is reinforced by the connection of the model (1) with the Cahn-Hilliard equation, as explained in [1].

While such a steadily propagating "patchwork" solution can be constructed theoretically for a given propagation law $V(\kappa)$ and speed c, it represents only one of a wide class of possible solutions. Many more questions remain. It is not at all clear, for example, that a steady solution of this type would generally evolve from small perturbations on an unstable flat interface, and if it did so what value of c would be selected (although the value $c = V_M$ would seem to have some special significance), or how finely different phases might be distributed. It is much more likely that any such solution is itself unstable in a variety of ways and a far wider range of dynamical behaviour would then occur. This article has outlined a number of ongoing avenues of research that are available for understanding the implications of non-monotonicity in a speed-curvature relationship through analytical investigation alone. Some further details can be found in [1].

Acknowledgments: The author is grateful to Oliver Kerr, David Crighton, Michael Berry, Stan Osher and Charlie Elliott for useful discussions and suggestions.

References

[1] J.W. DOLD AND D.G. CRIGHTON (1998) *Non-monotonic curvature-dependent propagation.* (submitted).

[2] O.S. KERR AND J.W. DOLD (1998) *Slowly varying flames in near stoichiometric mixtures with general Lewis numbers.* (submitted).

[3] G.I. SIVASHINSKY (1997) *Nonlinear analysis of hydrodynamic instability in laminar flames—I. Derivation of basic equations.* Acta Astronautica **4**, 1177–1206.

[4] F. BAI, C.M. ELLIOTT, A. GARDINER, A. SPENCE AND A.M. STUART (1995) *The viscous Cahn-Hilliard equation. Part I: computations.* Nonlinearity **8**, 131–160.

[5] J.A. SETHIAN (1985) *Curvature and the evolution of fronts.* Comm. in Math. Phys. **101**, 487–499.

[6] J.A. SETHIAN. "Level set methods." Cambridge University Press, 1996.

J. W. Dold
Mathematics Department
UMIST, Manchester
U.K.

M. GRASSELLI

Asymptotic Analysis of a Phase-Field Model with Memory

Abstract

We present some results about asymptotic parameter analysis of a phase-field model with memory based on a heat flux law of the Gurtin-Pipkin type.

1 Introduction

Consider a material which occupies a bounded domain $\Omega \subset \mathbb{R}^3$. This material undergoes a solid-liquid phase transition process caused by temperature variations. To describe this phenomenon in the framework of the phase-field theory, we first introduce a pair of state variables, namely, the (relative) temperature ϑ and the phase-field χ, that is a regularized phase proportion (see, *e.g.*, [5, 19]). Then we want to determine their evolution on a finite time interval $[0, T]$, $T > 0$, starting from suitable initial and boundary conditions. In order to do that, we can use the well-known system of equations (see, *e.g.*, [5, 18, 19, 23] and references therein)

$$\partial_t e + \nabla \cdot \mathbf{q} = r \tag{1.1}$$

$$\mu \chi_t - \nu \Delta \chi + \beta(\chi) \ni \gamma(\vartheta, \chi) \tag{1.2}$$

in the cylindrical domain $Q := \Omega \times (0, T)$, where e is the internal energy, $\nabla \cdot$ stands for the spatial divergence operator, $\mathbf{q}$ is the heat flux, r denotes the heat supply, β is a maximal monotone graph in $\mathbb{R}^2$, and γ is a given smooth function. Moreover, μ and ν are positive constants called time relaxation parameter and interfacial energy parameter.

We assume that the internal energy e is given by the constitutive assumption

$$e = \vartheta + \lambda \chi \tag{1.3}$$

λ being a positive constant proportional to the latent heat per unit mass. As is well known, the heat flux $\mathbf{q}$ is usually related to $\nabla \vartheta$ through the Fourier law. On the contrary, here we suppose that the heat flux depends on the whole past history of $\nabla \vartheta$ according to the model proposed by Gurtin and Pipkin (cf. [20]), *i.e.*,

$$\mathbf{q}(x, t) = -\int_{-\infty}^{t} k(t - s) \nabla \vartheta(x, s) ds \qquad (x, t) \in \Omega \times (-\infty, T) \tag{1.4}$$

where $k : (0, +\infty) \to \mathbb{R}$ is a smooth memory kernel such that $k(0) > 0$. We recall that constitutive law (1.4) entails that ϑ propagates at finite speed (see [21, 22] and their references for detailed discussions about (1.4)).

Taking (1.3) into account and assuming that the past history of ϑ is given up to the initial time $t = 0$, equation (1.1) can be rewritten this way

$$\partial_t(\vartheta + \lambda\chi) - \Delta(k * \vartheta) = g \qquad \text{in } Q \tag{1.5}$$

where Δ stands for the spatial Laplace operator, $*$ indicates the usual time convolution product over $(0,t)$ (*i.e.*, $(a * b)(t) = \int_0^t a(t - s)b(s)ds$, $t > 0$), and g incorporates a known term depending on the past history of ϑ up to $t = 0$. As we shall see, from the mathematical viewpoint, we are dealing with a hyperbolic-parabolic coupled system instead of a parabolic one as in the standard phase-field model based on the Fourier law.

Let us associate now with (1.2), (1.5) the set of initial and boundary conditions

$$(\vartheta + \lambda\chi)(\cdot, 0) = e_0, \quad \chi(\cdot, 0) = \chi_0 \qquad \text{in } \Omega \tag{1.6}$$

$$\partial_n(k * \vartheta) = h \qquad \text{on } \Sigma := \Gamma \times (0, T) \tag{1.7}$$

$$\partial_n\chi = 0 \qquad \text{on } \Sigma \tag{1.8}$$

being $\Gamma = \partial\Omega$ and denoting by ∂_n the outward normal derivative to Γ. Here e_0, χ_0, and h are given functions and h may also depend on the past history of ϑ up to $t = 0$. We have thus formulated an initial and boundary value problem which describes the evolution of ϑ and χ on $[0, T]$. This kind of problem was first analyzed in [1] by using a semigroup approach (see also [7, 14, 15], for further related results), taking $\beta(\chi) = \chi^3$ and γ as a linear combination of ϑ and χ. Then, by means of different techniques, several well-posedness results and regularity theorems were established in [8, 9] for more general β and γ (see also [3, 4, 17] for other phase-field models accounting for memory effects).

Some recent joint papers with Colli and Gilardi (cf. [10, 11, 12]) are devoted to an asymptotic analysis of problem (1.2), (1.5-8) with respect to the parameters μ, ν and the kernel k (see also [3, 7, 16, 18] for related results). More precisely, [10] is devoted to investigating the behavior of the solution $(\vartheta_\nu, \chi_\nu)$ to (1.2), (1.5-8) as ν goes to 0 (cf. also [3, 16, 18]). In this case, the formal asymptotic limit is a hyperbolic Stefan problem with phase relaxation introduced and studied by Visintin in the case of a Maxwell-Cattaneo heat flux law (see, *e.g.*, [25]). Regarding this problem, an existence result for weak solutions has been obtained in [13], but, as far as we know, the only available uniqueness theorem is in one spatial dimension for a Maxwell-Cattaneo law (see [24]). The main result of [10] shows that there is a subsequence of $\{(\vartheta_\nu, \chi_\nu)\}$ which suitably converges to a solution of the formal limit problem as $\nu \searrow 0$. In [11] we analyze the behavior of $\{(\vartheta_\mu, \chi_\mu)\}$ as $\mu \searrow 0$, $(\vartheta_\mu, \chi_\mu)$ being the solution to (1.2), (1.5-8) (cf. also [7]). There, the formal asymptotic limit is a quasi-stationary phase-field model with memory. This problem is shown to have a unique solution under some monotonicity assumptions and that allows us to prove the convergence of the whole sequence to the solution of the formal limit problem. In addition, by strengthening the hypotheses, some error estimates are obtained as well. The asymptotic analysis performed in [12] may be viewed as a sort of justification of models like (1.2), (1.5-8). In fact, we consider problem (1.2), (1.5-8) with k replaced by a memory kernel k_ε such that the sequence $\{k_\varepsilon\}$ suitably converges as $\varepsilon \searrow 0$ to $k_0\delta$, where $k_0 > 0$ and δ denotes the Dirac mass. Then we show that the corresponding solution $(\vartheta_\varepsilon, \chi_\varepsilon)$ converges as $\varepsilon \searrow 0$ to a solution of the standard parabolic phase-field model based on the Fourier heat conduction law $\mathbf{q} = -k_0\nabla\vartheta$. Moreover, provided that a uniqueness theorem for the limit problem holds, some error estimates are established. Hence, the standard parabolic phase-field model can be approximated by a phase-field model with memory.

In this paper, we present an overview of some of the results mentioned above and, in addition we give a slight generalization of [10, Thm. 2.6]. To be precise, Sections 3, 4, and 5 are devoted to the cases $\nu \searrow 0$, $\mu \searrow 0$, and $k \to k_0\delta$, respectively; while in Section 2, we recall, for the reader's convenience, some results on existence, uniqueness, and regularity which turn out to be useful in the sequel.

2 Existence, uniqueness, and regularity

Here we collect some results regarding the well-posedness of the initial and boundary value problem (1.2), (1.5-8) (see [8, 9, 10]). To introduce a convenient formulation of such a problem, we first need some notation. Define $W := H^2(\Omega)$, $V := H^1(\Omega)$, and $H := L^2(\Omega)$, identify H with its dual space H', and remark that $V \hookrightarrow H \hookrightarrow V'$ with dense and compact injections. Also, let $(\cdot, \cdot)$ stand both for the scalar product in H and for the duality pairing between V' and V, and let $\|\cdot\|$ denote the norm in either H or H^N. The linear and continuous operator $A : V \to V'$ specified by

$$(Av_1, v_2) := \int_\Omega \nabla v_1 \cdot \nabla v_2 \qquad v_1, v_2 \in V$$

is not injective in our framework, depending on the choice of V. In fact, as an unbounded operator from H to H, it reduces to $-\Delta$ with Neumann homogeneous boundary conditions. If X is a real Banach space and $v \in L^1(0, T; X)$, we put $(1 * v)(t) := \int_0^t v(s)ds$ for any $t \in [0, T]$, while z' and z_t indicate the total (or time, if, e.g., X is a function space on Ω) derivative of $z \in W^{1,1}(0, T; X)$.

As far as k, g, e_0 are concerned, we assume (cf. [10, Sec. 2])

$$k \in W^{2,1}(0, T), \quad k(0) > 0 \tag{2.1}$$

$$g \in L^1(0, T; H), \quad h \in W^{1,1}(0, T; L^2(\Gamma)) \tag{2.2}$$

$$e_0 \in H. \tag{2.3}$$

As maximal monotone graph from $\mathbb{R}$ to $\mathbb{R}$, β is the subdifferential of

$$j : \mathbb{R} \to (-\infty, +\infty] \quad \text{convex and lower semicontinuous function.} \tag{2.4}$$

In addition, we allow j to be non-negative and admit minimum 0 in 0, which yields

$$\beta = \partial j, \quad 0 = j(0) = \min j, \quad \text{whence} \quad 0 \in \beta(0). \tag{2.5}$$

Finally, regarding γ and χ_0, we suppose

$$\gamma \in C^{0,1}(\mathbb{R}^2) \tag{2.6}$$
$$\chi_0 \in H \quad \text{and} \quad j(\chi_0) \in L^1(\Omega). \tag{2.7}$$

Then, setting

$$(f(t), v) := (g(t), v) + \int_\Gamma h(t)v \quad \text{for } v \in V \text{ and a.a. } t \in (0, T) \tag{2.8}$$

and denoting by $D(\beta)$ the domain of β, the notion of weak solution to (1.2), (1.4-7) is made precise by the following variational formulation

Problem (P). *Find a pair (ϑ, χ) and an auxiliary element ξ satisfying*

$$\vartheta \in W^{1,1}(0, T; V') \cap C^0([0, T]; H), \quad k * \vartheta \in C^0([0, T]; V) \tag{2.9}$$
$$\chi \in H^1(0, T; V') \cap L^2(0, T; V), \quad \xi \in L^2(Q) \tag{2.10}$$
$$\chi \in D(\beta) \quad \text{and} \quad \xi \in \beta(\chi) \quad \text{a.e. in } Q \tag{2.11}$$
$$\partial_t(\vartheta + \lambda\chi) + A(k * \vartheta) = f \quad \text{in } V', \text{ a.e. in } (0, T) \tag{2.12}$$
$$\mu\chi_t + \nu A\chi + \xi = \gamma(\vartheta, \chi) \quad \text{in } V', \text{ a.e. in } (0, T) \tag{2.13}$$
$$(\vartheta + \lambda\chi)(0) = e_0, \quad \chi(0) = \chi_0. \tag{2.14}$$

Existence and uniqueness are ensured by (cf. [10, Thm. 2.1])

Theorem 2.1. *Let the assumptions (2.1-7) hold. Then Problem (P) admits a solution and the pair (ϑ, χ) is uniquely determined.*

Remark 2.2. From (2.10), via interpolation, it follows $\chi \in C^0([0, T]; H)$ and, consequently, both the initial conditions of (2.14) hold in H.

Note that (2.2) and (2.8) imply

$$f \in L^1(0, T; H) + W^{1,1}(0, T; V'). \tag{2.15}$$

This is the general assumption we need for showing Theorem 2.1. Therefore, from now on, let us argue in terms of f.

Some regularity results are contained in (see [10, Thm. 2.4])

Theorem 2.3. *Let (2.1-7) hold and let ϑ, χ, ξ solve Problem (P). If $\chi_0 \in V$, then*

$$\chi \in H^1(0, T; H) \cap C^0([0, T]; V) \cap L^2(0, T; W). \tag{2.16}$$

Consequently, equation (1.2) holds a.e. in Q and condition (1.8) is satisfied a.e. in Σ. Moreover, if $e_0 \in V$ and $f \in W^{1,1}(0,T;H) + W^{2,1}(0,T;V')$ with $f(0) \in H$, then

$$e = \vartheta + \lambda \chi \in W^{2,1}(0,T;V') \cap W^{1,\infty}(0,T;H) \cap L^{\infty}(0,T;V). \qquad (2.17)$$

Furthermore, if $\chi_0 \in W$, $\partial_n \chi_0 = 0$ on Γ, $\chi_0 \in D(\beta)$ a.e. in Ω, and $\beta^0(\chi_0) \in H$ (β^0 representing the minimal section of β), then

$$\chi \in W^{1,\infty}(0,T;H) \cap H^1(0,T;V) \cap L^{\infty}(0,T;W), \quad \xi \in L^{\infty}(0,T;H). \qquad (2.18)$$

Remark 2.4. Theorems 2.1 and 2.3 can be further generalized in several ways. For instance, the internal energy e (cf. (1.3)) may depend on the past histories of ϑ and χ (see [10]). Also, e can have a nonlinear and smooth dependence on χ (see [8]).

3 The case $\nu \searrow 0$.

Here we investigate the asymptotic behavior of Problem (P) as the coefficient ν tends to 0. In [10] we treated the case where γ linearly depends on ϑ only. Here we slightly generalize that result by letting

$$\gamma(\vartheta, \chi) = \lambda \vartheta + \sigma(\chi) \qquad (3.1)$$

where

$$\sigma \in C^{0,1}(\mathbb{R}), \quad \sigma(0) = 0. \qquad (3.2)$$

Indeed, we can prove (cf. [10, Thm. 2.6])

Theorem 3.1. Let (2.1-7) and (3.1-2) hold. For any $\nu > 0$, denote by $(\vartheta_\nu, \chi_\nu, \xi_\nu)$ the solution of Problem (P) corresponding to the initial data e_0 and $\chi_{0\nu}$ in (2.14), with $\chi_{0\nu} \in V$ fulfilling

$$\nu^{1/2} \|\chi_{0\nu}\|_V \leq C_1 \quad \text{and} \quad \int_\Omega j(\chi_{0\nu}) \leq C_2 \quad \forall \nu > 0 \qquad (3.3)$$

for some positive constants C_1, C_2, and

$$\chi_{0\nu} \to \chi_0 \quad \text{strongly in } H \text{ as } \nu \searrow 0. \qquad (3.4)$$

Then there exist ϑ, χ, and ξ such that, at least for a subsequence,

$$\vartheta_\nu \to \vartheta \quad \text{weakly star in } L^{\infty}(0,T;H) \text{ and strongly in } C^0([0,T];V') \qquad (3.5)$$

$$k * \vartheta_\nu \to k * \vartheta \quad \text{weakly star in } L^{\infty}(0,T;V) \text{ and strongly in } C^0([0,T];H) \qquad (3.6)$$

$$\chi_\nu \to \chi \quad \text{weakly in } H^1(0,T;H) \qquad (3.7)$$

$$\nu \chi_\nu \to 0 \quad \text{weakly in } L^2(0,T;W) \text{ and strongly in } L^{\infty}(0,T;V) \qquad (3.8)$$

$$\xi_\nu \to \xi \quad \text{weakly in } L^2(Q) \qquad (3.9)$$

277

as ν goes to 0. Moreover, any triplet (ϑ, χ, ξ) obtained as limit of subsequences of $(\vartheta_\nu, \chi_\nu, \xi_\nu)$ solves the limit problem specified by (2.9-12), (2.14), and

$$\chi \in H^1(0, T; H) \tag{3.10}$$

$$\mu\chi + \xi = \lambda u + \sigma(\chi) \quad \text{a.e. in } Q \tag{3.11}$$

Remark 3.2. The existence of $(\vartheta_\nu, \chi_\nu, \xi_\nu)$ is ensured by Theorem 2.1. Also, note that χ_ν fulfills (2.16), thanks to Theorem 2.3.

Remark 3.3. A collection of data $\chi_{0\nu}$ obeying (3.3-4) can be found, for instance, in [9, formula (3.5)]. Otherwise, compare to [2, pp. 281-282]. Let us stress that Theorem 3.1 entails, in particular, the existence of solutions to the phase relaxation problem (2.9-12), (2.14), (3.10-11). This fact generalizes the existence result of [13].

Proof. Following [10, Sec. 4], we introduce the freezing index $u_\nu := 1 * \vartheta_\nu$ and observe that, on account of (2.1) and (3.1-2), the pair (u_ν, χ_ν) solves

$$u_\nu'' + k(0)(Au_\nu + u_\nu) = f_1 + f_2 + k(0)u_\nu - k' * Au_\nu + \lambda\chi_\nu',$$
$$\text{in } V', \text{ a.e. in } (0, T) \tag{3.12}$$

$$\mu\chi_\nu' + \nu A\chi_\nu + \xi_\nu = \lambda u_\nu' + S\chi_\nu \quad \text{in } V', \text{ a.e. in } (0, T) \tag{3.13}$$

$$\chi_\nu \in D(\tilde{\beta}) \quad \text{and} \quad \xi_\nu \in \tilde{\beta}(\chi_\nu) \quad \text{a.e. in } Q \tag{3.14}$$

$$u_\nu(0) = 0, \quad u_\nu'(0) = \vartheta_0, \quad \chi_\nu(0) = \chi_{0\nu}. \tag{3.15}$$

Note that ξ_ν is not the same function as in (2.11) since $\tilde{\beta} : \mathbb{R} \to 2^{\mathbb{R}}$ is the maximal monotone graph defined by

$$\tilde{\beta}(r) = Sr - \sigma(r) + \beta(r) \quad r \in \mathbb{R} \tag{3.16}$$

where $S = \|\sigma'\|_{L^\infty(\mathbb{R})}$. Also, to get (3.12), $k(0)u_\nu$ has been added to both sides of (2.12) and f has been split into $f_1 \in L^1(0, T; H)$ and $f_2 \in W^{1,1}(0, T; V')$.

Testing (3.12) by u_ν' and (3.13) by χ_ν', the same argument used in [10, Sec. 4] leads us to find a constant C_3, depending only on $k(0), \mu, T, S, \|\vartheta_0\|, \|\chi_0\|, \|j(\chi_0)\|_{L^1(\Omega)}$, $\|f_1\|_{L^1(0,T;H)}, \|f_2\|_{W^{1,1}(0,T;V')}$, and $\|k'\|_{W^{1,1}(0,T)}$, such that

$$\|u_\nu\|^2_{W^{1,\infty}(0,T;H) \cap L^\infty(0,T;V)} + \|\chi_\nu\|^2_{H^1(0,T;H)} + \nu\|\chi_\nu\|^2_{L^\infty(0,T;V)} \leq C_3. \tag{3.17}$$

Then, multiplying (3.13) by ξ_ν and integrating in space and time, one gets (cf. [10, Sec. 4])

$$\|\xi_\nu\|^2_{L^2(Q)} \leq C_4 + \lambda^2 C_5 \tag{3.18}$$

In addition, on account of (2.1), by comparison to (3.12-13), we deduce

$$\|u_\nu'' - f_1\|_{L^2(0,T;V')} + \nu\|\chi_\nu\|_{L^2(0,T;W)} \leq C_6. \tag{3.19}$$

Here, the constants C_4, C_5, C_6 have the same dependences as C_3 does.

The uniform bounds (3.17-19) allow us to recover all the convergences (3.5-9) and, owing to (3.19), using the generalized Ascoli theorem, we can also deduce (cf. [10, Sec. 4])

$$u_\nu \to u = 1 * \vartheta \quad \text{strongly in } C^1([0, T; V']) \cap C^0([0, T]; V'). \tag{3.20}$$

To conclude the proof, we only need to show that (cf. (3.7), (3.9), and (3.14))

$$\xi \in \tilde{\beta}(\chi) \qquad \text{a.e. in } Q. \tag{3.21}$$

Observe first that we just need to prove the inequality (cf., e.g., [2, p. 42])

$$\limsup_{\nu \searrow 0} \int_0^T e^{-2(S/\mu)t} \int_\Omega \xi_\nu \chi_\nu \le \int_0^T e^{-2(S/\mu)t} \int_\Omega \xi \chi \tag{3.22}$$

recalling that the exponential weight gives an equivalent norm in $L^2(Q)$. Then, multiply equation (3.13) by $e^{-2(S/\mu)t}\chi_\nu$ and integrate in space and time. We obtain

$$\int_0^T e^{-2(S/\mu)t} \int_\Omega \xi_\nu \chi_\nu = \lambda \int_0^T e^{-2(S/\mu)t} \int_\Omega u'_\nu \chi_\nu$$
$$- \nu \int_0^T e^{-2(S/\mu)t} \|\nabla \chi_\nu(t)\|^2 \, dt - \int_0^T e^{-2(S/\mu)t} \int_\Omega (\mu \chi'_\nu - S\chi_\nu)\chi_\nu. \tag{3.23}$$

Observe that

$$\int_0^T e^{-2(S/\mu)t} \int_\Omega (\mu \chi'_\nu - S\chi_\nu)\chi_\nu = \frac{\mu}{2} e^{-2(S/\mu)T} \|\chi_\nu(T)\|^2 - \frac{\mu}{2} \|\chi_{0\nu}\|^2. \tag{3.24}$$

On the other hand, multiplying by $e^{-2(S/\mu)t}\chi$ the limiting equation (3.10) and integrating in space and time yield (cf. also (3.24))

$$\int_0^T e^{-2(S/\mu)t} \int_\Omega \xi \chi = \lambda \int_0^T e^{-2(S/\mu)t} \int_\Omega u'\chi - \frac{\mu}{2} e^{-2(S/\mu)T} \|\chi(T)\|^2 + \frac{\mu}{2} \|\chi_0\|^2. \tag{3.25}$$

In addition, note that, taking advantage of (3.7) and (3.20) via integration by parts in time, we can prove (cf. [10, Sec. 4])

$$\lim_{\nu \searrow 0} \int_0^T e^{-2(S/\mu)t} \int_\Omega u'_\nu \chi_\nu = \int_0^T e^{-2(S/\mu)t} \int_\Omega u'\chi. \tag{3.26}$$

Hence, on account of (3.23-26), we infer (3.21). In fact, we have

$$\limsup_{\nu \searrow 0} \int_0^T e^{-2(S/\mu)t} \int_\Omega \xi_\nu \chi_\nu$$
$$\le \lambda \int_0^T e^{-2(S/\mu)t} \int_\Omega u'\chi + \frac{\mu}{2} \|\chi_0\|^2 - \frac{\mu}{2} \limsup_{\nu \searrow 0} e^{-2(S/\mu)T} \|\chi_\nu(T)\|^2$$
$$\le \int_0^T e^{-2(S/\mu)t} \int_\Omega \xi \chi.$$

Remark 3.4. Recalling [10, Rem. 4.1], we note that γ may also linearly depend on the past history of ϑ through an integrable relaxation kernel.

4 The case $\mu \searrow 0$.

We observe first that, introducing the integrated enthalpy (see (1.3)) $w := 1 * e$ and the graph $\alpha : \mathbb{R} \to 2^{\mathbb{R}}$ defined by

$$\alpha(z) := \beta(z) - \gamma(z) + \lambda^2 z \qquad \forall z \in \mathbb{R}, \tag{4.1}$$

Problem (P) can be reformulated this way (cf. [11, Sec. 2]).

Problem (EP). *Find a pair (w, χ) and an auxiliary element ξ satisfying*

$$w \in W^{2,1}(0, T; V') \cap C^1([0, T]; H) \cap C^0([0, T]; V) \tag{4.2}$$

$$\chi \in H^1(0, T; V') \cap L^2(0, T; V), \quad \xi \in L^2(Q) \tag{4.3}$$

$$\chi \in D(\alpha) \quad \text{and} \quad \xi \in \alpha(\chi) \text{ a.e. in } Q \tag{4.4}$$

$$\partial_{tt} w + k(0) A w = f - k' * A w - \lambda k * A \chi \quad \text{in } V', \text{ a.e. in } (0, T) \tag{4.5}$$

$$\mu \chi_t + \nu A \chi + \xi = \lambda w_t \quad \text{in } V', \text{ a.e. in } (0, T) \tag{4.6}$$

$$w(0) = 0, \quad w_t(0) = e_0, \quad \chi(0) = \chi_0. \tag{4.7}$$

Of course, the unique solvability of Problem (EP) is ensured by Theorem 2.1.

Observe that setting $\mu = 0$ in (4.6) and taking (4.4) into account, we formally get the limiting elliptic differential inclusion

$$\nu A \chi + \alpha(\chi) \ni \lambda w_t.$$

Hence, in order to start investigating the asymptotic behavior of solutions to Problem (P) (or (EP)), it seems reasonable to provide some monotonicity conditions on the graph α. More precisely, we suppose

$$\alpha = \partial s : \mathbb{R} \to 2^{\mathbb{R}} \qquad \text{with} \qquad \alpha(0) \ni 0 \tag{4.8}$$

$$s : \mathbb{R} \to [0, +\infty] \text{ is proper, convex, lower semicontinuous and } s(0) = 0 \tag{4.9}$$

$$(\eta_1 - \eta_2)(z_1 - z_2) \geq c(z_1 - z_2)^2 \quad \forall z_i \in D(\alpha), \quad \forall \eta_i \in \alpha(z_i), \quad i = 1, 2 \tag{4.10}$$

for some $c > 0$, where $D(\alpha)$ is the effective domain of α. It is worth recalling that, from a physical viewpoint, the strong monotonicity condition (4.10) basically corresponds to requiring the latent heat be large enough.

Consider now three sequences $\{f_\mu\}, \{e_{0\mu}\}, \{\chi_{0\mu}\}$ such that

$$f_\mu \in L^1(0, T; H) + W^{1,1}(0, T; V') \tag{4.11}$$

$$e_{0\mu} \in H, \qquad \chi_{0\mu} \in V, \quad \text{and} \quad s(\chi_{0\mu}) \in L^1(\Omega) \tag{4.12}$$

$$f_\mu \to f \qquad \text{in} \quad L^1(0, T; H) + W^{1,1}(0, T; V') \tag{4.13}$$

$$e_{0\mu} \to e_0 \qquad \text{in} \quad H \tag{4.14}$$

$$\mu^{1/2} \chi_{0\mu} \to 0 \qquad \text{in} \quad H \tag{4.15}$$

$$\mu^{1/2} \|\chi_{0\mu}\|_V + \mu \|s(\chi_{0\mu})\|_{L^1(\Omega)} \leq C_7 \tag{4.16}$$

for some $C_7 > 0$ and any $\mu \in (0,1)$.

Taking advantage of the integrated enthalpy formulation, namely, Problem (EP), we can prove the following convergence results (see [11, Thm. 2.2])

Theorem 4.1. *Let (2.1) and (4.8-10) hold. For any $\mu > 0$, denote by $(\vartheta_\mu, \chi_\mu, \xi_\mu)$ the solution to Problem (P) which corresponds to the source term f_μ and the initial data $e_{0\mu}$ and $\chi_{0\mu}$ in (2.14) fulfilling (4.11-12). If the data f_μ, $e_{0\mu}$, and $\chi_{0\mu}$ satisfy (4.13-16) as well, then there exists a triplet (ϑ, χ, ξ) such that the convergences listed below hold.*

$$
\begin{array}{llr}
\vartheta_\mu \to \vartheta & strongly\ in\ L^2(0,T;H) & (4.17) \\
1 * \vartheta_\mu \to 1 * \vartheta & strongly\ in\ C^0([0,T];V) & (4.18) \\
\chi_\mu \to \chi & strongly\ in\ L^2(0,T;V) & (4.19) \\
\mu^{1/2}\chi_\mu \to 0 & strongly\ in\ L^2(0,T;H) & (4.20) \\
\chi_\mu \to \chi & weakly\ in\ L^2(0,T;W) & (4.21) \\
\mu\chi_\mu \to 0 & weakly\ in\ H^1(0,T;H) & (4.22) \\
\xi_\mu \to \xi & weakly\ in\ L^2(Q) & (4.23) \\
\mu^{1/2}\chi_\mu \to 0 & weakly\ star\ in\ L^\infty(0,T;V). & (4.24)
\end{array}
$$

In addition, the triplet (ϑ, χ, ξ) solves the problem

$$
\begin{array}{llr}
\partial_t(\vartheta + \lambda\chi) + A(k * \vartheta) = f & in\ V',\ a.e.\ in\ (0,T) & (4.25) \\
\nu A\chi + \xi = \lambda(\vartheta + \lambda\chi) & in\ V',\ a.e.\ in\ (0,T) & (4.26) \\
\chi \in D(\alpha)\ \ and\ \ \xi \in \alpha(\chi) & a.e.\ in\ Q & (4.27) \\
(\vartheta + \lambda\chi)(0) = e_0. & & (4.28)
\end{array}
$$

Remark 4.2. Theorem 4.1 ensures the existence of a solution to the formal limit problem (4.25-28). Indeed, one can just assume $f \in L^1(0,T;H) + W^{1,1}(0,T;V')$ and $e_0 \in H$, choosing then $f_\mu = f$, $e_{0\mu} = e_0$, and $\chi_{0\mu} = 0$ as approximating data. Moreover, in this case assumptions (2.1) and (4.8-10) also allow proving uniqueness (cf. [11, Thm. 2.3]). Hence, the whole sequence in Theorem 4.1 converges.

Remark 4.3. Theorem 4.1 is a reformulation of the original statement (see [11, Thm. 2.2]) in terms of the pair (ϑ, χ). Indeed, [11, Thm. 2.2] refers to Problem (EP) since this formulation appears to be more convenient from the mathematical viewpoint. In this context, it is worth recalling that $w_\mu \to w = 1 * (\vartheta + \lambda\chi)$ strongly in $C^1([0,T];H) \cap C^0([0,T];V)$ as $\mu \searrow 0$.

By strengthening the assumptions on the data sequences, some error estimates can be shown. More precisely, in addition to (4.11-16) let us introduce the further hypotheses

$$X_{0\mu} \in W : \quad \partial_n X_{0\mu} = 0 \quad \text{on } \Gamma \tag{4.29}$$

$$X_{0\mu} \in D(\alpha) \quad \text{a.e. in } \Omega \quad \text{and} \quad \alpha^0(X_{0\mu}) \in H \tag{4.30}$$

$$\|f_\mu\|_{W^{1,1}(0,T;H)+W^{2,1}(0,T;V')} + \|f_\mu(0)\| + \|e_{0\mu}\|_V \le C_8 \tag{4.31}$$

$$\exists \ \xi_{0\mu} \in \alpha(X_{0\mu}) \quad \text{a.e. in } \Omega \quad \text{such that}$$

$$\|X_{0\mu}\| + \mu^{-1/2}\|\nu A X_{0\mu} + \xi_{0\mu} - \lambda e_{0\mu}\| \le C_9 \tag{4.32}$$

for some C_8, $C_9 > 0$ and any $\mu \in (0,1)$, where $\alpha^0(z)$ denotes the element of $\alpha(z)$ having minimum modulus.

Remark 4.4. Note that assumptions (4.29-32) ensure that (ϑ_μ, X_μ) enjoys (2.16-18) (cf. Thm. 2.3). Also, these assumptions allow obtaining uniform bounds in stronger norms (cf. [11, Thm. 2.4 and Rem. 2.5]). Thanks to these bounds, we can deduce further convergences which are helpful in order to get error estimates.

The error estimates are given by (see [11, Thm. 2.6]).

Theorem 4.5. *Let the assumptions of Theorem 4.1 hold. If, in addition, (4.29-32) hold, then there are two positive constants C_{10} and C_{11} such that, for any $\mu \in (0,1)$,*

$$\|\vartheta_\mu - \vartheta\|_{L^2(0,T;H)} + \|1 * \vartheta_\mu - 1 * \vartheta\|_{C^0([0,T];V)}$$
$$+ \|X_\mu - X\|_{L^2(0,T;V)} \le C_{10}(\mu + \varepsilon_\mu) \tag{4.33}$$

$$\|\vartheta_\mu - \vartheta\|_{C^0([0,T];H)} + \|X_\mu - X\|_{C^0([0,T];V)} \le C_{11}(\mu^{1/2} + \varepsilon_\mu) \tag{4.34}$$

where

$$\varepsilon_\mu := \|e_{0\mu} - e_0\| + \|f_\mu - f\|_{L^1(0,T;H)+W^{1,1}(0,T;V')}$$

and (ϑ, X) is the unique solution to the limit problem (4.25-28).

5 The case $k \to k_0 \delta$.

Recalling the Introduction, a convenient approximation k_ε of $k_0 \delta$ is obtained by taking as relaxation kernel k in equation (1.5) the unique solution k_ε to the singular perturbation problem (see [12])

$$\varepsilon k'_\varepsilon + k_\varepsilon = \pi_\varepsilon \quad \text{in } (0,T) \tag{5.1}$$

$$k_\varepsilon(0) = \frac{k_0}{\varepsilon} \tag{5.2}$$

for some $\varepsilon > 0$, where π_ε is a given smooth function. Let us observe that from (5.1-2) we easily deduce

$$k_\varepsilon = k_0 \delta_\varepsilon + \delta_\varepsilon * \pi_\varepsilon \quad \text{with} \quad \delta_\varepsilon(t) = \varepsilon^{-1} exp[-t/\varepsilon] \quad t \in [0,T].$$

Therefore, k_ε is a smooth approximation of $k_0\delta$ provided that the sequence $\{\pi_\varepsilon\}$ suitably converges to 0 as ε goes to 0. In particular, if $\pi_\varepsilon \equiv 0$, note that the underlying heat flux constitutive assumption associated with k_ε is the Maxwell-Cattaneo's law (see [6]).

In this framework, the formal limit problem when ε goes to 0 is obtained by taking $k = k_0\delta$ in (1.5) and (1.7). This leads to the well-known standard parabolic phase-field model (cf., *e.g.*, [5, 18, 19, 23])

$$\partial_t (\vartheta + \lambda\chi) - k_0\Delta\vartheta = g \tag{5.3}$$

$$\mu\partial_t\chi - \nu\Delta\chi + \beta(\chi) \ni \gamma(\vartheta,\chi) \tag{5.4}$$

in Q, with initial conditions (1.6) and boundary conditions (cf. (1.7))

$$k_0\partial_n\vartheta = h, \quad \partial_n\chi = 0 \quad \text{on } \Sigma. \tag{5.5}$$

We give some asymptotic results which show that the solution $(\vartheta_\varepsilon, \chi_\varepsilon)$ to Problem (P) characterized by k_ε converges to the solution to a suitable variational formulation of (5.2-4), (1.6). More precisely, consider

$$f \in H^1(0,T;H) + W^{2,1}(0,T;V'), \quad f(0) \in H \tag{5.6}$$

and introduce two sequences of initial data $\{e_{0\varepsilon}\}$, $\{\chi_\varepsilon\}$ such that

$$e_{0\varepsilon} \in V, \quad \chi_{0\varepsilon} \in V, \quad \text{and} \quad j(\chi_{0\varepsilon}) \in L^1(\Omega).$$

In addition, let

$$\pi_\varepsilon \in W^{1,1}(0,T) \quad \forall \varepsilon > 0, \quad \pi_\varepsilon \to 0 \text{ in } W^{1,1}(0,T) \text{ as } \varepsilon \searrow 0. \tag{5.7}$$

Denote by $(\vartheta_\varepsilon, \chi_\varepsilon, \xi_\varepsilon)$ the solution to Problem (P) given by Theorem 2.1, where k, e_0, χ_0 have been substituted with k_ε, $e_{0\varepsilon}$, $\chi_{0\varepsilon}$, respectively. Note that, due to (5.5-6), Theorem 2.4 also implies that the regularity properties (2.16-17) hold.

Then we can state the first convergence theorem (cf. [12, Thm. 2.2])

Theorem 5.1. *Let (2.5-6) and (5.5-7) hold. Moreover, assume*

$$e_{0\varepsilon} \to e_0 \text{ in } V', \quad \chi_{0\varepsilon} \to \chi_0 \text{ in } V \tag{5.8}$$

as $\varepsilon \searrow 0$ and that, for any $\varepsilon \in (0,1)$,

$$\|j(\chi_{0\varepsilon})\|_{L^1(\Omega)} + \varepsilon^{1/2}\|e_{0\varepsilon}\|_V \le C_{12} \tag{5.9}$$

for some positive constant C_{12}. Then, there exists a triplet (w, χ, ξ) such that, possibly for a subsequence,

$$\vartheta_\varepsilon \to \vartheta \quad \text{strongly in } C^0([0,T];H) \tag{5.10}$$

$$1 * \vartheta_\varepsilon \to 1 * \vartheta \quad \text{strongly in } C^0([0,T];V) \tag{5.11}$$

$$\chi_\varepsilon \to \chi \quad \text{weakly star } H^1(0,T;H) \cap L^\infty(0,T;V) \cap L^2(0,T;W) \tag{5.12}$$

$$\chi_\varepsilon \to \chi \quad \text{strongly in } C^0([0,T];H) \cap L^2(0,T;V) \tag{5.13}$$

$$\xi_\varepsilon \to \xi \quad \text{weakly in } L^2(Q) \tag{5.14}$$

283

as $\varepsilon \searrow 0$. In addition, the triplet (ϑ, χ, ξ) solves the problem

$$\vartheta + \lambda \chi + A(1 * \vartheta) = e_0 + 1 * f \quad \text{in } V', \quad \text{a.e. in } (0, T) \tag{5.15}$$

$$\mu \partial_t \chi + \nu A \chi + \xi = \gamma(\vartheta, \chi) \quad \text{in } V', \quad \text{a.e. in } (0, T) \tag{5.16}$$

$$\xi \in \beta(\chi) \quad \text{a.e. in } Q \tag{5.17}$$

$$\chi(0) = \chi_0. \tag{5.18}$$

Remark 5.2. The solution to the limit problem (5.2-4), (1.6) given by Theorem 5.1 is quite weak (compare to (5.15-18)). To ensure its uniqueness (and thus the convergence of the whole sequences), we need stronger assumptions on e_0, that is $e_0 \in H$. Of course, in this case, the convergence hypothesis (5.8) has to be modified accordingly (cf. Theorem 5.3 below, also see [12, Thm. 2.3 and Rem. 2.4]).

Taking Remark 5.2 into account, we can now present a result regarding error estimates (cf. [12, Thms. 2.5 and 2.6], also see [12, Rem. 5.1]).

Theorem 5.3. Let (2.5-6) and (5.5-9) hold. Then, assume that

$$e_{0\varepsilon} \to e_0 \quad \text{in } H, \quad \text{as } \varepsilon \searrow 0 \tag{5.19}$$

and, for any $\varepsilon \in (0, 1)$,

$$\|\pi_\varepsilon\|_{W^{1,1}(0,T)} + \|e_{0\varepsilon} - e_0\|_{V'} + \|\chi_{0\varepsilon} - \chi_0\| \le C_{13} \varepsilon^{1/2} \tag{5.20}$$

for some $C_{13} > 0$. Then there is a positive constant C_{14} such that

$$\|\vartheta_\varepsilon - \vartheta\|_{L^2(0,T;H)} + \|1 * \vartheta_\varepsilon - 1 * \vartheta\|_{C^0([0,T];V)}$$
$$+ \|\chi_\varepsilon - \chi\|_{L^2(0,T;V) \cap C^0([0,T];H)} \le C_{14} \varepsilon^{1/2} \tag{5.21}$$

for any $\varepsilon \in (0, 1)$. If (5.19) and (5.20) are replaced, respectively, by

$$e_{0\varepsilon} \to e_0 \quad \text{in } V, \quad \text{as } \varepsilon \searrow 0 \tag{5.22}$$

$$\|\pi_\varepsilon\|_{W^{1,1}(0,T)} + \|e_{0\varepsilon} - e_0\|_{V'} + \|\chi_{0\varepsilon} - \chi_0\| \le C_{15} \varepsilon \tag{5.23}$$

for any $\varepsilon \in (0, 1)$ and for some $C_{15} > 0$, then there is a positive constant C_{16} such that, for any $\varepsilon \in (0, 1)$,

$$\|\varepsilon - \vartheta\|_{L^2(0,T;H)} + \|1 * \vartheta_\varepsilon - 1 * \vartheta\|_{C^0([0,T];V)}$$
$$+ \varepsilon^{1/2} \|\vartheta_\varepsilon - \vartheta\|_{C^0([0,T];H)} + \|\chi_\varepsilon - \chi\|_{L^2(0,T;V) \cap C^0([0,T];H)} \le C_{16} \varepsilon. \tag{5.24}$$

Remark 5.4. Hypothesis (5.5) can be generalized by taking a sequence $\{f_\varepsilon\}$ fulfilling (5.5) which suitably converges to a function $f \in L^1(0, T; H) + L^2(0, T; V')$. In particular, Theorem 5.1 still holds if f just belongs to $L^1(0, T; V')$ (see [12, Thm. 2.2]). In this case, note that the uniqueness of solutions to problem (5.15-18) is no longer guaranteed (cf. Rem. 5.2).

Remark 5.5. Theorems 5.1 and 5.3 still hold when e has a nonlinear and smooth dependence on λ (see [12]).

References

[1] S. AIZICOVICI AND V. BARBU, *Existence and asymptotic results for a system of integro-partial differential equations*, NoDEA Nonlinear Differential Equations Appl., **3** (1996), 1–18.

[2] V. BARBU, "Nonlinear semigroups and differential equations in Banach spaces," Noordhoff, Leyden, 1976.

[3] G. BONFANTI AND F. LUTEROTTI, *Regularity and convergence results for a phase-field model with memory*, Math. Methods Appl. Sci., to appear.

[4] G. BONFANTI AND F. LUTEROTTI, *Global solution to a phase-field model with memory and quadratic nonlinearity*, Adv. Math. Sci. Appl., to appear.

[5] G. CAGINALP, *An analysis of a phase field model of a free boundary*, Arch. Rational Mech. Anal., **92** (1986), 205–245.

[6] C. CATTANEO, *Sulla conduzione del calore*, Atti Sem. mat. Fis. Univ. Modena, **3** (1948), 83–101.

[7] L. CHIUSANO AND P. COLLI, *Positive kernel and zero time relaxation in a phase field model with memory*, Comm. Appl. Anal., to appear.

[8] P. COLLI, G. GILARDI, AND M. GRASSELLI, *Global smooth solution to the standard phase-field model with memory*, Advances Differential Equations, **2** (1997), 453–486.

[9] P. COLLI, G. GILARDI, AND M. GRASSELLI, *Well-posedness of the weak formulation for the phase-field model with memory*, Advances Differential Equations, **2** (1997), 487–508.

[10] P. COLLI, G. GILARDI, AND M. GRASSELLI, *Convergence of phase field to phase relaxation models with memory*, Ann. Univ. Ferrara Sez. VII (N.S.), **41** (1995), 1–14.

[11] P. COLLI, G. GILARDI, AND M. GRASSELLI, *Asymptotic analysis of a phase field model with memory for vanishing time relaxation*, Hiroshima Math. J., to appear.

[12] P. COLLI, G. GILARDI, AND M. GRASSELLI, *Asymptotic justification of the phase-field model with memory*, Comm. Appl. Anal., to appear.

[13] P. COLLI AND M. GRASSELLI, *An existence result for a hyperbolic phase transition problem with memory*, Appl. Math. Lett., **5** (1992), 99–102.

285

[14] P. COLLI AND P. LAURENÇOT, *Existence and stabilization of solutions to the phase-field model with memory*, J. Integral Equations Appl., to appear.

[15] P. COLLI AND PH. LAURENÇOT, *Uniqueness of weak solutions to the phase-field model with memory*, J. Math. Sci. Univ. Tokyo **5** (1998), 459–476.

[16] P. COLLI AND J. SPREKELS, *On a Penrose-Fife model with zero interfacial energy leading to a phase-field system of relaxed Stefan type*, Ann. Mat. Pura Appl. (4), **169** (1995), 269–289.

[17] P. COLLI AND J. SPREKELS, *Weak solution to some Penrose-Fife phase-field systems with temperature-dependent memory*, J. Differential Equations, **142** (1998), 54–77.

[18] A. DAMLAMIAN, N. KENMOCHI, AND N. SATO, *Subdifferential operator approach to a class of nonlinear systems for Stefan problems with phase relaxation*, Nonlinear Anal., **23** (1994), 115–142.

[19] G.J. FIX, *Phase field models for free boundary problems*, in "Free boundary problems: theory and applications. Vol. II," A. Fasano and M. Primicerio eds., pp. 580–589, Pitman Res. Notes Math. Ser., **79**, Longman, London, 1983.

[20] M.E. GURTIN AND A.C. PIPKIN, *A general theory of heat conduction with finite wave speeds*, Arch. Rational Mech. Anal., **31** (1968), 113–126.

[21] D.D. JOSEPH AND L. PREZIOSI, *Heat waves*, Rev. Modern Phys., **61** (1989), 41–73.

[22] D.D. JOSEPH AND L. PREZIOSI, *Addendum to the paper "Heat waves" [Rev. Mod. Phys. 61, 41(1989)]*, Rev. Modern Phys., **62** (1990), 375–391.

[23] N. KENMOCHI, *Systems of nonlinear PDEs arising from dynamical phase transitions*, in "Phase transitions and hysteresis", A. Visintin ed., 39–86, Lecture Notes in Math., **1584**, Springer-Verlag, Berlin, 1994.

[24] N. V. SHEMETOV, *Existence and stability results for the hyperbolic Stefan problem with phase relaxation*, Ann. Mat. Pura Appl. (4), **168** (1995), 301–316.

[25] A. VISINTIN, *Stefan problem with phase relaxation*, IMA J. Appl. Math., **34** (1985), 225–245.

Maurizio Graselli
Dipartimento di Matematica "F. Brioschi"
Politecnico di Milano
Via E. Bonardi 9, 20133 Milano, Italy

286

A. Visintin
Models of Nucleation and Growth

1 Introduction

In this paper we outline a model of phase evolution in two-phase (e.g., solid and liquid) systems. We account for not only smooth growth, but also for discontinuous evolution. An example of the latter mode is *nucleation,* namely, the formation of a new connected component of one of the two phases, for instance, of solid in the interior of an undercooled liquid.

At a fine *(mesoscopic)* length-scale, surface tension effects become relevant; under stationary conditions, equilibrium is represented by the classical *Gibbs-Thomson law,*

$$\theta = -c\kappa \qquad \text{on the solid-liquid interface } S. \tag{1}$$

Here κ is the mean curvature of S (assumed positive for solid balls), θ the relative temperature (namely, the difference between the actual absolute temperature τ and the constant value τ_E at which a planar solid–liquid interface is at equilibrium), c a positive coefficient. Denoting the signed velocity of advancement of the liquid by v, evolution can then be represented by *mean curvature flow:*

$$av = \theta + c\kappa \qquad \text{on } S\,(a > 0). \tag{2}$$

This law has been extensively analyzed, mainly with no forcing term θ and independently from its connection with phase transitions, at first by Brakke [5], Gage [14], Grayson [15], Huisken [18], Barles [4], Sethian [28], Osher and Sethian [25], Evans and Spruck [13], Chen, Giga and Goto [9], and then by many others. For instance, see the reviews of Evans, Soner and Souganidis [12], Ilmanen [19], and the volumes [6, 10].

Intense mathematical research has been devoted to the Stefan problem and its generalizations; for instance, see the recent monograph V. [31]. The Stefan problem with surface tension has been studied by Luckhaus [20], V. [30], Gurtin and Soner [17], and others. Nucleation has been dealt with by Gurtin [16] without accounting for surface energy. So-called *phase-field* models have been studied in several papers, cf., e.g., Caginalp [7], Penrose and Fife [26, 27] and references therein. Soner [29] has shown that phase transition with mean curvature flow can be asymptotically derived from the phase-field model.

Phase nucleation is the main concern of this paper. This author already dealt with this phenomenon in [32, 33], inserting the Gibbs-Thomson law (1.1) into a *two-scale* Stefan model; see also V. [31, Chaps. VIII, IX]. In [34, 35] he then proposed and studied a model which couples nonlinear mean curvature flow and nucleation, assuming a prescribed temperature evolution. Here we review and discuss those developments.

287

This is just a first step toward the analysis of a mesoscopic model of phase transitions, in which the evolution of temperature and phase will be coupled.

2   Macroscopic and Mesoscopic Free Energies

Macroscopic Free Energy. Let us consider a material capable of attaining two phases (solid and liquid, say), which occupies a bounded region $\Omega \subset \mathbb{R}^3$. We assume that it fulfills several simplifying hypotheses: incompressibility, homogeneity, anisotropy, and so on. We want to characterize the stationary phase configurations which can be attained in correspondence to a prescribed (non-necessarily uniform) temperature field. The latter can be maintained stationary by a suitably distributed heat source or sink, coupled with appropriate boundary conditions.

We shall consider two models characterized by different length-scales: a *macroscopic* scale (of about 10^{-1} cm, say), which is characteristic of laboratory measurements; a *mesoscopic* scale (of about 10^{-5}cm, say), which is intermediate between the macroscopic scale and that of molecular phenomena. We assume that at this finer scale thermodynamics locally retains its validity, and differential calculus can be used.

At the macroscopic length-scale, we shall denote by $\varphi(\in [0,1])$ the local liquid concentration, and define the *phase function* $\chi := 2\varphi - 1(\in [-1,1])$. Thus

$$
\begin{array}{lll}
\chi = 1 & \text{in the liquid phase,} & \\
\chi = -1 & \text{in the solid phase,} & (3) \\
\chi \in]-1,1[& \text{in the so-called } mushy \text{ } region;
\end{array}
$$

the latter representing a fine liquid–solid mixture.

We shall denote the relative temperature by θ, and assume that $\theta \in L^\infty(\Omega)$. The *macroscopic* linearized Helmholtz free energy then reads

$$
\tilde{F}_\theta(\chi) := -\frac{L}{2\tau_E} \int_\Omega \theta \chi \, dx \qquad \forall \chi \in L^1(\Omega) \text{ such that } |\chi| \le 1 \text{ a.e.,} \qquad (4)
$$

plus an additive function of θ. L is the density of latent heat, and τ_E the absolute temperature of equilibrium for a solid–liquid mixture; both are assumed to be positive constants. $\tilde{F}_\theta(\chi)$ is set equal to $+\infty$ for all $\chi \in L^1(\Omega)$ such that $|\chi| > 1$ in a positive measure set. Obviously, χ minimizes $\tilde{F}_\theta$ iff

$$
\begin{array}{lll}
\chi = 1 & \text{where } \theta > 0, & \\
\chi = -1 & \text{where } \theta < 0, & (5) \\
\chi \in [-1,1] & \text{where } \theta = 0.
\end{array}
$$

This model does not account for *undercooling* prior to solid nucleation; this can be quite relevant, e.g., up to 370 K for platinum [8, Chap. 3].

288

Mesoscopic Free Energy. At a sufficiently fine length-scale, surface tension becomes relevant, and liquid and solid parts can be distinguished in the mushy regions. One then deals with pure phases, liquid and solid, without any mesoscopic mushy region. This corresponds to replacing the convex constraint $|\chi| \leq 1$ by a *nonconvex* one: $|\chi| = 1$ pointwise in Ω.

The mesoscopic structure of two-phase systems is characterized by nonconvexity and surface tension. The latter is at the basis of the *Gibbs-Thomson law*.

We assume that the boundary Γ of Ω is of Lipschitz class, and introduce the perimeter functional $P : L^1(\Omega) \to \mathbf{R} \cup \{+\infty\}$:

$$
P(\chi) := \begin{cases} \dfrac{1}{2} \displaystyle\int_\Omega |\nabla \chi| & \text{if } |\chi| = 1 \text{ a.e. in } \Omega, \\[2ex] +\infty & \text{otherwise;} \end{cases} \tag{6}
$$

here $\int_\Omega |\nabla \chi| := \sup\{\int_\Omega \chi \nabla \cdot \vec{\eta} : \vec{\eta} \in C_0^1(\Omega)^N, |\vec{\eta}| \leq 1 \text{ in } \Omega\}$. So $P(\chi)$ equals the interface area, whenever χ is the characteristic function of some set.

Let us denote by σ the surface density of free energy at the solid–liquid interface, by σ_L (σ_S, resp.) the surface density of free energy relative to a surface separating the liquid phase (solid phase, resp.) from an external material. The functional

$$
\begin{aligned}
F_\theta(\chi) &:= -\frac{L}{2\tau_E} \int_\Omega \theta \chi dx + \sigma P(\chi) + \sigma_L \int_\Gamma \frac{1+\chi}{2} d\Gamma + \sigma_S \int_\Gamma \frac{1-\chi}{2} d\Gamma \\[2ex]
&= -\frac{L}{2\tau_E} \int_\Omega \theta \chi dx + \sigma P(\chi) + \frac{\sigma_L - \sigma_S}{2} \int_\Gamma \chi d\Gamma + \frac{\sigma_L + \sigma_S}{2} |\Gamma| \quad \forall \chi \in \text{Dom}(P)
\end{aligned} \tag{7}
$$

represents the total phase contribution to the free energy. Notice that the trace of χ on Γ is meaningful; moreover $F_\theta(\chi) < +\infty$ only if $|\chi| = 1$ a.e. in Ω, so that by minimizing F_θ the constraint "$|\chi| = 1$ a.e. in Ω" is automatically imposed.

Crystals are anisotropic, and indeed the isotropy assumption is more appropriate for liquid–vapour than for solid–liquid systems. However, a large part of our discussion might be extended to heterogeneous and anisotropic materials, provided that $\sigma \int_\Omega |\nabla \chi|$ were replaced by $\int_\Omega |\sigma \cdot \nabla \chi|$, where $\sigma = \sigma(x)$ is a positive definite 3×3 tensor.

Via the *direct method* of the calculus of variations, one can show that for any $\theta \in L^1(\Omega)$ the functional F_θ has an (absolute) minimizer, provided that

$$
|\sigma_L - \sigma_S| \leq \sigma. \tag{8}
$$

The latter condition is equivalent to the semicontinuity of F_θ, see Massari and Pepe [23]. If it were violated, it would be easy to exhibit counterexamples in which F_θ has no minimizer.

We stress that here the nonconvexity of the potential (or equivalently of the constraint on χ) is compensated by the presence of the perimeter term, because of the compactness of the inclusion $BV(\Omega) \subset L^1(\Omega)$. Loosely speaking, surface tension *sustains* nonconvexity.

Asymptotic Behaviour as $\sigma \to 0$. For a moment let us denote by F_θ^σ the functional defined in (2.5), displaying the dependence on σ.

For water $\sigma \simeq 1.8 \times 10^{-6}$ cal cm^{-2}, hence at the macroscopic length-scale σ can be regarded as vanishing. As $\sigma \to 0$ (whence $\sigma_L, \sigma_S \to 0$, by (2.6)) along a sequence, F_θ^σ Γ-*converges* (in the sense of De Giorgi [11]) to F_θ, cf. (2.2). That is:

(i) for any $\chi \in L^1(\Omega)$ and any sequence $\{\chi_{\sigma_n}\} \subset L^1(\Omega)$ such that $\chi_{\sigma_n} \to \chi$ strongly in $L^1(\Omega)$, we have $\liminf_{\sigma_n \to 0} F_\theta^{\sigma_n}(\chi_{\sigma_n}) \geq F_\theta(\chi)$;

(ii) for any $\chi \in L^1(\Omega)$, there exists a sequence $\{\chi_{\sigma_n}\} \subset L^1(\Omega)$ such that $\chi_{\sigma_n} \to \chi$ strongly in $L^1(\Omega)$ and $\lim_{\sigma_n \to 0} F_\theta^{\sigma_n}(\chi_{\sigma_n}) = F_\theta(\chi)$.

We then conclude that the macroscopic convex model can be retrieved as the limit of the mesoscopic nonconvex model as $\sigma \to 0$, as far as *absolute* minima are concerned.

A Double-Well Potential. An alternative approach can be used at an even finer length-scale. Let us consider a so-called *double-well* potential, e.g., $W_1(v) := \left(1 - v^2\right)^2$ for any $v \in \mathbb{R}$, fix two positive parameters a and ε, and represent the free energy by a functional of the form

$$F_\theta^\varepsilon(\chi) := \int_\Omega \left(\frac{\varepsilon a}{2}|\nabla\chi|^2 + \frac{1}{\varepsilon}W(\chi) - \frac{L}{2\tau_E}\theta\chi\right)dx + \frac{\sigma_L - \sigma_S}{2}\int_\Gamma \chi d\Gamma \qquad \forall\chi \in H^1(\Omega) \quad (9)$$

(plus a function of θ). If χ is nonuniform, the first two terms are in competition: $(\varepsilon a/2)|\nabla\chi|^2$ penalizes sharp variations of χ, whereas $(1/\varepsilon)W(\chi)$ penalizes deviations from $|\chi| = 1$. For small ε, any either absolute or relative minimizer of F_θ^ε attains values close to ± 1 in the whole Ω, except for thin transition layers which represent neighbourhoods of the phase interfaces. The coefficients a, ε are so small that the layer thickness is typically of order 10^{-7} cm. This length-scale is so close to that of molecular phenomena, that the use of a continuous model might be questioned.

A well-known result of Modica and Mortola [24] shows that as $\varepsilon \to 0$, F_θ^ε Γ-converges to the mesoscopic free energy functional F_θ^σ, provided that $a = 9\sigma^2/128$.

3 The Gibbs-Thomson Law

Absolute and Relative Minimizers. We shall say that $\chi \in L^1(\Omega)$ is a *relative minimizer* of F_θ if it is not an absolute minimizer, and there exists $K(\chi) > 0$ such that

$$F_\theta(\chi) \leq F_\theta(v) \qquad \forall v \in L^1(\Omega) \text{ such that } \|v - \chi\|_{L^1(\Omega)} \leq K(\chi). \tag{10}$$

Proposition 3.1 *(Existence of Relative Minimizers) (i) For any $\theta \in L^\infty(\Omega)$, if $\theta \leq 0$ and $\theta \not\equiv 0$ in Ω, then $\chi \equiv 1$ is a relative minimizer of F_θ in $L^1(\Omega)$ (and the same* holds for $\chi \equiv -1$, if $\theta \geq 0$ and $\theta \not\equiv 0$).

(ii) If the σ-term were dropped in (2.5), then F_θ would have no relative minimizer for any $\theta \in L^\infty(\Omega)$.

This is a simple consequence of the nonconvexity of the functional and of the presence of the (convex and coercive) higher order term $\sigma P(\chi)$, see V. [31, p. 168].

At constant temperature, (states represented by) absolute minimizers of the free energy persist for any time. On the other hand, relative minimizers may persist for some time, but eventually decay because of *fluctuations*. Those states are called *metastable*. By the same token, relative maximizers and saddle points decay instantaneously.

Theorem 3.2 *(Gibbs-Thomson Law and Contact Angle Condition) Let $\theta \in W^{1,1}(\Omega)$, $\chi \in L^1(\Omega)$, and the boundary S of the set $\Omega^* := \{x \in \Omega : \chi(x) = 1\}$ be of class C^1. Let us denote by $\vec{n}$ the unit normal vector to S oriented toward Ω^*, and set $\kappa := \nabla_S \cdot \vec{n}/2$ (= interface mean curvature). If χ is either an absolute or relative minimizer of F_θ in $L^1(\Omega)$, then $\kappa \in L^1(S)$ and*

$$\gamma_0 \theta = -\frac{2\sigma\tau_E}{L}\kappa \qquad a.e. \ on \ S. \tag{11}$$

Moreover, denoting by ω the angle between $\vec{n}$ and the outward normal vector to Γ, and endowing $S \cap \Gamma$ with the one-dimensional Hausdorff measure, we have

$$\cos\omega = \frac{\sigma_S - \sigma_L}{\sigma} \qquad a.e. \ on \ S \cap \Gamma. \tag{12}$$

The simple argument is based on a local representation of S as a Cartesian graph, see V. [31, p. 169]. Whenever $\theta \in L^p(\Omega)$ with $p > 3$, by a classical result of geometric measure theory due to Almgren [1], $\partial\Omega^*$ is a manifold of class $C^{1,(p-3)/2p}$, and the latter result can be applied.

For instance, for water at atmospheric pressure at about $0°\,\text{C}$ we have $\sigma \simeq 1.8 \times 10^{-6}$ cal cm^{-2}, $L \simeq 80$ cal cm^{-3}, $\tau_E \simeq 273$ K. Hence $2\sigma\tau_E/L \simeq 1.2 \times 10^{-5}$ cm K, and nucleation at an undercooling of few degrees corresponds to a critical radius of order 10^{-5} cm.

4   Mean Curvature Flow

Radial Setting. In order to get some understanding of nucleation, we first describe isothermal nucleation of a solid phase in the interior of a liquid at a uniform temperature $\theta < 0$. We just deal with *homogeneous* nucleation, which corresponds to a new solid phase not in contact with the container.

By the isoperimetric property of the sphere and by the uniformity of the temperature, we can confine ourselves to varying the phase in balls, the position of the center
being immaterial. The variation of the free energy $F_\theta(\chi)$ due to formation of a solid
ball of radius R ($\leq \tilde{R}$, say) only depends on R, and will be denoted by $\varphi(R)$. By
(2.5), we have

$$\varphi(R) = \frac{4\pi L}{3\tau_E}\theta R^3 + 4\pi\sigma R^2 \qquad \forall R \geq 0. \tag{13}$$

cf. Fig. 1. Hence $\varphi'(R_c) = 0$ iff $R_c := -2\sigma\tau_E/L\theta(> 0)$. The *critical radius* R_c
coincides with the value prescribed by the Gibbs-Thomson law (3.2). If $\tilde{R}$ is not too
small, $\varphi(\tilde{R}) < 0$; $R = \tilde{R}$, $R = R_c$, and $R = 0$ are then an absolute minimizer, a
relative maximizer, and a relative minimizer of φ in $[0, \tilde{R}]$, respectively.

In first approximation, it seems reasonable to assume that the dynamics is governed
by descent along the potential φ. By a suitable choice of the relaxation coefficient,
the *gradient flow* for φ yields the law of *motion by mean curvature*, with a forcing
term proportional to θ:

$$a\frac{dR}{dt} = -\frac{\varphi'(R)}{4\pi R^2} = -\frac{2\sigma}{R} - \frac{L\theta}{\tau_E} \qquad (a : \text{positive coefficient}). \tag{14}$$

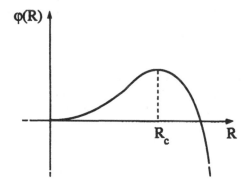

Figure 1.

General Setting. Let us denote by $\kappa(\cdot, t)$ the mean curvature of the interface $S_t :=$
$\partial\Omega^*(t)$ (with the convention that $\kappa > 0$ for solid balls). Defining the *signed distance*
from the solid–liquid interface

$$\rho(x, t) := \text{ess-dist}\,(x, \Omega \setminus \Omega^*(t)) - \text{ess-dist}\,(x, \Omega^*(t)) \qquad \forall(x, t), \tag{15}$$

$\partial \rho / \partial t$ equals the velocity of advancement of the liquid front. The *mean curvature flow* then reads

$$a \frac{\partial \rho}{\partial t} = 2\sigma \kappa + \frac{L}{\tau_E} \theta \qquad \text{on } S. \tag{16}$$

The gradient flow for the free energy functional $F_\theta^\varepsilon(\chi)$ reads $c \partial \chi / \partial t + F_\theta^{\varepsilon\,\prime}(\chi) = 0$, where F_θ^ε is the Fréchet derivative and c is a positive coefficient. This is equivalent to the *Allen-Cahn* (or *Landau-Ginzburg*) *equation*

$$c \frac{\partial \chi}{\partial t} - \varepsilon a \Delta \chi + \frac{4}{\varepsilon} \chi (\chi^2 - 1) = \frac{L\theta}{2\tau_E} \qquad \text{in } Q. \tag{17}$$

5 Nucleation and Growth

By (4.3), it is clear that phase nucleation cannot occur by mean curvature flow, but only as a fluctuation takes R from 0 to some value larger than R_c. If we regard this as a *Poisson process*, in a small time interval δt nucleation occurs with probability proportional to $\delta t \exp[-\varphi(R_c(\theta))/k\tau]$. So nucleation is stochastic and discontinuous, whereas mean curvature flow is deterministic and continuous.

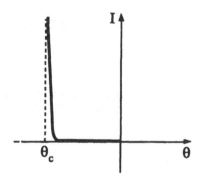

Figure 2.

By the classical theory of nucleation, for a metal and at the beginning of the process, the *nucleation rate* I (:= number of nuclei formed per unit volume per unit time) varies from almost zero to a very large value in a very small range of temperatures, which can be identified with a single value $\theta_c (< 0)$, see Fig. 2. Therefore, for a prescribed time-scale, nucleation can be assumed to occur at a *critical temperature* θ_c.

Nuclei then grow by mean curvature flow, until either they impinge on other nuclei, or undercooling is eliminated by release of latent heat.

Nucleation and Hysteresis. Here we outline a rather simplified model. For the sake of simplicity, we assume that the thresholds for solid and liquid nucleation are symmetric, equal to $\pm c$. For a moment let us neglect the dependence on x, and set $\chi_\pm(t) := \chi(t \pm 0)$ for any t.

The undercooling and superheating required to trigger nucleation are schematically outlined in Fig. 3. There the arrows represent the dynamics; notice the *bistable* zone for $-c \le \theta \le c$. (The oblique dashed line, which separates the two half-planes in which $\theta + c\chi$ has different sign, is only drawn to help the reader in interpreting the system (5.1), see below.)

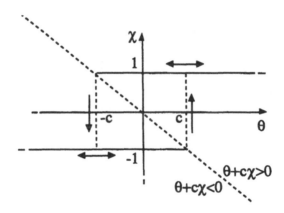

Figure 3.

Let us set $\text{sign}(x) := \{-1\}$ if $x < 0$, $\text{sign}(x) := \{1\}$ if $x > 0$ and $\text{sign}(0) := [-1, 1]$. In analytic terms, the χ vs. θ dependence outlined in Fig. 3 can be written as follows:

$$\begin{cases} \chi_+ \in \text{sign}(\theta + c\chi_-) \\ |\chi| = 1 \qquad \forall t; \end{cases} \tag{18}$$

or also, setting $\psi(v) := 0$ if $|v| = 1$ and $\psi(v) := +\infty$ otherwise,

$$\psi(\chi_+) - \psi(v) \le (\theta + c\chi_-)(\chi_+ - v) \qquad \forall v \in \mathbf{R}, \forall t. \tag{19}$$

Let us now reinsert the x-dependence. Assuming that $\theta(\cdot, t) \in L^\infty(\Omega)$ for any t, we represent nucleation by the variational inequality

$$P(\chi_+) - P(v) \le \frac{L}{2\tau_E} \int_\Omega (\theta + c\chi_-)(\chi_+ - v) dx \qquad \forall v \in L^\infty(\Omega), \forall t. \tag{20}$$

294

Nonlinear mean curvature flow. At this point the exigence arises of matching nucleation and mean curvature flow. We then propose to modify (4.4) as follows:

$$\alpha\left(\frac{\partial\rho}{\partial t}\right) = 2\sigma\kappa + \frac{L}{\tau_E}\theta \qquad \text{on } S_t, \forall t \in I_c, \tag{21}$$

where $\alpha : \mathbf{R} \to \mathbf{R}$ is nonconstant, bounded and monotone. For instance, see Fig. 4,

$$\alpha(\xi) := \begin{cases} -Lc/\tau_E & \text{if } \xi < -Lc/a\tau_E, \\ a\xi & \text{if } -Lc/a\tau_E \leq \xi \leq Lc/a\tau_E, \\ Lc/\tau_E & \text{if } \xi > Lc/a\tau_E. \end{cases} \tag{22}$$

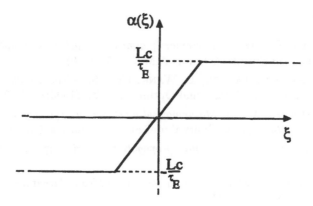

Figure 4.

Let us briefly compare the nonlinear law (5.4) and the standard mean curvature flow (4.4). The boundedness of the function α has two relevant consequences: it entails that the mean curvature is uniformly bounded, whenever the same holds for θ; this has a regularizing effect in space. But it also entails a loss of regularity for the velocity $\partial\rho/\partial t$, and in fact evolution can be discontinuous. Thus here space singularities do not appear on the moving surface, since (roughly speaking) they can be avoided by means of jumps of the phase.

295

Evolution Modes. We then distinguish two modes of evolution.

(i) *Mean Curvature Flow.* This is smooth in time, and occurs at almost any instant (with respect to the ordinary Lebesgue measure). This is represented by a law on the moving surface, (5.4).

(ii) *Singular Evolution.* This can appear in several forms: as phase nucleation, or as phase annihilation, or as merging of two separate phases, or as splitting of a single phase, or as other changes in the phase topology. This is represented by the variational inequality (5.3) which acts on the whole domain Ω (not just the moving surface). According to this law, the phase volume does not need to vary continuously in time. This seems to be consistent with the classical physical description of nucleation at a mesoscopic length-scale.

(5.3) and (5.4) provide complementary and consistent information. They can be coupled in a weak formulation, for which existence of a solution has been proved, see V. [35]. Of course, the next step in this program is to couple (5.3) and (5.4) with the energy conservation equation.

References

[1] F. ALMGREN, *Existence and regularity almost everywhere of elliptic variational problems with constraints.* Memoirs A.M.S. **165** (1976)

[2] F. ALMGREN, J.E. TAYLOR, L. WANG, *Curvature-driven flows: a variational approach.* S.I.A.M. J. Control and Optimization **31** (1993) 387–437

[3] F. ALMGREN, L. WANG, *Mathematical existence of crystal growth with Gibbs-Thomson curvature effects.* Journal of Geometric Analysis (to appear)

[4] G. BARLES, *Remark on a flame propagation model.* Rapport I.N.R.I.A. **464**, 1985

[5] K.A. BRAKKE, *The Motion of a Surface by its Mean Curvature.* Princeton University Press, Princeton 1978

[6] G. BUTTAZZO, A. VISINTIN (EDS.), *Motion by Mean Curvature and Related Topics.* De Gruyter, Berlin 1994

[7] G. CAGINALP, *An analysis of a phase field model of a free boundary.* Arch. Rational Mech. Anal. **92** (1986) 205–245

[8] B. CHALMERS, *Principles of Solidification.* Wiley, New York 1964

[9] Y.G. CHEN, Y. GIGA, S. GOTO, *Uniqueness and existence of viscosity solutions of generalized solutions of mean curvature flow equation.* J. Differential Geom. **33** (1991) 749–786

[10] A. DAMLAMIAN, J. SPRUCK, A. VISINTIN (EDS.), *Curvature Flow and Related Problems.* Gakkōtosho Scientific, Tokyo 1995

[11] E. DE GIORGI, T. FRANZONI, *Su un tipo di convergenza variazionale.* Atti Accad. Naz. Lincei Cl. Sci. Mat. Fis. Natur. **58** (1975) 842–850

[12] L.C. EVANS, M. SONER, P.E. SOUGANIDIS, *Phase transitions and generalized motion by mean curvature.* Comm. Pure Appl. Math. **45** (1992) 1097–1123

[13] L.C. EVANS, J. SPRUCK, *Motion of level sets by mean curvature I.* J. Diff. Geom. **33** (1991) 635–681

[14] M. GAGE, *An isoperimetric inequality with applications to curve shortening.* Duke Math. J. **50** (1983) 1225–1229

[15] M. GRAYSON, *The heat equation shrinks embedded plane curves to points.* J. Diff. Geometry **26** (1987) 285–314

[16] M.E. GURTIN, *Thermodynamics and supercritical Stefan equations with nucleations.* Quart. Appl. Math. **LII** (1994) 133–155

[17] M.E. GURTIN, H.M. SONER, *Some remarks on the Stefan problem with surface structure.* Quart. Appl. Math. **L** (1992) 291–303

[18] G. HUISKEN, *Flow by mean curvature of convex surfaces into spheres.* J. Differ. Geom. **20** (1984) 237–266

[19] T. ILMANEN, *Elliptic regularization and partial regularity for motion by mean curvature.* Memoirs A.M.S. **520** (1994)

[20] S. LUCKHAUS, *Solutions of the two-phase Stefan problem with the Gibbs-Thomson law for the melting temperature.* Euro. J. Appl. Math. **1** (1990) 101–111

[21] S. LUCKHAUS, *Solidification of alloys and the Gibbs-Thomson law.* Preprint 1994

[22] S. LUCKHAUS, T. STURZENHECKER, *Implicit time discretization for the mean curvature flow equation.* Calc. Var. **3** (1995) 253–271

[23] U. MASSARI, L.PEPE, *Su di una impostazione parametrica del problema dei capillari.* Ann. Univ. Ferrara **20** (1974) 21–31

[24] L. MODICA, S. MORTOLA, *Un esempio di Γ-convergenza.* Boll. Un. Mat. Ital. **5** (1977) 285–299

[25] S. OSHER, J.A. SETHIAN, *Fronts propagating with curvature dependent speed: algorithms based on Hamilton-Jacobi formulations.* J. Comput. Physics **79** (1988) 12–49

[26] O. PENROSE, P.C. FIFE, *Thermodynamically consistent models of phase-field type for the kinetics of phase transitions.* Physica D **43** (1990) 44–62

[27] O. PENROSE, P.C. FIFE, *On the relation between the standard phase-field model and a "thermodynamically consistent" phase-field model.* Physica D **69** (1993) 107–113

[28] J.A. SETHIAN, *Curvature and evolution of fronts.* Comm. Math. Phys. **101** (1985) 487–499

[29] M. SONER, *Convergence of the phase-field equation to the Mullins-Sekerka problem with kinetic undercooling.* Arch. Rational Mech. Anal. **131** (1995) 139–197

297

[30] A. VISINTIN, The Stefan problem with surface tension. In: *Mathematical Models of Phase Change Problems* (J.F. Rodrigues, ed.). Birkhäuser, Basel 1989, pp. 191–213

[31] A. VISINTIN, *Models of Phase Transitions*. Birkhäuser, Boston 1996

[32] A. VISINTIN, Two-scale Stefan problem with surface tension. In: *Nonlinear Analysis and Applications* (N. Kenmochi, M. Niezgódka, P. Strzelecki, eds.). Gakkotosho, Tokyo 1996, pp. 405–424

[33] A. VISINTIN, *Two-scale model of phase transitions*. Physica D **106** (1997), 66-80

[34] A. VISINTIN, *Motion by mean curvature flow and nucleation*. C.R. Acad. Sc. Paris, Serie I **325** (1997), 55-60

[35] A. VISINTIN, *Nucleation and mean curvature flow*. Communications in P.D.E.s **23** (1998), 17–35

Augusto Visintin
Dipartimento di Matematica
Università di Trento
Trento, Italy

Part 5.

Computational Methods and Numerical Analysis

B. COCKBURN* P.-A. GREMAUD† AND X. YANG
New Results in Numerical Conservation Laws

Abstract

We study the notions of consistency and accuracy of a large family of methods (Finite Volume Methods) for numerically solving conservation laws. Two types of difficulties arise: first, the solutions have low regularity (shock waves); second, when working on nonuniform non-Cartesian grids, consistency is lost. It is shown that even when consistency is lost due to irregularities of the grid, not only are the methods still convergent, but no loss in the optimal convergence rate occurs (supraconvergence). In other words, the optimal rate, here $\mathcal{O}(\Delta^{1/2})$, is preserved provided (i) the algorithm is in conservation form, (ii) the numerical flux is consistent (not to be confused with the consistency of the scheme). This is one of the first supraconvergence results for low regularity (BV) problems.

1 Introduction

This paper is devoted to the analysis of Finite Volume schemes for conservation laws. More precisely, we study the delicate interplay between the *accuracy* of a given method on the one hand, and the properties of the underlying meshes on the other hand. The grid enters into play, not only through its *size*, but also its *(non)uniformity* as well as its *topology*. It is part of the usual folklore of this field that, if general grids are used, and unless the numerical flux takes somehow into account the "irregularities" of the grids, a *loss of consistency* occurs. To our knowledge, a full analysis of the situation for nonlinear hyperbolic problems was first offered in [3].

Given a general conservation law in $\mathbf{R}^N$

$$v_t + \nabla \cdot f(v) = 0, \tag{1}$$

we want to approximate the weak entropy solution v. If $\mathcal{T}_h$ stands for a partition (mesh) of $\mathbf{R}^N$, K for a generic element of $\mathcal{T}_h$ (cell, no restriction of size or shape) and e for an edge of K ($e = \bar{K} \cap \bar{K}'$, K, K' neighboring cells), we consider

$$u_K^{n+1} = u_K^n - \frac{\Delta t}{|K|} \sum_{e \in \partial K} |e| f_{e,K}(u_K^n, u_{K_e}^n), \tag{2}$$

where Δt is the time step and $f_{e,K}(\cdot, \cdot)$ the numerical flux.

---

*Partially supported by the National Science Foundation (Grant DMS-9407952) and by the University of Minnesota Supercomputer Institute.

†Partially supported by the Army Research Office through grant DAAH04-95-1-0419.

The study of schemes of type (2) can also be viewed as a first step toward the analysis of higher order methods, which are often constructed by considering corrections to (2).

2 Conservativity, regularity and consistency

Very few results about the justification (to say nothing about the accuracy) of approximating v, solution of (1), by u, solution of (2)

$$u(t,x) = u_K^n, \qquad \text{for } (t,x) \in [t^n, t^{n+1}) \times K,$$

are available, even in the scalar case ($u \in \mathbb{R}$). Why is that? Three properties play, or appear to play, an important role: *conservativity, regularity of the approximate solution* and *consistency*. Let us examine each of those properties.

The first issue is easily dealt with. In the case, for instance, of compactly supported data, one observes

$$\frac{d}{dt} \int_{\mathbb{R}^N} v(t,x)\,dx = 0.$$

Such a property is also desirable at the discrete level, i.e.

$$\sum_{K \in \mathcal{T}_h} u_K^n |K| = \sum_{K \in \mathcal{T}_h} u_K^0 |K|.$$

This implies some restriction on the numerical flux $f_{e,K}$

$$f_{e,K}(u_K, u_{K_e}) + f_{e,K_e}(u_{K_e}, u_K) = 0, \tag{3}$$

where K and K_e share the face e. It is not difficult to convince oneself that nonconservative schemes will in general fail to converge to v, see, e.g., [5].

The issue of the regularity of the approximate solution u is more technical. For historical reasons [8], it was long believed that establishing some regularity properties of u was a necessary first step toward *proving* convergence. However, establishing the "required" properties, typically that the total variation of the approximate solution does not increase in time, proved to be a daunting task. This was done in [8] for monotone schemes on uniform Cartesian grids and in [10] for monotone schemes on nonuniform Cartesian grids. In [1-3], it was shown that *no regularity properties of u are necessary*, at least not explicitly. This led to the first, to our knowledge, optimal result in the case of general nonuniform non-Cartesian grids [3] or, in other words, in the case of methods that do not require a dimensional splitting.

Let us now turn to the last property mentioned above, consistency. This is by far the most delicate point. First, and obviously, the numerical flux $f_{e,K}$ should be consistent with the nonlinearity f (or else the wrong problem is solved), i.e.,

$$f_{e,K}(u,u) = f(u) \cdot n_{e,K}, \tag{4}$$

302

where $n_{e,K}$ is the outward unit normal at the edge e of the finite volume K. Thus, *consistency of the numerical flux* (4) is required. This is a necessary, but by no means sufficient, condition to ensure *consistency of the numerical scheme*.

From here on, by consistency, we mean consistency of the numerical scheme. A first observation is that, for problems with low regularity like conservation laws (BV regularity), the very notion of consistency is only *formal*. Indeed, it is based on the notion of truncation error. Formally, one considers u solution of (2) as a smooth function, performs a Taylor expansion locally about the barycenter x_K of a generic cell K and gets expressions (model equations) of the type

$$u_t(t, x_K) + \nabla \cdot f(u(t, x_K)) = TE(t, x_K),$$

where TE is by definition the truncation error. If, when $\Delta x = \max\{$ diameter$(K), K \in \mathcal{T}_h\}$ goes to zero so does TE, then the scheme is said to be consistent.

Note that in the present case of conservation laws, the question is further complicated by the fact that in order to converge not only to a weak solution of (1), but rather to the weak entropy solution of (1), it appears that TE should not only go to zero as Δx decreases, but should do so in a way compatible with the entropy condition. For monotone methods on uniform Cartesian grids, this is easily seen to hold true. Indeed, one has

$$TE = \nabla \cdot (\nu \nabla u) + \text{h.o.t.},$$

where, under a suitable CFL condition, $\nu = \mathcal{O}(\Delta x)$ is the numerical viscosity of the algorithm under consideration, and where h.o.t. stands for higher order terms [1]. It is equally easy to see that on general meshes, the same monotone schemes have a truncation error of the following type [2, 3]:

$$TE = \mathcal{O}(1) + \nabla \cdot (\nu \nabla u) + \text{h.o.t.},$$

where the terms $\mathcal{O}(1)$ appear when the irregularities of the meshes are not compensated through fine tuning of the numerical flux. Such corrections are almost never considered in practice.

Our main point is the following: if the numerical method is in conservation form (3), and if the numerical flux is consistent (4), no *order* of convergence is lost, even though consistency (of the scheme) has been lost due to irregularities of the grid, see [3] for precise statements. We emphasize that the results in [3] are *not* formal, and take fully into account the possibility of shock formation, for instance. To sum this up, the notion of "consistency of the numerical scheme" appears to be irrelevant. Such a phenomenon, known as *supraconvergence*, has been studied in various contexts [4, 6, 9] and/or at the formal level [11]. To the authors' knowledge, [3] is the first rigorous explanation of supraconvergence for low regularity nonlinear hyperbolic problems.

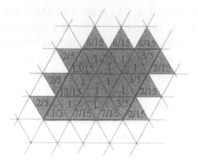

Figure 1. Exact numerical solution for a monotone example with increase in the total variation.

3 Stability and Accuracy

In many respects, uniform Cartesian grids present, *theoretically*, advantages over less regular meshes. What kind of implications, if any, does this have in practice? We look first at the impact of using general grids on the *stability* of the method, and then on its *accuracy*.

Let us consider again the class of monotone schemes, and examine the influence of the *topology* of the grid on the stability. It is known that on Cartesian grids, the total variation of the approximate solution does not increase in time (see [8] and [10] for proofs in the case of uniform and nonuniform grids, respectively). This property is lost if non-Cartesian grids are used, as can be seen from the following trivial example.

In Figure 1, (1) is solved with $f(v) = (0, b)v$ on a uniform non-Cartesian grid. The initial condition is 1 inside the highlighted parallelogram and 0 outside. As a way of illustration, the classical Lax-Friedrichs numerical flux is used

$$f_{e,K}(u, v) = \frac{1}{2}(f(u) + f(v)) \cdot n_{e,K} + C(u - v).$$

The problem is solved with $\dfrac{b\Delta t}{\Delta x} = 1/5$ and $C = \dfrac{1}{10}\dfrac{\Delta x}{\Delta t}$, where Δx is again the diameter of the cells. The support of the numerical solution after one time step, as well as its exact values are displayed. As can be readily checked, and in spite of the monotonicity of the scheme, the total variation is increasing

$$TV(u^0) = 12\,|e| \qquad TV(u^1) = \frac{226}{15}\,|e| \approx 15.0667\,|e|, \qquad |e| = \text{edge} = \frac{\sqrt{3}}{2}\Delta x.$$

How do the irregularities of the grid affect the accuracy of the numerical solutions? Here, the accuracy is measured through the error $v - u$ in the $L^\infty(0, T; L^1(\mathbf{R}^N))$-norm.

304

To fix the ideas, we restrict our attention to the following kind of numerical flux:

$$f_{e,K}(v,w) = f_{cent;e,K}(v,w) - f_{visc;e,K}(v,w),$$
$$f_{cent;e,K}(v,w) = a_{e,K}\, f(v) \cdot n_{e,K} + b_{e,K}\, f(w) \cdot n_{e,K},$$
$$f_{visc;e,K}(v,w) = \alpha_e\,(N_e(w) - N_e(v)),$$

where N_e determines the dissipativity of the scheme. The following conditions correspond, respectively, to consistency (of $f_{e,K}$) and conservativity

$$a_{e,K} + b_{e,K} = 1, \qquad a_{e,K_e} = b_{e,K}\ \&\ b_{e,K_e} = a_{e,K}. \tag{5}$$

The error analysis of the terms brought in by not having a highly regular grid can be done separately on the centered part, $f_{cent;e,K}$, and viscous part, $f_{visc;e,K}$, of the numerical flux. The "centered terms" are found to be zero if and only if

$$\delta_e \equiv x_e - x_K - b_{e,K_e}(x_{K_e} - x_K) = 0, \quad \text{for any } e.$$

Note that if, say, $b_{e,K_e} = 1/2$, then δ_e measures the difference between x_e and the average $(x_K + x_{K_e})/2$. The "viscous terms" can be analyzed in the same way, and are found to be zero if and only if the barycenter x_K is equal to a specific convex combination of the barycenters of the neighboring finite volumes

$$x_K = \frac{\sum_{e \in \partial K} |e|\,\alpha_e\, N'_e(c)\, x_{K_e}}{\sum_{e \in \partial K} |e|\,\alpha_e\, N'_e(c)}.$$

On a general grid, those conditions are not satisfied. A simplicial mesh with all the bad properties (i.e., "large" δ_e and x_K not in the convex hull of the barycenters of the neighboring cells) is displayed in Figure 2.

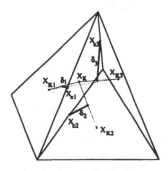

Figure 2. Example of a bad mesh.

The effect of the irregularities of the grid are then naturally measured by how far away one is from satisfying the above two conditions. More precisely, one has

$$
\| u(t^N) - v(t^N) \|_{L^1(\mathbf{R}^d)} \le 2 \| u_0 - v_0 \|_{L^1(\mathbf{R}^d)} + C_1 \Delta x^{1/2}
$$
$$
+ \frac{C_2}{\Delta x^{1/2}} \big\{ \| \delta \|_{\ell^\infty(\mathcal{E}_{\Delta x})/\mathbf{R}^d} + \| \alpha \|_{\ell^\infty(\mathcal{E}_{\Delta x})} \big\} + \text{h.o.t}, \qquad (6)
$$

where

$$
\| \delta \|_{\ell^\infty(\mathcal{E}_{\Delta x})/\mathbf{R}^d} = \inf_{\hat{\delta} \in \mathbf{R}^d} \sup_{e \in \mathcal{E}_{\Delta x}} \max_{1 \le j \le d} | (\delta_e - \hat{\delta})_j |
$$
$$
\| \alpha \|_{\ell^\infty(\mathcal{E}_{\Delta x})} = \sup_{e \in \mathcal{E}_{\Delta x}} | \alpha_e | \max_{1 \le j \le d} | (x_{K_e} - x_K)_j |,
$$

and where $\mathcal{E}_{\Delta x}$ stands for the set of all the edges. We refer the reader to [3, Theorem 2.3] for a demonstration as well as the explicit description of the constants C_1 and C_2. We now analyze this result and conclude by a series of remarks.

1. The constant C_2 is found to depend to the total variation of the numerical solution u. If we assume this total variation to be bounded, the previous result shows that even though the method is formally inconsistent, not only is it still convergent (supraconvergence), but the optimal order of convergence $\mathcal{O}(\Delta x^{1/2})$ is preserved. Indeed, on any grid, both $\| \delta \|_{\ell^\infty(\mathcal{E}_{\Delta x})/\mathbf{R}^d}$ and $\| \alpha \|_{\ell^\infty(\mathcal{E}_{\Delta x})}$ are at least of order Δx.

2. Technically, what makes the supraconvergence result possible is that the above quantity δ_e is entirely edge based. Indeed, thanks to consistency (of $f_{e,K}$) and conservativity (5), K and K_e can be interchanged in δ_e. Therefore, *consistency (of the numerical flux) and conservativity (a global property) are enough to overcome apparent losses of consistency of the method (a local property).*

3. A similar analysis can be done independently of any assumption on the approximate solution u, [3, Theorem 2.1]. However, in that case, the estimates involve relative variations of the coefficient δ_e (and a corresponding term for α), rather than merely their values. Therefore, unless those relative variations are zero (consistent schemes, possible only on Cartesian grids or affine transformations of those) or go to zero faster than (an adequate power of) Δx (*p*-consistent schemes [3]), the order of convergence appears to be reduced (for *p*-consistent schemes), or lost (for inconsistent schemes).

4. In conclusion, the use of highly irregular grids does not influence the asymptotic rate of convergence, provided that the numerical flux is consistent and the method conservative. However, and evidently for a given finite Δx, changing the regularity of the mesh does change the size of the error. An indication of how much this affects the accuracy is given by the estimate (6). More precise estimates would require a local analysis.

5. Finally, the ideas behind the proofs are strongly related to the seminal works of Kružkov [7] and Kuznetsov [8]. There is, however, one major difference. In [8], the approach is essentially *a posteriori*, thus the need to establish some regularity properties of the approximate solution. In [1-3], the dual point of view is taken, leading to an essentially *a priori* approach, where the regularity of the numerical solution does not play any explicit role.

References

[1] B. COCKBURN AND P.-A. GREMAUD, A priori error estimates for numerical methods for scalar conservation laws. Part I: The general approach, *Math. Comp.* 65 (1996), 533–573.

[2] B. COCKBURN AND P.-A. GREMAUD, A priori error estimates for numerical methods for scalar conservation laws. Part II: Flux-splitting monotone schemes on irregular Cartesian grids *Math. Comp.* 66 (1997), 547–572.

[3] B. COCKBURN, P.-A. GREMAUD AND XIANGRONG YANG, A priori error estimates for numerical methods for scalar conservation laws. Part III: Multi-dimensional flux-splitting monotone schemes on non-Cartesian grids, *SIAM J. Numer. Anal*, submitted.

[4] B. GARCÍA-ARCHILLA, A supraconvergent scheme for the Korteweg-de-Vries equation, *Numer. Math.* 61 (1992), 291–310.

[5] T. Y. HOU AND P. G. LEFLOCH, Why nonconservative schemes converge to wrong solutions: error analysis., *Math. Comp.* 62 (1994), 497–530.

[6] H.-O. KREISS, T.A. MANTEUFFEL, B. SWARTZ, B. WENDROFF, AND A.B. WHITE, JR., Supra-convergent schemes on irregular grids, *Math. Comp.* 47 (1986), 537–554.

[7] S.N. KRUŽKOV, First order quasilinear equations in several independent variables, *Math. USSR Sbornik* 10 (1970), 217–243.

[8] N.N. KUZNETSOV, Accuracy of some approximate methods for computing the weak solutions of a first-order quasi-linear equation, *USSR Comp. Math. and Math. Phys.* 16 (1976), 105–119.

[9] T.A. MANTEUFFEL AND A.B. WHITE, JR., The numerical solution of second-order boundary value problems on nonuniform meshes, *Math. Comp.* 47 (1986), 511–535.

[10] R. SANDERS, On Convergence of monotone finite difference schemes with variable spatial differencing, *Math. Comp.* 40 (1983), 91–106.

[11] B. WENDROFF, Supraconvergence in two dimensions, *Los Alamos National Laboratory report LA-UR-95-3068*, 1995.

Bernardo Cockburn
School of Mathematics
University of Minnesota
127 Vincent Hall, Minneapolis
MN 55455, USA

Pierre-Alain Gremaud
Center for Research in Scientific Computation and
Department of Mathematics
North Carolina State University
Raleigh, NC 27695-8205, USA

Xiangrong Yang
School of Mathematics
University of Minnesota
127 Vincent Hall, Minneapolis
MN 55455, USA

J.I. DÍAZ AND E. SCHIAVI

On a Degenerate System in Glaciology Giving Rise to a Free Boundary

1 Introduction

The mechanism whereby large ice sheets can surge periodically was recently studied by Fowler and Johnson ([4]) and Fowler ([3]). They proposed a two-dimensional ice sheet simplified model that includes basal ice sliding dependent on the basal water pressure and which consists in the following system:

$$h_t - \left[(\delta + Q)^S h^{R+1}|h_x|^{R-1} h_x\right]_x = a(t, x)$$

$$Q_x - (\delta + Q)^S \left[h^{R+1}|h_x|^{R+1} - \mu h^R |h_x|^R \xi^{-1/2}\right] - \gamma + \lambda h^{-1} \geq 0, \quad Q \geq 0$$

$$\left(Q_x - (\delta + Q)^S \left[h^{R+1}|h_x|^{R+1} - \mu h^R |h_x|^R \xi^{-1/2}\right] - \gamma + \lambda h^{-1}\right) Q = 0$$

$$\xi_x = (\delta + Q)^S h^R |h_x|^R$$

over $\Omega \times (0, T)$ with $T > 0$, $\Omega = (0, 1)$ (the scaled spatial domain), where R and S are some positive numbers satisfying $R > 1$ and $0 < S < 1$. The unknown variables are *the ice depth* $h(t, x)$, *the accumulated ice velocity* $\xi(t, x)$, and *the basal water flow* $Q(t, x)$. The *complementary formulation* for Q has been introduced in order to deal with the *cold* $(Q = 0)$ and *temperate* $(Q > 0)$ transition at the base. The points of $\Omega \times (0, T)$ separating those two zones are the free boundaries of the problem. Prescribing suitable boundary conditions for h, ξ, Q and an initial condition for h, Fowler and Schiavi ([5]) solved numerically the system by using a fully implicit backward finite difference scheme for h and an improved Euler method for Q and ξ. Numerical computations indicated a series of *surges* and showed that a front propagates backward during a surge. Moreover, their analysis suggested that the problem, as formulated, does not have smooth solutions. The main goal of this work is to present the mathematical analysis of the implicit backward scheme system showing the existence of a weak solution for the discretized system. Actually, this time discretized solution corresponds to the notion of *mild solution* of the evolution system as used in semigroup theory (see, e.g., Benilan ([1])).

2 The implicit discretized system

We define $p = R+1$, $m = (2R+1)/R$. We start by considering the following initial and boundary value problem: given h_0, h_D, ξ_D, Q_D and a strictly positive *accumulation rate* function $a(t, x)$, find three functions, h, Q and ξ satisfying

$$(S) \begin{cases} \partial_t h - \left[(\delta + Q)^S | (h^m)_x |^{p-2} (h^m)_x \right]_x = a \quad \text{in } \Omega \times (0, T), \\[2mm] \partial_x Q + \beta(Q) \ni (\delta + Q)^S h^p |h_x|^p - \mu \xi_x \xi^{-1/2} + \gamma - \lambda h^{-1} \quad \text{in } \Omega \times (0, T), \\[2mm] \partial_x \xi = (\delta + Q)^S h^{p-1} |h_x|^{p-1} \quad \text{in } \Omega \times (0, T), \\[2mm] h(t, 1) = h_D(t, 1), \quad t \in (0, T), \\[2mm] h_x(t, 0) = 0, Q(t, 0) = Q_D(t), \xi(t, 0) = \xi_D(t) \quad t \in (0, T), \\[2mm] h(0, x) = h_0(x) \quad \text{on } \Omega. \end{cases}$$

Here β denotes the maximal monotone graph defined by

$$\beta(r) = \emptyset \quad \text{if} \quad r < 0, \quad \beta(0) = (-\infty, 0], \quad \beta(r) = 0 \quad \text{if} \quad r > 0.$$

The coefficients γ, μ, λ are $O(1)$ dimensionless parameters. The positive constant δ, $0 < \delta \ll 1$ represents the ice shearing component in the flow when $Q = 0$. Given a positive integer number N and letting $k = T/N$, the time step of the discretization, we denote by $I_n = I_{n,k} = (t_{n-1}, t_n) = ((n-1)k, nk)$, $(n = 1, ...N, t_n = nk)$ the associated sub-intervals of $(0, T)$. Let $V \doteq V_{hm} \times V_\xi \times V_Q$ the Banach space defined by $V_{hm} \doteq \{\phi \in W^{1,p}(\Omega) : \phi(1) = 0\}$, $V_\xi \doteq \{\psi \in W^{1,p'}(\Omega) : \psi(0) = 0\}$, $V_Q \doteq \{\eta \in W^{1,1}(\Omega) : \eta(0) = 0\}$. We shall assume the following hypothesis on the data of the problem:

$$Q_D, \xi_D, h_D \in C[0, T], h_0 \in C[0, 1], h_D(0) = h_0(1) \tag{1}$$

$$\left. \begin{array}{l} M_D > h_D > m_D > 0, M_0 > h_0 > m_0 > 0, a > 0, Q_D \geq 0 \text{ and } \xi_D > 0, \\ \text{for some constants } M_D > m_D > 0 \text{ and } M_0 > m_0 > 0. \end{array} \right\} \tag{2}$$

It is useful to introduce the following notation:

$$A = (\delta + Q)^S, \ B = h^p |h_x|^p, \ C = \mu \xi_x \xi^{-1/2}, \ D = \lambda h^{-1}, \ E = h^{p-1} |h_x|^{p-1}. \tag{3}$$

After defining the piecewise constant in time approximations of the data in the usual manner, we consider the elliptic discretized system

310

$$(S_{k,n}) \begin{cases} \partial_t^{-k} h_{k,n} - \left[A_{k,n} |(h_k^m)_x|^{p-2} (h_k^m)_x \right]_x = a_{k,n} & \text{in } \Omega, \\[2mm] \partial_x Q_{k,n} + \beta(Q_{k,n}) \ni (\delta + Q_{k,n})^S B_{k,n} - C_{k,n} + \gamma - D_{k,n} & \text{in } \Omega, \\[2mm] \partial_x \xi_{k,n} = A_{k,n} E_{k,n} & \text{in } \Omega, \\[2mm] h_{k,n}(t,1) = h_{D_{k,n}}(t,1) & t \in (0,T), \\[2mm] (h_{k,n})_x(t,0) = 0, Q_{k,n}(t,0) = Q_{D_{k,n}}(t), \xi_{k,n}(t,0) = \xi_{D_{k,n}}(t) & t \in (0,T), \\[2mm] h_{k,0}(x) = h_0(x) & \text{on } \Omega, \end{cases}$$

where

$$\partial_t^{-k} h_{k,n}(t,.) \doteq \frac{h_{k,n}(.) - h_{k,n-1}(.)}{k} \quad \forall n = 1,..N$$

and $A_{k,n}$, $B_{k,n}$, $C_{k,n}$, $D_{k,n}$ and $E_{k,n}$ are defined as in (3) replacing h, ξ and Q by $h_{k,n}$, $\xi_{k,n}$ and $Q_{k,n}$.

Definition 2.1 *Given a, h_D, Q_D, ξ_D, h_0 satisfying hypothesis (1), (2) and $a_{k,n}$, $h_{D_{k,n}}$, $Q_{D_{k,n}}$, $\xi_{D_{k,n}}$ the associated discretized functions, we say that $(h_{k,n}^m, \xi_{k,n}, Q_{k,n})$ is a weak solution of $(S_{k,n})$ if*

$$(h_{k,n}^m(t,.), \xi_{k,n}(t,.), Q_{k,n}(t,.)) \in [h_D^m + V_{h^m}] \times [\xi_D + V_\xi] \times [Q_D + V_Q], \quad a.e. \ t \in (0,T),$$

there exists $b_{k,n} \in L^1(\Omega)$, with $b_{k,n}(x) \in \beta(Q_{k,n}(x))$ a.e. $x \in (0,1)$, and the following conditions hold:

$$\int_0^1 \partial_t^{-k} h_{k,n}^m(t)\phi + \int_0^1 (\delta + Q_{k,n})^S |(h_{k,n}^m)_x|^{p-2}(h_{k,n}^m)_x \phi_x = \int_0^1 a\,\phi$$

$$\int_0^1 \xi_{k,n}\psi_x + \int_0^1 (\delta + Q_{k,n})^S h_{k,n}^{p-1} |(h_{k,n})_x|^{p-1} \psi = \xi_{k,n}(1,t)\psi(1,t)$$

$$\int_0^1 Q_{k,n}\eta_x + \int_0^1 \left[(\delta + Q_{k,n})^S h_{k,n}^p |(h_{k,n})_x|^p + \gamma \right] \eta = Q_{k,n}(1,t)\eta(1,t) +$$

$$+ \mu \int_0^1 (\xi_{k,n})_x (\xi_{k,n})^{-1/2}\eta + \lambda \int_0^1 h_{k,n}^{-1}\eta + \int_0^1 b_{k,n}\eta$$

for all test functions ϕ, ψ, $\eta \in V_h \times V_\xi \times V_Q$.

3 Existence of weak solutions via an iterative scheme

In order to prove the existence of a weak solution of $(S_{k,n})$, we shall use an iterative process which allows decoupling the system into three separate problems: $P(h^j_{k,n})$, $P(\xi^j_{k,n})$ and $P(Q^j_{k,n})$. Later we shall obtain *a priori* estimates which allow us to prove the convergence of such iterative schemes. The decoupled problem is the following: For each j we shall find three functions $(h^j_{k,n})^m$, $(\xi^j_{k,n})$ and $(Q^j_{k,n})$ satisfying

$$(S^j_{k,n}) \begin{cases} \partial_t^{-k} h^j_{k,n} - \left[A^{j-1}_{k,n} |\partial_x[(h^j_{k,n})^m]|^{p-2} \partial_x[(h^j_{k,n})^m] \right]_x = a_{k,n} & \text{on } \Omega, \\[2mm] \partial_x \xi^j_{k,n} = A^{j-1}_{k,n} E^j_{k,n} & \text{in } \Omega, \\[2mm] \partial_x Q^j_{k,n} + \beta(Q^j_{k,n}) \ni (\delta + Q^j_{k,n})^S B^j_{k,n} - C^j_{k,n} + \gamma - D^j_{k,n} & \text{in } \Omega, \\[2mm] h^j_{k,n}(t,1) = h_{D_{k,n}}(t,1) & t \in (0,T), \\[2mm] (h^j_{k,n})_x(t,0) = 0, Q^j_{k,n}(t,0) = Q_{D_{k,n}}(t), \xi^j_{k,n}(t,0) = \xi_{D_{k,n}}(t) & t \in (0,T), \\[2mm] h_{k,0}(x) = h_0(x) & \text{in } \Omega. \end{cases}$$

In order to study this system we study separately three problems:
First step: Problem $P(h^j_{k,1})$. We introduce the change of unknown $w \doteq (h^j_{k,n})^m$. By defining $w_D \doteq (h_{D_{k,n}})^m$, $\hat{A} \doteq k A^{j-1}_{k,n}$ and $f \doteq k a_{k,n} + h_{k,n-1}$, for each j-step of the iterative process function, w must satisfy

$$P(h^j_{k,n}) \begin{cases} -\partial_x \left(\hat{A} |w_x|^{p-2} w_x \right) + w^{1/m} = f & \text{in } \Omega, \\[2mm] w_x(0) = 0, w(1) = w_D. \end{cases}$$

Let $V_w := V_{h^m}$. As usual, given $\hat{A} \in L^\infty(\Omega)$, $\hat{A} > 0$ and $f \in L^\infty(\Omega)$, we say that w is a *bounded weak solution* of $P(h^j_{k,n})$ if $w \in w_D + V_w$ and it satisfies

$$-\int_\Omega \hat{A} |w_x|^{p-2} w_x \, \phi_x + \int_\Omega w^{1/m} \phi = \int_\Omega f \phi, \qquad \forall \phi \in V_w.$$

The existence of a unique approximate solution $w = (h^j_{k,n})^m \in w_D + V_w$ is a well-known result in the literature (see, e.g., the exposition made in Díaz ([2])). Notice that, in particular, $B^j_{k,n} \in L^1(\Omega)$, $E^j_{k,n} \in L^{p'}(\Omega)$. We can get some *a priori* estimates on w (i.e., on $(h^j_{k,n})^m$). First, by using the comparison principle and suitable super- and subsolutions built thanks to the assumptions on the data, we obtain

Lemma 3.1 *Let $h^j_{k,n}$ be a weak bounded solution of problem $P(h^j_{k,n})$. Then there exist two real positive numbers m^*, M^* (depending only on the data of the problem)*

such that

$$0 < m^* \leq h_{k,n}^j(t,x) \leq M^* < +\infty \qquad on \quad \Omega.$$

In particular, $D_{k,n}^j > 0$, a.e. $x \in \Omega$, $D_{k,n}^j \in L^\infty(\Omega)$ and $D_{k,n}^j$ is uniformly bounded in $L^\infty(\Omega)$ with respect to j.

A second *a priori* estimate can be obtained by using an energy method.

Lemma 3.2 *Function $(h_{k,n}^1)^m - (h_{D_{k,n}})^m$ is uniformly bounded in the energy space V_w. In particular, $B_{k,n}^j$ and $E_{k,n}^j$ are uniformly bounded in their respective spaces.*

Second step: Problem $P(\xi_{k,n}^j)$. Let $h_{k,n}^j$ be the weak solution of problem $P(h_{k,n}^j)$. Then $A_{k,n}^{j-1} \in L^\infty(\Omega)$, $E_{k,n}^j \in L^{p'}(\Omega)$ and $A_{k,n}^{j-1}E_{k,n}^j \doteq (\delta + Q_{k,r}^{j-1})^S (h_{k,r}^j)^{p-1} |(h_{k,r}^j)_x|^{p-1} \in L^{p'}(\Omega)$. We consider the problem

$$P(\xi_{k,n}^j) \begin{cases} (\xi_{k,n}^j)_x & = A_{k,n}^{j-1}E_{k,n}^j \quad in\ \Omega, \\ \xi_{k,n}^j(0) & = \xi_{D_{k,n}}. \end{cases}$$

Definition 3.1 *We shall say that $\xi_{k,n}^j \in W^{1,p'}(\Omega)$ is a weak solution of problem $P(\xi_{k,n}^j)$ if $\xi_{k,n}^j - \xi_{D_{k,n}} \in V_\xi$ and*

$$\int_0^1 \xi_{k,n}^j \psi_x + \int_0^1 A_{k,n}^{j-1}E_{k,n}^j \psi = \xi_{D_{k,n}}(1)\psi(1), \qquad \forall\ \psi \in V_\xi.$$

It is straightforward to show the existence of a unique function $\xi_{k,n}^j$ weak solution of problem $P(\xi_{k,n}^j)$. Since $(\xi_{k,n}^j)_x(x) \geq 0$, a.e. $x \in \Omega$, a direct integration leads to the following result:

Lemma 3.3 *$\xi_{k,n}^j(x) > 0$, $\forall x \in \Omega$. In particular, $C_{k,n}^j \geq 0$, a.e. $x \in \Omega$.*

Third step: Problem $P(Q_{k,n}^j)$. Let $h_{k,n}^j$ and $\xi_{k,n}^j$ be the weak solutions of $P(h_{k,n}^j)$ and $P(\xi_{k,n}^j)$, respectively. We consider $B_{k,n}^j$, $C_{k,n}^j$ y $D_{k,n}^j$ defined by (3). Notice that $D_{k,n}^j \in C([0,1])$, $C_{k,n}^j \in L^{p'}(\Omega)$, but $B_{k,n}^j$ is, in general, merely in $L^1(\Omega)$. We introduce the problem

$$P(Q_{k,n}^j) \begin{cases} \partial_x Q_{k,n}^j + \beta(Q_{k,n}^j) & \ni\ (\delta + Q_{k,n}^j)^S B_{k,n}^j + \gamma - C_{k,n}^j - D_{k,n}^j \quad in\ \Omega, \\ Q_{k,n}^j(0) & = Q_{D_{k,n}}. \end{cases}$$

Definition 3.2 *We shall say that $Q_{k,n}^j \in Q_{D_{k,n}} + V_Q \subset W^{1,1}(\Omega)$ is a weak solution if there exist a function $z \in L^1(\Omega)$ such that $z(x) \in \beta(Q_{k,n}^j(x))$, a.e. $x \in \Omega$ and*

$$
\int_0^1 Q_{k,n}^j \eta_x + \int_0^1 (\delta + Q_{k,n}^j)^S B_{k,n}^j \eta + \gamma \int_0^1 \eta =
$$
$$
= \int_0^1 C_{k,n}^j \eta + \int_0^1 D_{k,n}^j \eta + \int_0^1 z\eta + Q_{k,n}^j(1)\eta \tag{4}
$$

for each $\eta \in V_Q$.

We have

Theorem 3.1 *There exists a unique weak solution of $P(Q_{k,n}^j)$.*

Proof. We approximate the maximal monotone graph β and function $b(Q) \doteq (\delta + Q)^S$ by some sequences of Lipschitz functions generating some approximating regularized problems of solutions $(Q_{k,n}^j)_\epsilon \in W^{1,1}(\Omega)$ (the existence of solutions of such problems is consequence of a Banach fixed point argument). Moreover, we get that $\|(Q_{k,n}^j)_\epsilon\|_{L^\infty(\Omega)} \leq C$ and $\|(\partial_x Q_{k,n}^j)_\epsilon\|_{L^1(\Omega)} \leq C$. Passing to the limit in the weak formulation of the regularizing problems it is possible to show that $(Q_{k,n}^j)_\epsilon \to Q_{k,n}^j$ as $\epsilon \to \infty$ strongly in $L^2(\Omega)$ with $Q_{k,n}^j$ solution of $P(Q_{k,n}^j)$.

4 Convergence

We already obtained the *a priori* estimates

$$\|(h_{k,n}^j)_x\|_{L^p(\Omega)} \leq C, \quad \|(\xi_{k,n}^j)_x\|_{L^{p'}(\Omega)} \leq C, \quad \|Q_{k,n}^j\|_{L^\infty(\Omega)} \leq C, \quad \|(Q_{k,n}^j)_x\|_{L^1(\Omega)} \leq C;$$

uniformly in j, $\forall k, n$ fixed and $\forall t \in I_{k,n}$. So

$$\|A_{k,n}^{j-1}\|_{L^\infty(\Omega)} \leq C, \qquad \|B_{k,n}^j\|_{L^1(\Omega)} \leq C, \qquad \|C_{k,n}^j\|_{L^{p'}(\Omega)} \leq C,$$

$$\|D_{k,n}^j\|_{L^\infty(\Omega)} \leq C, \qquad \|E_{k,n}^j\|_{L^{p'}(\Omega)} \leq C.$$

By applying Poincaré inequality, and Sobolev and Lebesgue theorems we get the following result:

Lemma 4.1 *Let $\{(h_{k,n}^j)^m\}$, $\{\xi_{k,n}^j\}$ and $\{Q_{k,n}^j\}$ be the sequences of solutions of problems $P(h_{k,n}^j)$, $P(\xi_{k,n}^j)$ and $P(Q_{k,n}^j)$, respectively. Then, $\forall k, n$ fixed and $\forall t \in I_{k,n}$ we have*

$$h_{k,n}^j \to h_{k,n}, \quad \xi_{k,n}^j \to \xi_{k,n}, \quad Q_{k,n}^j \to Q_{k,n} \text{ strongly in } L^q(\Omega), \ \forall q \geq 1, \text{ , when } j \to \infty.$$

Moreover, $h_{k,n}^j \to h_{k,n}$, $\xi_{k,n}^j \to \xi_{k,n}$ strongly in $C^0([0,1])$.

314

An analysis of system $(S_{k,n}^j)$ reveals that the difficult term, in order to pass to the limit in the weak formulation, is the product $(\delta + Q^j)^S |h_z^j|^p$ representing the frictional heating due to viscous dissipation. Nevertheless, we have

Lemma 4.2 $(\delta + Q^{j-1})^S |h_z^j|^p \to (\delta + Q)^S |h_z|^p$, strongly in $L^1(\Omega)$, when $j \to \infty$.

Proof. We consider $w^j = (h^j)^m$ with h^j solution of problem $P(h_{k,n}^j)$. Without loss of generality we can suppose that $w_D \equiv 0$. Multiplying by $(h^j)^m$ and integrating by parts, we have

$$\int_\Omega (\delta + Q^{j-1})^S |w_x^j|^p dx = -\frac{1}{k} \int_\Omega (w^j)^{\frac{1}{m}+1} + \int_\Omega f^j w^j \, dx$$

with $f^j = f_{k,n}^j = a_{k,n} + \frac{1}{k}(w_{k,n-1}^j)^{1/m} \in L^\infty(\Omega)$, $\|f^j\|_{L^\infty(\Omega)} \leq C$, uniformly in j. Using Lebesgue theorem

$$\int_\Omega (w^j)^{\frac{1}{m}+1} dx \quad \to \quad \int_\Omega w^{\frac{1}{m}+1} dx, \text{ when } j \to \infty,$$

and we deduce that

$$\int_\Omega f^j w^j \, dx \quad \to \quad \int_\Omega f w \, dx, \text{ when } j \to \infty \tag{5}$$

(remember that $f^j \to f$, $w^j \to w$ strongly in $L^2(\Omega)$). Then we deduce that

$$\int_\Omega (\delta + Q^{j-1})^S |w_x^j|^p dx \quad \to \quad -\frac{1}{k} \int_\Omega w^{\frac{1}{m}+1} dx + \int_\Omega f w \, dx, \text{ when } j \to \infty.$$

But multiplying in $P(h_{k,n}^j)$ by w and integrating in Ω, we get

$$\int_\Omega (\delta + Q)^S |w_x|^p \, dx = -\frac{1}{k} \int_\Omega w^{\frac{1}{m}+1} + \int_\Omega f w \, dx.$$

Hence

$$\int_\Omega (\delta + Q^{j-1})^S |w_x^j|^p dx \to \int_\Omega (\delta + Q)^S |w_x|^p \, dx.$$

As a consequence, we get the strong convergence of $(h_j^m)_x$ to $(h^m)_x$ in $L^p(\Omega)$.

Lemma 4.3 We have that $w_x^j \to w_x$ strongly in $L^p(\Omega)$ and

$$(\delta + Q^j)^S |h_z^j|^p \to (\delta + Q)^S |h_z|^p$$

strongly in $L^1(\Omega)$.

Proof. Subtracting the equations verified by w_j and w, and multiplying by $w_j - w$ we get

$$I_j \doteq \int_\Omega (\delta + Q^{j-1})^S \left[|w_x^j|^{p-2} w_x^j - |w_x|^{p-2} w_x \right] (w_x^j - w_x)\, dx$$

$$= \int_\Omega \left[(\delta + Q)^S - (\delta + Q^{j-1})^S \right] \left(|w_x|^{p-2} w_x \right) (w_x^j - w_x)\, dx -$$

$$- \frac{1}{k} \int_\Omega \left((w^j)^{\frac{1}{m}} - w^{\frac{1}{m}} \right) (w^j - w)\, dx + \int_0^1 (f^j - f)(w^j - w).$$

By the previous lemma and (5) we deduce that $I_j \to 0$ if $j \to \infty$. Finally, as $p > 2$ (remember that $p = R + 1$ and $R > 1$), Q^j are uniformly bounded in $L^\infty(\Omega)$. Moreover, it is well-known (see, e.g., Díaz ([2], Lemma 4.10)) that there exists $C > 0$ (independently of j) such that

$$C \int_\Omega |w_x^j - w_x|^p \le \int_\Omega (\delta + Q^{j-1})^S \left[|w_x^j|^{p-2} w_x^j - |w_x|^{p-2} w_x \right] (w^j - w)\, dx.$$

So, $w_x^j \to w_x$, strongly in $L^p(\Omega)$. Moreover, since $\{(\delta + Q^j)^S\}$ is uniformly bounded in $L^\infty(\Omega)$ and $|h_x^j|^p \to |h_x|^p$ strongly in $L^1(\Omega)$, we obtain the second conclusion. Thus, we have proved

Theorem 4.1 *The sequence* $(h_{k,n}^j, \xi_{k,n}^j, Q_{k,n}^j)$ *converges, when* $j \to \infty$, *to* $(h_{k,n}, \xi_{k,n}, Q_{k,n})$ *solution of* $(S_{k,n})$.

References

[1] BENILAN, PH., *Equations d'evolution dans un espace de Banach quelconque et applications*, Thesis, Univ. Paris XI, Orsay, 1972.

[2] DÍAZ, J.I., *Nonlinear Partial Differential Equations and Free Boundaries*. Ed. Pitman, London, 1985.

[3] FOWLER, A.C., Glaciers and ice sheets, in *The Mathematics of Models for Climatology and Environment*, J.I. Díaz ed., NATO ASI Series I 48, Springer, Berlin, 1997, pp.301-336.

[4] FOWLER; A. AND JOHNSON, C., Ice-sheet surging and ice-stream formation, *Ann. Glaciol.*, **23**, 68-73.

[5] FOWLER, A.C. AND SCHIAVI, E., A theory of ice sheet surges. To appear in *Journal of Glaciology*.

J.I. Díaz
Universidad Complutense de Madrid
Facultad C.C. Matemáticas
Departamento Matemática Aplicada
Avenida Complutense s/n, 20040
Spain

E. Schiavi
Universidad Autonoma de Madrid
Facultad C.C. Económicas
Departamento análisis económico: economía cuantitativa
Cantoblanco, Madrid
Spain

G. C. GEORGIOU AND A. G. BOUDOUVIS

Singular Finite Element Solutions of the Axisymmetric Extrudate-Swell Problem

We solve the axisymmetric, creeping Newtonian extrudate-swell problem for the case of zero surface tension. Both the standard and the singular finite element methods are used and the convergence of the numerical solutions with mesh refinement is studied. The numerical results show that the singular finite elements accelerate the convergence of the free surface considerably; they perform well when coarse or moderately refined meshes are used.

1   Introduction

In this work, we revisit the singular finite element method (SFEM) developed by Georgiou et al. for solving Newtonian flow problems with boundary stress singularities [1, 2]. In the SFEM, special elements incorporating the radial form of the local solution by means of singular basis functions are employed in a small region around the singularity, while standard elements are used in the rest of the domain. The idea of incorporating the form of the local singularity solution into the numerical scheme was borrowed from analogous methods used in fracture mechanics (see, e.g., [1] and references therein). The basic motive behind using singular methods is to improve the accuracy and the rate of convergence of the solution with mesh refinement, which are rather unsatisfactory with standard numerical methods, especially in the neighborhood of the singularity. The poor performance of the standard FEM is attributed to the fact that the calculated pressure and stresses cannot be infinite at the singular point, as required by the local asymptotic solution, and are thus tainted by spurious oscillations. This difficulty is overcome with the SFEM.

Georgiou et al. applied the SFEM to the planar Newtonian extrudate-swell problem which describes the extrusion of a viscous fluid through a die into an inviscid medium [2]. This is a well-known free surface problem; at low Reynolds numbers, the fluid swells as it comes out of the die. Another important characteristic of this flow is the presence of a stress singularity at the exit of the die, resulting from the sudden change in the boundary condition from the wall of the die to the free surface of the extrudate. The extrudate-swell problem is extremely important in polymer processing and has thus been the focus of a plethora of experimental and numerical studies in the last twenty-five years [3].

The singular finite element calculations for the planar Newtonian extrudate-swell problem have revealed that the spurious stress oscillations that characterize the stresses in the standard finite element solution are eliminated [2]. Similar observations have been made when solving the planar Newtonian stick-slip and 2:1 expansion problems [1, 2]. The former problem is the special case of the extrudate-swell problem in the

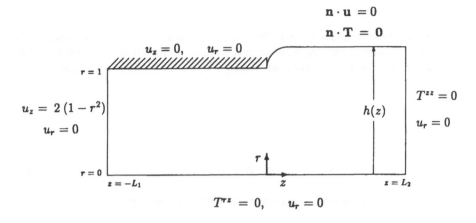

Figure 1: Geometry and boundary conditions for the extrudate-swell problem.

limit of infinite surface tension, in which the free surface becomes completely flat. In the case of the planar extrudate-swell problem, the convergence of the free surface profile with mesh refinement is considerably accelerated by using the singular finite elements [2].

The two main drawbacks of the SFEM have also been addressed in Ref. [1]. First, extensive mesh refinement is not possible with the SFEM. As the mesh is refined, the singular elements become smaller in size, and, consequently, the size of the region over which the singularity is given special attention is reduced. Second, the method can be implemented only if the radial form of the local solution is known, at least approximately. This implies that the method is not applicable to many important problems such as most viscoelastic flow problems in which the inaccuracies, stemming from the failure to approximate satisfactorily the stress behavior near the singularity, are, in general, more severe.

In this paper, we solve the round Newtonian extrudate-swell problem at zero Reynolds number (creeping flow) and zero surface tension, using both the standard and the singular finite element methods. We systematically study the convergence of the numerical solutions with mesh refinement. Our objective is to compare the performance of the two methods and to obtain accurate estimates of the position of the free surface and the extrudate-swell ratio. These results can be quite useful in testing other numerical methods proposed in the literature.

2   Governing Equations

The flow geometry and the dimensionless governing equations and boundary conditions for the steady-state axisymmetric extrudate-swell problem are depicted in Figure 1. The scaling parameter for lengths is the radius R, the velocity vector $\mathbf{u}$ is

319

scaled by the mean velocity U, and, finally, the pressure p and the stress tensor $\mathbf{T}$ are measured in units of $\eta U/R$, where η is the viscosity. For creeping, incompressible Newtonian flow with zero gravity, the continuity and momentum equations may be written as

$$\nabla \cdot \mathbf{u} = 0 \tag{1}$$

and

$$\nabla \cdot \mathbf{T} = \mathbf{0}. \tag{2}$$

The Newtonian stress tensor for incompressible fluid is given by

$$\mathbf{T} = -p\,\mathbf{I} + [\nabla \mathbf{u} + (\nabla \mathbf{u})^T], \tag{3}$$

where $\mathbf{I}$ is the unit tensor and the superscript T denotes the transpose of a tensor.

The boundary conditions are depicted in Figure 1. Along the axis of symmetry, the usual symmetry conditions apply. Along the wall, both velocity components u_z and u_r are zero (there is neither slip nor penetration). The kinematic condition for the free surface,

$$\mathbf{n} \cdot \mathbf{u} = 0, \tag{4}$$

where $\mathbf{n}$ is the unit normal vector pointing outward from the free surface, provides an additional equation needed for the calculation of the unknown position of the free surface h. For zero surface tension, the momentum balance on the free surface requires the tangential and normal stress components to vanish,

$$\mathbf{n} \cdot \mathbf{T} = \mathbf{0}. \tag{5}$$

The inflow and outflow boundaries are taken at finite distances L_1 and L_2 sufficiently far from the exit so that the flow can be considered fully developed at the inlet and fully translational at the outflow plane. Hence, the radial velocity is zero at both planes, whereas the axial velocity is parabolic at the inlet. The normal stress at the outlet is taken equal to zero.

2.1 Radial form of the singularity

A prerequisite for the construction of the singular finite elements is the knowledge of the radial form of the singularity. In the case of steady plane flow near a corner of angle α formed by a rigid boundary and a flat free surface (Figure 2), the local solution is obtained using separation of variables in polar coordinates (r, θ) centered

320

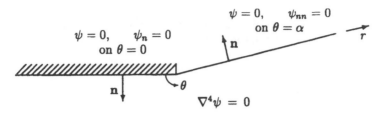

Figure 2: Local analysis of the singularity.

at the singular point [4, 5]. The streamfunction Ψ is expressed as an eigenfunction expansion:

$$\Psi(r, \theta) = \sum_{i=1}^{\infty} r^{\lambda_i + 1} f_i(\theta) , \qquad (6)$$

where the eigenvalues λ_i and the form of the functions $f_i(\theta)$ are determined by imposing the boundary conditions. Michael [4] showed that for vanishingly small surface tension on a planar free surface, the contact angle α must be equal to π, i.e., that the slope of the free surface is zero at the singular point. Trogdon and Joseph solved the stick-slip problem with the Wiener-Hopf method and used a singular perturbation analysis about $Ca=0$ to obtain the shape of the free surface at high surface tension [6]. Their analysis predicts two possible values for the contact angle: one is $\alpha=\pi$ and the other is some arbitrary value, which is rejected since it leads to discontinuities in streamlines near the exit [7]. Schultz and Gervasio extended Michael's analysis to finite surface tension and showed that either the separation angle is zero or the mean curvature is infinite [8]. The latter result has been confirmed by the numerical calculations of Salamon et al. [7]; they used a finite element analysis combining quasi-orthogonal mesh generation with local adaptive mesh refinement. Sturges calculated the first two eigenvalues for various values of the angle between the wall and the free surface [9] (in the case of zero surface tension). For angles in the interval $[\pi, 3\pi/2]$, the first eigenvalue is real and decreases from $1/2$ to $1/3$. In the case of $\alpha=\pi$, i.e., for the stick-slip flow, the separated solutions consist of two sets corresponding to symmetrical and antisymmetrical flows:

$$\psi = r^{\lambda+1} a_\lambda [\cos (\lambda + 1)\theta - \cos (\lambda - 1)\theta] , \qquad \text{for } \lambda = \frac{1}{2}, \frac{3}{2}, \frac{5}{2} \cdots , \qquad (7)$$

and

$$\psi = r^{\lambda+1} b_\lambda [(\lambda - 1) \sin (\lambda + 1)\theta - (\lambda + 1) \sin (\lambda - 1)\theta] , \quad \text{for } \lambda = 2, 3, 4, \cdots . \qquad (8)$$

Solutions (12) and (13) are exact in the case of the planar stick-slip problem [1]. The leading term of the antisymmetrical set ($\lambda=1/2$) leads to the inverse square root singularity of the velocity derivatives and the pressure. We observe that, at a fixed angle θ, the velocity components vary as

$$u_x \ (\text{or } u_y) \sim A_1 r^{1/2} + A_2 r + A_3 r^{3/2} + A_4 r^2 + \dots , \tag{9}$$

and the pressure varies as

$$p \sim B_1 r^{-1/2} + B_2 + B_3 r^{1/2} + B_4 r + \dots , \tag{10}$$

where A_i and B_i are constants.

In contrast to the polynomial basis functions used with ordinary elements, the basis functions for u and p over the singular elements are constructed so that they embody the leading terms of equations (14) and (15), respectively, in the radial direction. Toward this end, it is very convenient to have triangular and not rectangular singular elements with one vertex at the singular point. From equation (15), it is also clear that the pressure is infinite at the singular point. This fact is taken care of by the singular basis functions; thus, no pressure node is placed at the singular point. In other words, the need to calculate an infinite quantity is eliminated. This is one of the most important features of the singular finite elements, resulting in the elimination of the Gibbs-type oscillations of the pressure. More details about the construction and the complete expressions of the basis functions for u and p may be found in [1], where the planar stick-slip problem is solved.

Based on Michael's result that the slope of the free surface is zero at the exit [4], Georgiou et al. employed the elements constructed for the stick-slip problem to solve the planar extrudate-swell problem [2], assuming that the radial form of the local solution is approximately the same. With this assumption, the need for finding the angle of separation and constructing appropriate singular basis functions is circumvented. In fact, it was shown that the numerical results are not sensitive to small variations of the singularity powers used for the planar extrudate-swell problem. This assumption is supported by the numerical results of Salamon et al. for finite capillary numbers, which reveal that the values of λ are in the range $1.50 \ll \lambda \ll 1.55$ and correspond to singular albeit integrable stresses [7]. Georgiou et al. obtained accurate oscillation-free results with rather coarse meshes, not only for the limiting case of zero Reynolds number and zero surface tension but also for small and moderate values of the Reynolds number and for a wide range of capillary numbers [2]. Using the SFEM for non-zero Reynolds number flows is justified by the fact that the local solution remains unchanged around the singular point where viscous effects dominate. Moreover, as surface tension increases the flow approaches the stick-slip limit, the local solution of which was used for designing the singular elements.

Although Michael's analysis [4] does not apply to the axisymmetric extrudate-swell problem, we will use the same elements for solving the axisymmetric extrudate-swell

Table 1: Data for the ordinary and singular meshes ($L_2=20$).

Ordinary elements			Singular elements		
Mesh	Degrees of freedom	Size of corner elements	Mesh	Degrees of freedom	Radius of corner elements
OM1	1506	0.2	SM1	1654	0.24
OM2	2807	0.1	SM2	2955	0.12
OM3	4660	0.05	SM3	4808	0.06
OM4	7528	0.025	SM4	7676	0.03
OM5	12642	0.01	SM5	12790	0.012
OM6	17076	0.005	SM6	17224	0.006
OM7	30866	0.0025	SM7	31034	0.003

problem. In fact, the analytical solution for the axisymmetric stick-slip problem obtained by Sturges reveals that the velocity components u_z and u_r follow equations (14) and (15), respectively [9]. Thus, it is reasonable to assume that the radial form of the singularity in the axisymmetric extrudate-swell problem is the same. We will verify in Section 4 that the accuracy and the rate of convergence achieved with the singular elements are quite satisfactory.

3   Finite Element Formulation

The flow domain Ω is discretized by means of eight singular elements around the singular point and standard rectangular elements elsewhere. Let Φ^i and Ψ^i denote the basis functions used for the finite element approximations of the velocity vector and the pressure, respectively. Over the ordinary elements, Φ^i are biquadratic (P^2-C^0) and Ψ^i are bilinear (P^1-C^0). (These are the most common approximations used for Newtonian flow.) The ordinary elements contain nine velocity and four pressure nodes and are mapped onto a rectangular element in (ξ,ζ) coordinates by means of biquadratic shape functions. The singular elements are collapsed quadrilaterals with 13 velocity nodes and 8 pressure nodes mapped onto a 15-node rectangular element in the computational domain (ξ,ζ) by means of fourth-order polynomial shape functions in the ξ and second-order in the ζ direction. In the coordinate ξ, which corresponds to the radial direction, Φ^i and Ψ^i are constructed so that they embody the behavior of the leading four terms of equations (14) and (15), respectively. In the coordinate ζ, Φ^i and Ψ^i are quadratic and bilinear, respectively, so that the singular elements are compatible with the adjacent ordinary elements. More details about the singular functions are given in [1, 2]. Finally, to approximate the unknown position of the free surface h, we use quadratic basis functions $\mathcal{X}^i$. Thus, for the approximations of $\mathbf{u}$, p

323

and h, we have

$$\mathbf{u} = \sum_j^{N_u} \mathbf{u}^j \, \Phi^j \,, \tag{11}$$

$$p = \sum_j^{N_p} p^j \, \Psi^j \,, \tag{12}$$

$$h = \sum_j^{N_h} h^j \, \mathcal{X}^j \,, \tag{13}$$

where $\mathbf{u}^j$, p^j and h^j are the values of the unknowns at the jth node; N_u, N_p and N_h are the numbers of velocity, pressure and free-surface nodes, respectively.

We use the standard Galerkin method to weight the momentum, the continuity and the kinematic equations. The discretized equations are as follows:

$$\int_\Omega \nabla \mathbf{T} \cdot \nabla \Phi^i \, dV = 0 \,, \qquad\qquad i = 1, 2, \cdots, N_u \,, \tag{14}$$

$$\int_\Omega \nabla \cdot \mathbf{u} \, \Psi^i \, dV = 0 \,, \qquad\qquad i = 1, 2, \cdots, N_p \,, \tag{15}$$

$$\int_{\partial \Omega_F} \mathbf{n} \cdot \mathbf{u} \, W^i \, dS = 0 \,, \qquad\qquad i = 1, 2, \cdots, N_h \,, \tag{16}$$

where $\partial \Omega_F$ is the free surface. The nonlinear system of equations (19)-(21) is solved using the Newton method and a standard frontal subroutine. The mesh is updated at each iteration by the newly found free-surface location values. The nodes of the singular elements are at a constant distance from the singular point, rotating around the singular point according to the shape and the position of the free surface [2]. The standard 3×3 and a modified 5×3 Gaussian quadratures are used for the numerical integration over the ordinary and the singular elements, respectively [1].

4 Results and Discussion

For our convergence studies, we constructed seven ordinary element meshes, OM1 through OM7. We generated from them the singular element meshes, SM1 through SM7, by replacing eight elements around the singular point by eight singular and eight ordinary elements. All meshes extended up to five radii upstream ($L_1=5$). We considered three different values for the length L_2 of the domain downstream the exit, i.e., $L_2=5$, 20 and 40. In Table 1, useful data about the ordinary and singular meshes with $L_2=20$ are tabulated.

For zero Reynolds number, the value $L_2=5$ for the length of the extrudate is adequate to capture the swelling of the extrudate. The numerical results obtained with the three

324

Table 2: Calculated extrudate-swell ratios for zero surface tension (L_2=5); stars indicate that the corresponding solutions are tainted by wiggles near the exit.

Mesh	Ordinary elements	Singular elements
1	1.1594	1.1260
2	1.1435	1.1263
3	1.1356	1.1265
4	1.1314	1.1265
5	1.1287	1.1265
6	1.1276	1.1264*
7	1.1271	1.1260*

values of L_2 are the same up to five significant digits. The values of the extrudate-swell ratio h_f (the final jet radius) calculated with both the ordinary and singular elements are tabulated in Table 2. As pointed out by Tanner [3], swelling is generally reduced as the number of degrees of freedom is increased (i.e., as the mesh is refined), when using ordinary elements. From Table 2, we observe that this is not the case when the singular finite elements are used. We also observe that the singular element solution converges much faster than its ordinary element counterpart. Comparing the values of h_f given in Table 2, we note that the solution obtained with SM1 is more accurate than that obtained with the finest ordinary element mesh OM7. The acceleration of the convergence of the free surface with mesh refinement has been shown by Georgiou et al. for the planar problem [2].

As already mentioned, the performance of the SFEM deteriorates when the singular elements are small. Indeed, with the last two singular meshes (SM6 and SM7), oscillations appear on the free surface. The inaccuracies propagate downstream and they thus affect the calculated value of h_f. Note also that with more refined meshes, the SFEM diverges. We concluded that the most accurate results are obtained with mesh SM5. Hence, the converged value of h_f for zero Reynolds number and zero surface tension is 1.1265. Tanner provided a selection of h_f values from the literature and estimated the extrapolated value h_f=1.127±0.003 for an infinite number of degrees of freedom [3].

5   Conclusions

The standard and the singular finite element methods have been applied to the creeping, axisymmetric extrudate-swell problem. The convergence of the two methods with mesh refinement has been studied for zero surface tension. It has been found that the singular finite elements accelerate the convergence of the free surface considerably.

References

[1] G.C. GEORGIOU, L.G. OLSON, W.W. SCHULTZ AND S. SAGAN, A singular finite element for Stokes flow: the stick-slip problem, *Int. J. Numer. Methods Fluids*, **9**, 1353-1367 (1989).

[2] G.C. GEORGIOU, L. G. OLSON AND W.W. SCHULTZ, Singular finite elements for the sudden-expansion and the die-swell problems, *Int. J. Numer. Methods Fluids*, **10**, 357-371 (1990).

[3] R.I. TANNER, *Engineering Rheology*, Clarendon Press, Oxford (1988).

[4] D.H. MICHAEL, The separation of a viscous liquid at a straight edge, *Mathematica*, **5**, 82-84 (1958).

[5] H.K. MOFFATT, Viscous and resistive eddies near a sharp corner, *J. Fluid Mech.*, **18**, 1-18 (1964).

[6] S.A. TROGDON AND D.D. JOSEPH, The stick-slip problem for a round jet: II. Small surface tension, *Rheol. Acta*, **20**, 660 (1981).

[7] T.R. SALAMON, D.E. BORNSIDE, R.C. ARMSTRONG AND R.A. BROWN, The role of surface tension in the dominant balance in the die well singularity, *Phys. Fluids* **7(10)**, 2328 (1995).

[8] W.W. SCHULTZ AND C. GERVASIO, A study of the singularity in the die swell problem, *Quart. J. Mech. Appl. Math.* **43**, 407-425 (1990).

[9] L.D. STURGES, Die swell: the separation of the free surface, *J. Non-Newtonian Fluid Mech.* **6**, 155-159 (1979).

Georgios C. Georgiou
Department of Mathematics and Statistics
University of Cyprus
Kallipoleos 75, P.O. Box 537
1678 Nicosia, CYPRUS

Andreas G. Boudouvis
Department of Chemical Engineering
National Technical University of Athens
Zografou Campus, Athens 15780, GREECE

O. KLEIN

A Phase-Field System with Space-Dependent Relaxation Coefficient

1 Introduction

In [PF], Penrose and Fife derived a thermodynamically consistent type of phase-field systems as models for phase transitions like the melting of ice.

In this work, a special system of this type is considered, which is extended by a space-dependent relaxation coefficient, which may vanish on a set of positive measures.

A time-discrete scheme for this system will be presented, which is used to derive existence and uniqueness results.

Moreover, it will be shown that, as the time step size and some parameters tend to zero, the corresponding solution to the scheme converges to the weak solution to a singular Stefan problem in which the heat flux is proportional to the gradient of the inverse temperature.

Finally, numerical results will be presented.

The results in this work are part of the author's thesis [Kle-2].

2 The system of Penrose-Fife type

For positive numbers ε and δ and initial data $\theta_{0\varepsilon}$, $\chi_{0\varepsilon}$ in $L^2(\Omega)$, we consider a phase-field system (**PF**) of Penrose-Fife type.

(**PF**): Find (θ, u, χ) fulfilling

$$\theta \in H^1(0, T; L^2(\Omega)), \quad u \in L^2(0, T; H^2(\Omega)) \cap L^\infty(\Omega_T), \tag{2.1}$$

$$\chi \in H^1(0, T; H^1(\Omega)) \cap L^\infty(0, T; H^2(\Omega)), \tag{2.2}$$

$$\theta > 0, \quad u = \frac{1}{\theta}, \quad \chi \in D(\beta) \quad \text{a.e. in } \Omega_T, \tag{2.3}$$

$$c_0\theta_t + L\chi_t + \kappa\Delta u = g \quad \text{a.e. in } \Omega_T, \tag{2.4}$$

$$(\delta + \eta)\chi_t - \varepsilon\Delta\chi + \beta(\chi) - \rho\chi \ni L(u_C - u) - \frac{\rho}{2} \quad \text{a.e. in } \Omega_T, \tag{2.5}$$

$$\kappa\frac{\partial u}{\partial n} + \gamma u = \zeta, \quad \frac{\partial \chi}{\partial n} = 0 \quad \text{a.e. in } \Gamma \times (0, T), \tag{2.6}$$

$$\theta(\cdot, 0) = \theta_{0\varepsilon}, \quad \chi(\cdot, 0) = \chi_{0\varepsilon} \quad \text{a.e. in } \Omega. \tag{2.7}$$

Here, $\Omega \subset \mathbf{R}^N$ with $N \in \{2, 3\}$ denotes a bounded domain with smooth boundary Γ, $T > 0$ a final time and $\Omega_T := \Omega \times (0, T)$. Moreover, c_0, κ, L, u_C are positive constants, β denotes the subdifferential $\partial I_{[0,1]}$ of the indicator function of the interval

327

$[0, 1]$, ρ is a non-negative constant, and γ, ζ are given functions on $\Gamma \times (0, T)$. The space-dependent relaxation coefficient η is used to model anisotropy in space and may vanish on a set of positive measures.

Now, for a positive number ε and initial data $\theta_{0\varepsilon}$, $\chi_{0\varepsilon}$ in $L^2(\Omega)$, the following degenerate Penrose-Fife system (DPF) is considered.

(DPF): Find (θ, u, χ) fulfilling (2.1)-(2.7) with $\delta = 0$.

3 The time-discrete scheme

For $\varepsilon > 0$, $\delta > 0$, and $h = T/K$ with $K \in \mathbf{N}$, we consider a time-discrete scheme (D) for the Penrose-Fife system (PF).

(D): For $1 \leq m \leq K$, find $\theta_m \in L^2(\Omega)$, $u_m, \chi_m \in H^2(\Omega)$, such that

$$0 < u_m, \quad \theta_m = \frac{1}{u_m}, \quad \chi_m \in D(\beta) \qquad \text{a.e. in } \Omega, \tag{3.1}$$

$$c_0 \frac{\theta_m - \theta_{m-1}}{h} + L\frac{\chi_m - \chi_{m-1}}{h} + \kappa \Delta u_m = g_m \quad \text{a.e. in } \Omega, \tag{3.2}$$

$$(\delta + \eta)\frac{\chi_m - \chi_{m-1}}{h} - \varepsilon \Delta \chi_m + \beta(\chi_m)$$
$$- \rho \chi_m \ni L(u_C - u_m) - \frac{\rho}{2} \quad \text{a.e. in } \Omega, \tag{3.3}$$

$$-\kappa \frac{\partial u_m}{\partial n} = \gamma_m u_m - \zeta_m, \quad \frac{\partial \chi_m}{\partial n} = 0 \quad \text{a.e. in } \Gamma, \tag{3.4}$$

with

$$\theta_0 := \theta_{0\varepsilon}, \qquad u_0 := \frac{1}{\theta_{0\varepsilon}}, \qquad \chi_0 := \chi_{0\varepsilon}. \tag{3.5}$$

Here, for $1 \leq m \leq K$, g_m, γ_m, and ζ_m are defined by

$$g_m(x) := \frac{1}{h} \int_{(m-1)h}^{mh} g(x, t)\,dt, \quad \forall x \in \Omega,$$

$$\gamma_m(\sigma) := \frac{1}{h} \int_{(m-1)h}^{mh} \gamma(\sigma, t)\,dt, \quad \zeta_m(\sigma) := \frac{1}{h} \int_{(m-1)h}^{mh} \zeta(\sigma, t)\,dt, \quad \forall \sigma \in \partial\Omega.$$

For dealing with existence and convergence results, the following assumptions will be used:

(A1): It holds

$$g \in H^1(0, T; L^\infty(\Omega)), \quad \eta \in L^\infty(\Omega), \quad \eta \geq 0 \quad \text{a.e. in } \Omega, \tag{3.6}$$

$$\gamma \in W^{1,\infty}(0, T; L^\infty(\Gamma)) \cap L^\infty(0, T; C^1(\Gamma)), \tag{3.7}$$

$$\zeta \in H^1(0, T; L^2(\Gamma)) \cap L^\infty(\Gamma \times (0, T)) \cap L^\infty(0, T; H^{\frac{1}{2}}(\Gamma)), \tag{3.8}$$

$$\gamma \geq c_\gamma, \zeta \geq c_\zeta \quad \text{a.e. in } \Gamma \times (0, T), \tag{3.9}$$

where c_γ, c_ζ are positive constants.

(A2): The initial data $\theta_{0\varepsilon}, \chi_{0\varepsilon}$ fulfill

$$\theta_{0\varepsilon}, \frac{1}{\theta_{0\varepsilon}} \in H^1(\Omega) \cap L^\infty(\Omega), \quad \chi_{0\varepsilon} \in H^2(\Omega), \tag{3.10}$$

$$\theta_{0\varepsilon} > 0, \quad \chi_{0\varepsilon} \in D(\beta) \quad \text{a.e. in } \Omega, \quad \frac{\partial \chi_{0\varepsilon}}{\partial n} = 0 \quad \text{a.e. in } \Gamma. \tag{3.11}$$

For the time-discrete scheme, we have the following result:

Theorem 1 *The time-discrete problem (D) has a unique solution, if (A1), (A2), and $\rho h < \delta$ hold.*

4 Existence and convergence results for the Penrose-Fife system

The solution to the scheme **(D)** is used to construct an approximate solution $\left(\widehat{\theta}^{h\delta\varepsilon}, \widehat{u}^{h\delta\varepsilon}, \widehat{\chi}^{h\delta\varepsilon}\right)$ to the Penrose-Fife system **(PF)** by piecewise linear-in-time interpolation.

For the degenerate Penrose-Fife system, the following existence and approximation results hold:

Theorem 2 *Assume that (A1), (A2), $\varepsilon > 0$, $\eta \not\equiv 0$ hold and that $\rho \geq 0$ is sufficiently small. Moreover, assume that there is some $\alpha > 0$ such that*

$$-\varepsilon \Delta \chi_{0\varepsilon} + \beta(\chi_{0\varepsilon}) - \rho \chi_{0\varepsilon} \ni L\left(u_C - \frac{1}{\theta_{0\varepsilon}}\right) - \frac{\rho}{2} \quad \text{a.e. in } \{x \in \Omega : \eta(x) \leq \alpha\}. \tag{4.1}$$

Then the degenerate Penrose-Fife system (DPF) has one and only one solution (θ, u, χ).

As $h \searrow 0$ and $\delta \searrow 0$, such that $3\rho h < \delta$, we have

$$\widehat{\theta}^{h\delta\varepsilon} \longrightarrow \theta \quad \text{weakly in} \quad H^1(0, T; L^2(\Omega)), \tag{4.2}$$

$$\text{weakly star in} \quad L^\infty(0, T; H^1(\Omega)) \cap L^\infty(\Omega_T), \tag{4.3}$$

$$c_0 \widehat{\theta}_t^{h\delta\varepsilon} + L\widehat{\chi}_t^{h\delta\varepsilon} \longrightarrow c_0 \theta_t + L\chi_t$$

$$\text{weakly star in} \quad L^\infty(0, T; H^1(\Omega)^*), \tag{4.4}$$

$$\widehat{u}^{h\delta\varepsilon} \longrightarrow u \quad \text{weakly in} \quad H^1(0, T; L^2(\Omega)), \tag{4.5}$$

$$\text{weakly star in} \quad L^\infty(0, T; H^1(\Omega)) \cap L^\infty(\Omega_T), \tag{4.6}$$

$$\text{weakly in} \quad L^2(t_*, T; H^2(\Omega)) \,\forall\, 0 < t_* < T, \tag{4.7}$$

$$\widehat{\chi}^{h\delta\varepsilon} \longrightarrow \chi \quad \text{weakly in} \quad H^1(0,T;H^1(\Omega)), \tag{4.8}$$

$$\text{weakly star in} \quad L^\infty(0,T;H^2(\Omega)), \tag{4.9}$$

$$\eta\widehat{\chi}_t^{h\delta\varepsilon} \longrightarrow \eta\chi_t \quad \text{weakly star in} \quad L^\infty(0,T;L^2(\Omega)), \tag{4.10}$$

and there is a positive constant C such that

$$\left\|\widehat{\theta}^{h\delta\varepsilon} - \theta\right\|_{L^2(0,T;L^2(\Omega))} + \left\|\widehat{u}^{h\delta\varepsilon} - u\right\|_{L^2(0,T;L^2(\Omega))} + \left\|\widehat{\chi}^{h\delta\varepsilon} - \chi\right\|_{L^2(0,T;H^1(\Omega))}$$
$$+ \left\|\sqrt{\eta}\left(\widehat{\chi}^{h\delta\varepsilon} - \chi\right)\right\|_{C([0,T];L^2(\Omega))} + \left\|c_0\widehat{\theta}^{h\delta\varepsilon} + L\widehat{\chi}^{h\delta\varepsilon} - (c_0\theta + L\chi)\right\|_{C([0,T];H^1(\Omega)^*)}$$
$$\leq C\left(\sqrt{h} + \sqrt{\delta}\right). \tag{4.11}$$

For the non-degenerate Penrose-Fife system, we have the following result:

Theorem 3 *Assume that (A1), (A2), $\delta > 0$, and $\varepsilon > 0$ hold.*
Then there is a unique solution (θ, u, χ) to the Penrose-Fife system (PF).
As $h \searrow 0$, we have the convergences (4.2)-(4.10) and

$$\widehat{\theta}_t^{h\delta\varepsilon} \longrightarrow \theta_t \quad \text{weakly star in} \quad L^\infty(0,T;H^1(\Omega)^*), \tag{4.12}$$

$$\widehat{\chi}_t^{h\delta\varepsilon} \longrightarrow \chi_t \quad \text{weakly star in} \quad L^\infty(0,T;L^2(\Omega)), \tag{4.13}$$

and there is a positive constant C such that

$$\left\|\widehat{\theta}^{h\delta\varepsilon} - \theta\right\|_{L^2(0,T;L^2(\Omega)) \cap C([0,T];H^1(\Omega)^*)} + \left\|\widehat{u}^{h\delta\varepsilon} - u\right\|_{L^2(0,T;L^2(\Omega))}$$
$$+ \left\|\widehat{\chi}^{h\delta\varepsilon} - \chi\right\|_{C([0,T];L^2(\Omega)) \cap L^2(0,T;H^1(\Omega))} \leq C\sqrt{h}. \tag{4.14}$$

5 The Stefan problem

For an initial value $e_* \in L^2(\Omega)$, we consider the following weak formulation of the Stefan problem with a heat flux equal to the gradient of κ/θ.

(S): Find (θ, u, χ) fulfilling

$$\theta \in L^\infty(0,T;L^2(\Omega)), \quad u \in L^\infty(0,T;H^1(\Omega)), \quad \chi \in L^\infty(\Omega_T), \tag{5.1}$$

$$c_0\theta + L\chi \in H^1(0,T;H^1(\Omega)^*), \tag{5.2}$$

$$\theta > 0, \quad u = \frac{1}{\theta}, \quad \chi \in D(\beta) \quad \text{a.e. in } \Omega_T, \tag{5.3}$$

$$\left(\frac{\partial}{\partial t}(c_0\theta + L\chi)(\cdot,t), v\right)_{H^1(\Omega)^* \times H^1(\Omega)} = \int_\Gamma (\gamma u - \varsigma)(\cdot,t)v \, d\sigma$$
$$+ \kappa \int_\Omega \nabla u(\cdot,t) \cdot \nabla v \, dx + \int_\Omega g(\cdot,t)v \, dx, \, \forall v \in H^1(\Omega), \text{ for a.e. } t \in (0,T), \tag{5.4}$$

$$\beta(\chi) \ni L(u_C - u) \quad \text{a.e. in } \Omega_T, \tag{5.5}$$

$$(c_0\theta + L\chi)(\cdot, 0) = e_* \quad \text{in} \quad H^1(\Omega)^*. \tag{5.6}$$

For the initial value, we assume:

(A3): There are two functions θ_*, χ_* in $H^1(\Omega)$ and two positive constants a, b, such that

$$e_* = c_0\theta_* + L\chi_* \quad \text{a.e. in } \Omega, \quad u_* := \frac{1}{\theta_*} \in H^1(\Omega), \tag{5.7}$$

$$a \le u_* \le b, \quad 0 \le \chi_* \le 1, \quad L(u_C - u_*) \in \beta(\chi_*) \quad \text{a.e. in } \Omega. \tag{5.8}$$

For $\delta > 0$, $\varepsilon > 0$, let $(\theta^{\delta\varepsilon}, u^{\delta\varepsilon}, \chi^{\delta\varepsilon})$ be the solution to the Penrose-Fife system **(PF)** with $\eta \equiv 0$, $\rho = 0$, and initial values $\theta_{0\varepsilon}, \chi_{0\varepsilon}$ defined by

$$\frac{1}{\theta_{0\varepsilon}} - \frac{a}{2}\chi_{0\varepsilon} = u_* - \frac{a}{2}\chi_*, \quad \text{a.e. in } \Omega,$$

$$-\varepsilon\Delta\chi_{0\varepsilon} + \beta(\chi_{0\varepsilon}) \ni L\left(u_C - \frac{1}{\theta_{0\varepsilon}}\right) \quad \text{a.e. in } \Omega,$$

$$\chi_{0\varepsilon} \in H^2(\Omega), \quad \frac{\partial\chi_{0\varepsilon}}{\partial n} = 0 \quad \text{a.e. in } \Gamma.$$

For $h = \frac{T}{K}$ with $K \in \mathbf{N}$, the solution to the time-discrete scheme **(D)** leads to an approximation $\left(\hat{\theta}^{h\delta\varepsilon}, \hat{u}^{h\delta\varepsilon}, \hat{\chi}^{h\delta\varepsilon}\right)$ of $(\theta^{\delta\varepsilon}, u^{\delta\varepsilon}, \chi^{\delta\varepsilon})$. The following theorem yields that this approximation is also an approximation of the solution to the Stefan problem.

Theorem 4 *Assume that (A1) and (A3) hold. Then the Stefan problem (S) has a unique solution (θ, u, χ), and, as h, δ, and ε tend to 0, we have*

$$\hat{\theta}^{h\delta\varepsilon} \longrightarrow \theta \quad \textit{weakly star in} \quad L^\infty(0, T; L^2(\Omega)), \tag{5.9}$$

$$\hat{u}^{h\delta\varepsilon} \longrightarrow u \quad \textit{weakly star in} \quad H^1(0, T; L^{\frac{3}{2}}(\Omega)) \cap L^\infty(0, T; H^1(\Omega)), \tag{5.10}$$

$$\hat{\chi}^{h\delta\varepsilon} \longrightarrow \chi \quad \textit{weakly star in} \quad L^\infty(\Omega_T), \tag{5.11}$$

and there is a positive constant C such that

$$\left\|\hat{\theta}^{h\delta\varepsilon} - \theta\right\|_{L^2(0,T;L^1(\Omega))} + \left\|c_0\hat{\theta}^{h\delta\varepsilon} + L\hat{\chi}^{h\delta\varepsilon} - (c_0\theta + L\chi)\right\|_{C([0,T];H^1(\Omega)^*)}$$

$$+ \left\|\hat{u}^{h\delta\varepsilon} - u\right\|_{L^2(0,T;L^{\frac{3}{2}}(\Omega))} \le C\left(\sqrt{h} + \delta^{\frac{1}{4}} + \varepsilon^{\frac{1}{4}}\right). \tag{5.12}$$

Remark For $h \searrow 0$, the error estimates (5.12) and (4.14) lead to an error estimate for the approximation of the solution to the Stefan problem by the solutions $(\theta^{\delta\varepsilon}, u^{\delta\varepsilon}, \chi^{\delta\varepsilon})$ to the Penrose-Fife system, see [Kle-1, Kle-2]. In [CS], Colli and Sprekels have derived for $(\theta^{\delta\varepsilon}, u^{\delta\varepsilon}, \chi^{\delta\varepsilon})$ convergences similar to (5.9)-(5.11), but they have not shown an error estimate.

6 Numerical results

The system in **(D)** is solved numerically by using a finite element method derived from a method for some classes of nonlinear scalar problems (see [Glo, Kor]).

The implementation is based on KASKADE, an adaptive FEM Code derived at the Konrad-Zuse-Zentrum Berlin, see [BER].

Now, a numerical simulation of the anisotropic growth of ice in an undercooled liquid is presented.

The initial value $\theta_{0\epsilon}$ of the temperature is a constant below the melting temperature, and the initial value $\chi_{0\epsilon}$ for the order parameter corresponds to a germ of ice surrounded by an undercooled liquid, where the phases are separated by a transition zone.

The problem is considered on the disk $\Omega = \{(x, y) : x^2 + y^2 < 2\}$ and a space-dependent relaxation coefficient with a sixfold symmetry is used:

$\delta + \eta(x, y) = \cos(6\alpha(x, y)) + 1.1$, where $\alpha(x, y)$ is the polar angle corresponding to the point (x, y).

In Figure 1, the evolution of the phase boundary is shown, i.e., the $\frac{1}{2}$ level sets of

Figure 1: Evolution of the phase boundary. The thick line corresponds to $t = 40$.

χ are presented at time intervals of 4. The thicker one of these lines represents the phase boundary at $t = 40$. Figure 2 depicts the mesh that is generated to approximate the solution at this time.

References

[BER]   R. BECK, B. ERDMANN, AND R. ROITZSCH. KASKADE Users's Guide, Version 3.x. Technical Report TR 95-11, Konrad-Zuse-Zentrum, Berlin, 1995.

[CS]   P. COLLI AND J. SPREKELS. Stefan problems and the Penrose-Fife phase field model. To appear in Adv. Math. Sci. Appl.

[Glo]   R. GLOWINSKI. *Numerical Methods for Nonlinear Variational Problems.* Springer, 1984.

[Kle-1] O. KLEIN. A semidiscrete scheme for a Penrose-Fife system and some Stefan problems in $\mathbf{R}^3$. *Adv. Math. Sci. Appl.*, 7(1):491-523, 1997.

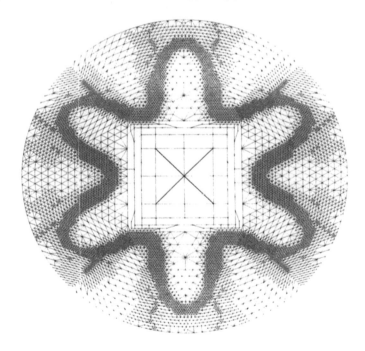

Figure 2: Generated mesh at t=40.

333

[Kle-2] O. KLEIN. *Existence and approximation results for phase-field systems of Penrose-Fife type and Stefan problems*. PhD thesis, Humboldt University, Berlin, 1997.

[Kor] R. KORNHUBER. *Adaptive Monotone Multigrid Methods for Nonlinear Variational Problems*. Advances in Numerical Mathematics. Wiley-Teubner, 1997.

[PF] O. PENROSE AND P. FIFE. Thermodynamically consistent models of phase-field type for the kinetics of phase transitions. *Physica D*, 43:44-62, 1990.

Olaf Klein
Weierstrass Institute for Applied Analysis and Stochastics
Mohrenstr. 39
D–10117 Berlin
Germany

R. Kornhuber[*]
Monotone Iterations for Elliptic Variational Inequalities

Abstract

A wide range of free boundary problems occurring in engineering and industry can be rewritten as a minimization problem for a strictly convex, piecewise smooth but non-differentiable energy functional. The fast solution of related discretized problems is a very delicate question, because usual Newton techniques cannot be applied. We propose a new approach based on convex minimization and constrained Newton type linearization. While convex minimization provides global convergence of the overall iteration, the subsequent constrained Newton type linearization is intended to accelerate the convergence speed. We present a general convergence theory and discuss several applications.

1 Introduction

We consider the minimization problem

$$u_j \in S_j : \qquad \mathcal{J}(u_j) + \phi_j(u_j) \le \mathcal{J}(v) + \phi_j(v) \qquad \forall v \in S_j \qquad (1.1)$$

on a finite-dimensional space S_j. The discrete problem (1.1) is typically resulting from the discretization of a related continuous analogue. The functional $\mathcal{J}$,

$$\mathcal{J}(v) = \tfrac{1}{2}a(v, v) - \ell(v), \qquad (1.2)$$

is induced by a continuous, symmetric and positive definite bilinear form $a(\cdot, \cdot)$ and by a linear functional ℓ. S_j is equipped with the energy norm $\| \cdot \| = a(\cdot, \cdot)^{1/2}$. The functional $\phi_j : S_j \to \mathbf{R} \cup \{+\infty\}$ is convex, lower semicontinuous and proper, i.e., $\phi_j(v) > -\infty$ and

$$\mathcal{K}_j = \{v \in S_j \mid \phi_j(v) < +\infty\} \ne \emptyset.$$

It is well-known that (1.1) then admits a unique solution $u_j \in S_j$.

Minimization problems of the form (1.1) with piecewise smooth nonlinearity ϕ_j arise in a large number of practical applications [3, 6, 8, 9, 11, 18]. As a consequence, there is a considerable interest in fast solvers motivating a variety of solution concepts [1, 2, 10, 13].

---

[*]The author gratefully acknowledges the hospitality of P. Deuflhard and his staff at the Konrad–Zuse–Center Berlin during the preparation of this manuscript. The work was supported by a Konrad–Zuse–Fellowship.

Algorithms from convex minimization, such as nonlinear Gauß-Seidel relaxation or steepest descent type methods, typically rely on local information about the objective function $\mathcal{J} + \phi_j$. This usually leads to rapidly deteriorating convergence rates when proceeding to larger spaces $\mathcal{S}_j$ or, equivalently, to more refined grids.

Because the energy $\mathcal{J} + \phi_j$ is not differentiable, classical Newton multigrid methods [12] cannot be applied without preceding regularization. Unfortunately, reasonable convergence speed may then have to be paid by unacceptable discretization errors and vice versa.

Extending recent *monotone multigrid methods* [14, 15, 16], we propose a new approach to the fast solution of (1.1). *Monotone iterations* are two-stage methods consisting of a globally convergent descent method and a subsequent *constrained* Newton linearization. The first substep is intended to fix the discrete free boundary, i.e., to deal with the non-smoothness of the problem, while the second substep is intended to increase the convergence speed once the discrete free boundary is (more or less) known. Note that this combination also has the flavor of an active set method. *Monotonically decreasing energy* is crucial for the global convergence of the overall iteration.

2 Monotone Iterations

Assume that $\mathcal{M}_j : \mathcal{S}_j \to \mathcal{S}_j$ satisfies

$$\mathcal{J}(\mathcal{M}_j(w)) + \phi_j(\mathcal{M}_j(w)) < \infty$$
$$\mathcal{J}(\mathcal{M}_j(w)) + \phi_j(\mathcal{M}_j(w)) \le \mathcal{J}(w) + \phi_j(w) \qquad \forall w \in \mathcal{S}_j, \qquad (2.1)$$

where

$$\mathcal{J}(\mathcal{M}_j(w)) + \phi_j(\mathcal{M}_j(w)) = \mathcal{J}(w) + \phi_j(w) \Leftrightarrow w = u_j. \qquad (2.2)$$

In addition, we require that

$$\limsup_{\nu \to \infty} \left(\mathcal{J}(\mathcal{M}_j(w^\nu)) + \phi_j(\mathcal{M}_j(w^\nu)) \right) \le \mathcal{J}(\mathcal{M}_j(\lim_{\nu \to \infty} w^\nu)) + \phi_j(\mathcal{M}_j(\lim_{\nu \to \infty} w^\nu)) \qquad (2.3)$$

holds for each convergent sequence $(w^\nu)_{\nu \ge 0} \subset \mathcal{K}_j$.

We shall see that the above conditions are sufficient for global convergence of the iteration $u_j^{\nu+1} = \mathcal{M}_j(u_j^\nu)$. However, the convergence speed may be unacceptably low. As a possible remedy, we introduce slightly more general *monotone iterations*

$$\bar{u}_j^\nu = \mathcal{M}_j(u_j^\nu)$$
$$u_j^{\nu+1} = \mathcal{C}_j(\bar{u}_j^\nu) \qquad (2.4)$$

where the additional substep $\mathcal{C}_j$ is intended to accelerate the convergence speed. Note that classical multigrid methods for self-adjoint linear problems can be interpreted in a similar way. Adopting multigrid terminology, $\mathcal{M}_j$ is called *fine grid smoother*, $\bar{u}_j^\nu$ is the *smoothed iterate* and $\mathcal{C}_j$ is called *coarse grid correction*.

We are now ready to state our basic convergence theorem.

Theorem 2.1 *Let ϕ_j be upper semicontinuous (and therefore continuous) on $\mathcal{K}_j$. Assume that the smoother $\mathcal{M}_j$ satisfies conditions (2.1) - (2.3) and that the coarse grid correction C_j has the monotonicity property*

$$\mathcal{J}(C_j(w)) + \phi_j(C_j(w)) \le \mathcal{J}(w) + \phi_j(w) \qquad \forall w \in \mathcal{S}_j. \tag{2.5}$$

Then the monotone iteration (2.4) is globally convergent.

Proof. For notational convenience, we introduce the abbreviation $\bar{\mathcal{J}} = \mathcal{J} + \phi_j$. Let us first show that the sequence of iterates $(u_j^\nu)_{\nu \ge 0}$ is bounded. As ϕ_j is convex, lower semicontinuous and proper, there are constants $c, C \in \mathbb{R}$, such that

$$\phi_j(v) \ge c\|v\| + C \qquad \forall v \in \mathcal{S}_j$$

(cf., e.g., [7]). As a consequence, we have

$$\bar{\mathcal{J}}(v) \ge \tfrac{1}{2}\|v\|^2 + (c - \|\ell\|)\|v\| + C \qquad \forall v \in \mathcal{S}_j, \tag{2.6}$$

so that $\|v\| \to \infty$ implies $\bar{\mathcal{J}}(v) \to \infty$. Hence, $(u_j^\nu)_{\nu \ge 0}$ must be bounded, because

$$\bar{\mathcal{J}}(u_j^\nu) \le \bar{\mathcal{J}}(\bar{u}_j^1) < \infty \qquad \forall \nu \ge 1$$

follows from (2.1) and (2.5).

Now, let $u_j^{\nu_k}$, $k \ge 0$, be an arbitrary, convergent subsequence of u_j^ν with the limit $u^* \in \mathcal{S}_j$,

$$\lim_{k \to \infty} u_j^{\nu_k} = u^*. \tag{2.7}$$

Such a subsequence exists, because u_j^ν is bounded and $\mathcal{S}_j$ has finite dimension. Note that $u^* \in \mathcal{K}_j$, because $(u_j^{\nu_k})_{k \ge 1} \subset \mathcal{U}_j := \{v \in \mathcal{S}_j | \ \bar{\mathcal{J}}(v) \le \bar{\mathcal{J}}(\bar{u}_j^1)\} \subset \mathcal{K}_j$ and the sublevel set $\mathcal{U}_j$ is closed. We now want to prove $u_j^* = u_j$. In the light of (2.2), it is sufficient to show

$$\bar{\mathcal{J}}(\mathcal{M}_j(u^*)) = \bar{\mathcal{J}}(u^*). \tag{2.8}$$

Observe that (2.1) and (2.5) imply

$$\bar{\mathcal{J}}(u_j^{\nu_k+1}) \le \bar{\mathcal{J}}(u_j^{\nu_k+1}) \le \bar{\mathcal{J}}(\mathcal{M}_j(u_j^{\nu_k})) \le \bar{\mathcal{J}}(u_j^{\nu_k}).$$

By virtue of the continuity of $\bar{\mathcal{J}}$ on $\mathcal{K}_j$, this leads to

$$\lim_{k \to \infty} \bar{\mathcal{J}}(\mathcal{M}_j(u_j^{\nu_k})) = \bar{\mathcal{J}}(u_j^*).$$

Now the equality (2.8) is an immediate consequence of conditions (2.1) and (2.3).

As $(u_j^{\nu_k})_{k \ge 0}$ was an arbitrary convergent subsequence, the whole sequence u_j^ν must converge to u_j. This completes the proof. □

As a by-product, we also get the convergence of the smoothed iterates

$$\lim_{k \to \infty} \bar{u}_j^\nu = u_j. \tag{2.9}$$

We emphasize that the coarse grid correction *alone* does not need to be convergent. This gives considerable flexibility in constructing C_j.

3 Fine Grid Smoother

All descent methods from convex minimization are natural candidates for the fine grid smoother $\mathcal{M}_j$.

EXAMPLE 3.1 (Nodal type nonlinearity)

Let S_j be the space of linear finite elements with respect to a triangulation $\mathcal{T}_j$ of a bounded polygonal domain Ω. The set of vertices of all triangles $t \in \mathcal{T}_j$ is called $\mathcal{N}_j$, $n_j = \#\mathcal{N}_j$ and

$$\Lambda_j = \left(\lambda^{(j)}_{p_1}, \ldots, \lambda^{(j)}_{p_{n_j}} \right)$$

denotes the nodal basis of S_j, ordered in a suitable way. Now assume that ϕ_j can be written as

$$\phi_j(v) = \sum_{p \in \mathcal{N}_j} \Phi_p(v(p)) \, h_p \tag{3.1}$$

with convex, lower semicontinuous and proper functions $\Phi_p : \mathbb{R} \to \mathbb{R} \cup \{+\infty\}$ and weights $h_p \in \mathbb{R}$. Then ϕ_j is convex, lower semicontinuous, proper, and continuous on $\mathcal{K}_j = \{v \in S_j | \, v(p) \in \text{dom } \Phi_p, p \in \mathcal{N}_j\}$.

The nonlinear Gauß–Seidel relaxation $\mathcal{M}_j^{\mathrm{GS}}$ (cf., e.g., [10, 16]) for the iterative solution of (1.1) reads as follows. Starting with a given iterate $w_0^\nu := u_j^\nu \in S_j$, we compute local corrections $v_l^\nu \in V_l := \text{span}\{\lambda^{(j)}_{p_l}\}$ from the n_j local subproblems

$$
\begin{aligned}
v_l^\nu \in V_l : \quad & \mathcal{J}(w_{l-1}^\nu + v_l^\nu) + \Phi_{p_l}(u_j^\nu(p_l) + v_l^\nu(p_l)) h_{p_l} \\
& \leq \mathcal{J}(w_{l-1}^\nu + v) + \Phi_{p_l}(u_j^\nu(p_l) + v(p_l)) h_{p_l}, \quad \forall v \in V_l,
\end{aligned}
\tag{3.2}
$$

setting $w_l^\nu = w_{l-1}^\nu + v_l^\nu$, $l = 1, \ldots, n_j$. Finally, we define $\mathcal{M}_j^{\mathrm{GS}}(u_j^\nu) := w_{n_j}^\nu$.

It is not difficult to show that $\mathcal{M}_j^{\mathrm{GS}}$ satisfies conditions (2.1) - (2.3).

If the nonlinearity ϕ_j does not have the form (3.1), then nonlinear Gauß–Seidel relaxation is no longer applicable, because (2.2) may be violated. In this case, other descent algorithms, such as bundle methods, should be used (cf., e.g., [13]).

4 Coarse Grid Correction

In most practical applications the functional ϕ_j is piecewise smooth. Then, for given $\bar{u}_j^\nu = \mathcal{M}_j u_j^\nu$, we can find a closed convex subset $\mathcal{K}_{\bar{u}_j^\nu} \subset S_j$ and a *smooth functional* $\phi_{\bar{u}_j^\nu} : S_j \to \mathbb{R}$, such that

$$\bar{u}_j^\nu \in \mathcal{K}_{\bar{u}_j^\nu}$$
$$\phi_{\bar{u}_j^\nu}(w) = \phi_j(w) + \text{const.} \qquad \forall w \in \mathcal{K}_{\bar{u}_j^\nu}.$$

Roughly speaking, all $w \in \mathcal{K}_{\bar{u}_j^\nu}$ must have the same phases as $\bar{u}_j^\nu$.

Let us consider the *constrained minimization* of the *smooth energy* $\mathcal{J} + \phi_{\bar{u}_j^\nu}$

$$u_j^* \in \mathcal{K}_{\bar{u}_j^\nu}: \quad \mathcal{J}(u_j^*) + \phi_{\bar{u}_j^\nu}(u_j^*) \leq \mathcal{J}(v) + \phi_{\bar{u}_j^\nu}(v) \quad \forall v \in \mathcal{K}_{\bar{u}_j^\nu}. \tag{4.1}$$

As a consequence of (2.9), we get $\text{dist}(u_j, \mathcal{K}_{\bar{u}_j^\nu}) \to 0$ as $\nu \to \infty$. Hence, the solutions u_j^* of (4.1) tend to u_j. Moreover, there is some hope that $u_j \in \mathcal{K}_{\bar{u}_j^\nu}$ holds for $\nu \geq \nu_0$ with ν_0 sufficiently large (see example 4.1 below). In this case, we even get $u_j^* = u_j$ $\forall \nu \geq \nu_0$. As a consequence, a monotone iteration (2.4) with coarse grid correction defined by $\mathcal{C}_j(\bar{u}_j^\nu) = u_j^*$ would produce the exact solution after a finite number of steps.

Of course, we cannot expect to solve (4.1) exactly. The main advantage of (4.1) is that Newton type linearization can be applied to the smooth energy $\mathcal{J} + \phi_{\bar{u}_j^\nu}$. More precisely, we approximate $\mathcal{J} + \phi_{\bar{u}_j^\nu}$ by the quadratic energy functional $\mathcal{J}_{\bar{u}_j^\nu}$,

$$\mathcal{J}_{\bar{u}_j^\nu}(w) = \tfrac{1}{2}a_{\bar{u}_j^\nu}(w, w) - \ell_{\bar{u}_j^\nu}(w) \approx \mathcal{J}(w) + \phi_{\bar{u}_j^\nu}(w) + \text{const.}, \quad w \in \mathcal{K}_{\bar{u}_j^\nu},$$

where the bilinear form

$$a_{\bar{u}_j^\nu}(w, w) = a(w, w) + \phi''_{\bar{u}_j^\nu}(\bar{u}_j^\nu)(w, w) \tag{4.2}$$

and the linear functional

$$\ell_{\bar{u}_j^\nu}(w) = \ell(w) - \phi'_{\bar{u}_j^\nu}(\bar{u}_j^\nu)(w) + \phi''_{\bar{u}_j^\nu}(\bar{u}_j^\nu)(\bar{u}_j^\nu, w)$$

are obtained by Taylor's expansion

$$\phi_{\bar{u}_j^\nu}(w) \approx \phi_{\bar{u}_j^\nu}(\bar{u}_j^\nu) + \phi'_{\bar{u}_j^\nu}(\bar{u}_j^\nu)(w - \bar{u}_j^\nu) + \tfrac{1}{2}\phi''_{\bar{u}_j^\nu}(\bar{u}_j^\nu)(w - \bar{u}_j^\nu, w - \bar{u}_j^\nu).$$

The resulting *linearized constrained problem*

$$w_j^* \in \mathcal{K}_{\bar{u}_j^\nu}: \quad \mathcal{J}_{\bar{u}_j^\nu}(w_j^*) \leq \mathcal{J}_{\bar{u}_j^\nu}(v) \quad \forall v \in \mathcal{K}_{\bar{u}_j^\nu} \tag{4.3}$$

can be regarded as a generalization of classical Newton linearization in case of smooth functionals ϕ_j. Indeed, if ϕ_j is twice differentiable on S_j, we can take $\mathcal{K}_{\bar{u}_j^\nu} = S_j$ and (4.3) becomes a linear system.

Let $\tilde{w}_j$ be an approximate solution of (4.3). Then, we define

$$u_j^{\nu+1} = \mathcal{C}_j(\bar{u}_j^\nu) := \bar{u}_j^\nu + \omega(\tilde{w}_j - \bar{u}_j^\nu), \tag{4.4}$$

where the damping parameter ω has to be chosen such that the monotonicity (2.5) holds. Here, we refer to well-known affine invariant damping strategies [4, 5].

If the exact solution $\tilde{w}_j = w_j^*$ of (4.3) is inserted in (4.4) and $u_j \in \mathcal{K}_{\bar{u}_j^\nu}$ holds for all $\nu \geq \nu_0$, then we can expect that the resulting monotone iteration (2.4) is converging quadratically for $\nu \geq \nu_0$. In practice, an approximation $\tilde{w}_j = \mathcal{MG}(\bar{u}_j^\nu)$ of w_j^* is obtained by one step of a suitable iterative scheme $\mathcal{MG}$. Here, multigrid typically comes into play. As for classical Newton multigrid methods we can expect that the convergence rates of $\mathcal{MG}$ asymptotically, i.e., for large ν, dominate the convergence speed of the overall monotone iteration (2.4). Hence, (asymptotically) fast solvers for (4.3) should produce (asymptotically) fast monotone iterations.

EXAMPLE 4.1 (Nodal type nonlinearity)

Let $\mathcal{T}_j$ be resulting from j refinements of an intentionally coarse triangulation $\mathcal{T}_0$ of a bounded polygonal domain Ω. In this way, we obtain a sequence of triangulations $\mathcal{T}_0, \ldots, \mathcal{T}_j$ and corresponding nested spaces $\mathcal{S}_0 \subset \cdots \subset \mathcal{S}_j$ of piecewise linear finite element functions. We assume for convenience that the triangulations are uniformly refined. Collecting all nodal basis functions from all refinement levels, we obtain the multilevel nodal basis $\Lambda_{\mathcal{S}}$,

$$\Lambda_{\mathcal{S}} = \left(\lambda_{p_1}^{(j)}, \lambda_{p_2}^{(j)} \ldots, \lambda_{p_{n_j}}^{(j)}, \ldots, \lambda_{p_1}^{(0)}, \ldots, \lambda_{p_{n_0}}^{(0)} \right),$$

with $m_{\mathcal{S}} = n_j + \cdots + n_0$ elements. As usual, the ordering $\lambda_l := \lambda_{p_l}^{(k_l)}$, $l = 1, \ldots, m_{\mathcal{S}}$, is taken from fine to coarse. Now assume, for example, that ϕ_j is given by (3.1) with

$$\Phi_p(z) = z^{1+\frac{1}{2}} \ \forall z \geq 0, \qquad \Phi_p(z) = +\infty \ \forall z < 0 \qquad \forall p \in \mathcal{N}_j.$$

Then we can choose

$$\mathcal{K}_{\bar{u}_j^\nu} = \{v \in \mathcal{S}_j | \ \tfrac{1}{2}\bar{u}_j^\nu(p) \leq v(p) \ \forall p \in \mathcal{N}_j^\circ(\bar{u}_j^\nu), \ v(p) = 0 \ \forall p \in \mathcal{N}_j^\bullet(\bar{u}_j^\nu)\},$$

where we have set

$$\mathcal{N}_j^\bullet(\bar{u}_j^\nu) = \{p \in \mathcal{N}_j | \ \bar{u}_j^\nu(p) = 0\}, \qquad \mathcal{N}_j^\circ(\bar{u}_j^\nu) = \mathcal{N}_j \setminus \mathcal{N}_j^\bullet(\bar{u}_j^\nu).$$

For the approximate solution of the linearized constrained problem (4.3), we can use the multilevel relaxation $\mathcal{MG}$ defined as follows. Starting with a given smoothed iterate $w_0^\nu = \bar{u}_j^\nu \in \mathcal{K}_j = \{v \in \mathcal{S}_j | \ v(p) \geq 0 \ \forall p \in \mathcal{N}_j\}$, we compute local corrections $v_l^\nu \in V_l := \mathrm{span}\{\lambda_l\}$ from the $m_{\mathcal{S}}$ local subproblems

$$v_l^\nu \in V_l \cap \mathcal{K}_{\bar{u}_j^\nu} : \qquad \mathcal{J}_{\bar{u}_j^\nu}(w_{l-1}^\nu + v_l^\nu) \leq \mathcal{J}_{\bar{u}_j^\nu}(w_{l-1}^\nu + v) \qquad \forall v \in V_l \cap \mathcal{K}_{\bar{u}_j^\nu}, \qquad (4.5)$$

setting $w_l^\nu = w_{l-1}^\nu + v_l^\nu$, $l = 1, \ldots, m_{\mathcal{S}}$. Finally, we define $\bar{w}_j = \mathcal{MG}(\bar{u}_j^\nu) := w_{m_{\mathcal{S}}}^\nu$. In the linear self-adjoint case, i.e., for $\mathcal{K}_{\bar{u}_j^\nu} = \mathcal{S}_j$ or, equivalently, for smooth ϕ_j, this is just the classical multigrid method with canonical restrictions and prolongations and Gauß-Seidel smoother. In practice, the local subproblems (4.5) are modified a bit in order to allow an implementation with optimal order of complexity [14, 16, 17]. In some cases, it may be more appropriate to use local damping parameters ω_l associated with each local correction v_l^ν instead of the global parameter ω in the correction step (4.4) [17]. Monotone iterations (2.4) involving such variants of multilevel relaxation are called *monotone multigrid methods.*

If our original problem (1.1) is non-degenerate and nonlinear Gauß-Seidel relaxation $\mathcal{M}_j^{GS}$ is used as fine grid smoother (cf. example 3.1), then it can be shown that $u_j \in \mathcal{K}_{\bar{u}_j^\nu}$ holds for sufficiently large ν. Moreover, the linearized constrained problem (4.3) asymptotically, i.e., for large ν, reduces to the linear problem

$$w_j^* \in S_j^\circ : \qquad a_{u_j}(w_j^*, v) = \ell_{u_j}(v) \qquad \forall v \in S_j^\circ \tag{4.6}$$

on the reduced space

$$S_j^\circ = \{v \in S_j| \ v(p) = 0 \ \forall p \in \mathcal{N}_j^\bullet(u_j)\}.$$

Multilevel relaxations automatically reduce to linear multigrid methods for (4.6) with multigrid convergence rates. Hence, we get asymptotic multigrid convergence rates for the resulting monotone multigrid methods. In our numerical experiments, we observed that the asymptotic behavior starts almost immediately, if nested iteration is used [17].

EXAMPLE 4.2 (Gradient type nonlinearity)

Let S_j and T_j be defined as in the preceding example. Assume that ϕ_j is given by

$$\phi_j(v) = \sum_{t \in T_j} |\nabla v(t)| \, h_t$$

with $|\cdot|$ denoting the Euclidean norm and suitable weights h_t. Then we can choose

$$\mathcal{K}_{\bar{u}_j^\nu} = \{v \in S_j| \ \tfrac{1}{2}|\nabla \bar{u}_j^\nu(t)| \leq |\nabla v(t)| \ \forall t \in T_j^\circ(\bar{u}_j^\nu), \ \nabla v(t) = 0 \ \forall t \in T_j^\bullet(\bar{u}_j^\nu)\},$$

where we have set

$$T_j^\bullet(\bar{u}_j^\nu) = \{t \in T_j| \ \nabla \bar{u}_j^\nu(t) = 0\}, \qquad T_j^\circ(\bar{u}_j^\nu) = T_j \setminus T_j^\bullet(\bar{u}_j^\nu).$$

Assume the fine grid smoother $\mathcal{M}_j$ has properties (2.1) - (2.3) and additionally guarantees $u_j \in \mathcal{K}_{\bar{u}_j^\nu}$ for sufficiently large ν. Then, similar to the previous example, (4.3) asymptotically reduces to the linear problem (4.6) with reduced space S_j° now given by

$$S_j^\circ = \{v \in S_j| \ \nabla v(t) = 0 \ \forall t \in T_j^\bullet(u_j)\}.$$

Again, multilevel relaxations for (4.3) asymptotically reduce to linear multigrid methods for this problem.

The actual construction of a fine grid smoother $\mathcal{M}_j$ with the desired properties is the subject of current research.

References

[1] D.P. BERTSEKAS. *Constrained optimization and Lagrange multiplier methods.* Academic Press, New York, 1984.

[2] R.W. COTTLE, J.S. PANG, AND R.E. STONE. *The Linear Complementary Problem.* Academic Press, Boston, 1992.

[3] J. CRANK. *Free and Moving Boundary Problems.* Oxford University Press, Oxford, 1988.

[4] P. DEUFLHARD. A modified Newton method for the solution of ill-conditioned systems of nonlinear equations with applications to multiple shooting. *Numer. Math.,* 22:289 – 315, 1974.

[5] P. DEUFLHARD AND M. WEISER. Global inexact Newton multilevel FEM for nonlinear elliptic problems. In W. Hackbusch and G. Wittum, editors, *Multigrid methods V,* Lecture Notes in Computational Science and Engineering, Springer, Berlin, Heidelberg, 1998.

[6] G. DUVAUT AND J.L. LIONS. *Les inéquations en mécanique et en physique.* Dunaud, Paris, 1972.

[7] I. EKELAND AND R. TEMAM. *Convex Analysis and Variational Problems.* North–Holland, Amsterdam, 1976.

[8] C.M. ELLIOTT AND J.R. OCKENDON. *Weak and Variational Methods for Moving Boundary Problems,* Volume 53 of *Research Notes in Mathematics.* Pitman, London, 1982.

[9] A. FRIEDMAN. *Variational Principles and Free Boundary Problems.* Wiley, New York, 1982.

[10] R. GLOWINSKI. *Numerical Methods for Nonlinear Variational Problems.* Springer, New York, 1984.

[11] R. GLOWINSKI, J.L. LIONS, AND R. TRÉMOLIÈRES. *Numerical Analysis of Variational Inequalities.* North–Holland, Amsterdam, 1981.

[12] W. HACKBUSCH. *Multi-Grid Methods and Applications.* Springer, Berlin, 1985.

[13] J.B. HIRIART-URRUTY AND C. LEMARÉCHAL. *Convex Analysis and Minimization Algorithms I,II.* Springer, Berlin, Heidelberg, New York, 1991.

[14] R. KORNHUBER. Monotone multigrid methods for elliptic variational inequalities I. *Numer. Math.,* 69:167 – 184, 1994.

[15] R. KORNHUBER. Monotone multigrid methods for elliptic variational inequalities II. *Numer. Math.*, 72:481 – 499, 1996.

[16] R. KORNHUBER. *Adaptive Monotone Multigrid Methods for Nonlinear Variational Problems*. Teubner, Stuttgart, 1997.

[17] R. KORNHUBER. Globally convergent multigrid methods for porous medium type equations. To appear.

[18] J.F. RODRIGUES. *Obstacle Problems in Mathematical Physics*. Number 134 in Mathematical Studies. North–Holland, Amsterdam, 1987.

Ralf Kornhuber
Freie Universitt Berlin
Institut fr Mathematik I
Arnimallee 2-6
D-14195 Berlin-Dahlem
Germany

343

R.H. Nochetto,* A. Schmidt†and C. Verdi‡

Adaptive Solution of Parabolic Free Boundary Problems with Error Control

Abstract

We derive *a posteriori* error estimates in natural energy norms with weights. They are useful in localizing the error extraction in a region of interest, and form the basis of an adaptive procedure. We use them to study the evolution of a persistent corner singularity and elucidate the issue of critical angle for instantaneous smoothing.

1 Introduction

The presence of interfaces, and associated lack of regularity, is responsible for global numerical pollution effects for parabolic free boundary problems. A cure consists of equidistributing discretization errors in adequate norms by means of highly graded meshes and varying time steps. Their construction relies on *a posteriori* error estimates, which are a fundamental component for the design of reliable and efficient adaptive algorithms for PDEs. These issues have been recently tackled in [6], [7], [8], [9], and are briefly discussed here.

We consider for simplicity the classical two-phase Stefan problem for an ideal material with constant thermal properties and unit latent heat

$$\partial_t u - \Delta \beta(u) = f \quad \text{in } Q = \Omega \times (0, T), \quad u(\cdot, 0) = u_0(\cdot) \quad \text{in } \Omega, \tag{1}$$

where $\beta(s) = \min(s, 0) + \max(s - 1, 0)$. A discrete solution U of (1) satisfies

$$\partial_t U - \Delta \beta(U) = f - \mathcal{R} \quad \text{in } Q, \quad U(\cdot, 0) = U_0(\cdot) \quad \text{in } \Omega, \tag{2}$$

where $\mathcal{R}$, a distribution with singular components and oscillatory behavior, is the so-called *parabolic residual*. Its size is to be determined in negative norms which entail averaging and thus better quantify oscillations.

In §2 we show how to represent the errors $e_u = u - U$ and $e_{\beta(u)} = \beta(u) - \beta(U)$ in terms of $\mathcal{R}$, and in §§3 and 4 we derive error estimates in *weighted norms*

$$\|e_u(\cdot, T)\|_{H^{-1}_{-\omega}(\Omega)} + \|e_{\beta(u)}\|_{L^2_{-\omega}(Q)} \leq \mathcal{E}(u_0, f, T, \Omega, \omega; U, h, \tau), \tag{3}$$

with a computable right-hand side; hereafter h stands for the mesh size and τ for the time step. The weight ω is space dependent and serves to localize the errors in regions

---

*Partially supported by NSF Grant DMS-9623394 and NSF SCREMS 9628467.
†Partially supported by EU Grant HCM "Phase Transitions and Surface Tension".
‡Partially supported by MURST and CNR Contract 95.00735.01.

of interest. These formulas are the basis of the adaptive algorithm of [6], [7], [8], [9]. We present in §5 an application to a persistent corner singularity, and elucidate the question of critical angle for instantaneous smoothing. We conclude in §6 with some remarks on the hyperbolic structure of (1) along with the essential role of adaptivity in this study.

2   Error Representation Formulas

Upon subtracting (2) from (1), the error e_u satisfies the equation

$$\partial_t e_u - \Delta(b e_u) = \mathcal{R} \tag{4}$$

in the sense of distributions, where

$$0 \le b(x,t) = \int_0^1 \beta'(su(x,t) + (1-s)U(x,t))\, ds \le 1$$

is discontinuous. To express e_u in suitable norms we resort to *parabolic duality*. We multiply (4) by a smooth test function ϕ and integrate by parts

$$\langle e_u(\cdot,T), \phi(\cdot,T)\rangle - \int_0^T \langle e_u, \partial_t \phi + b\Delta\phi\rangle = \langle u_0 - U^0, \phi(\cdot,0)\rangle + \mathcal{R}(\phi), \tag{5}$$

where $\langle \cdot, \cdot \rangle$ stands for the usual $L^2(\Omega)$ inner product. We are thus led to study the backward parabolic problem in *nondivergence* form [3], with vanishing diffusion coefficient b and regularization parameter $\delta \downarrow 0$,

$$\partial_t \phi + (b + \delta)\Delta\phi = -b^{1/2}\chi \quad \text{in } Q, \quad \phi(\cdot,T) = \rho \quad \text{in } \Omega. \tag{6}$$

Let $\omega \in C^{0,1}(\bar{\Omega})$ be a weight satisfying $\omega(x) \ge \omega_0 > 0$ and

$$\frac{|\nabla\omega(x)|}{\omega(x)} \le W \quad \forall\, x \in \Omega. \tag{7}$$

We introduce the following weighted norms over a generic domain Ξ

$$\|v\|_{L^2_{\pm\omega}(\Xi)} = \left(\int_\Xi v^2 \omega^{\pm 2}\right)^{1/2}, \quad \|v\|_{H^{-1}_{-\omega}(\Xi)} = \sup_{\eta \in H^1_0(\Xi)} \|\nabla\eta\|^{-1}_{L^2_\omega(\Xi)} \int_\Xi v\eta.$$

These norms will serve to localize the error estimates, and thereby force extra refinement, in regions of interest. The theory of nonlinear strictly parabolic problems [2] yields the existence of a unique solution ϕ, which satisfies

$$\max_{0 \le t \le T} \|\nabla\phi(\cdot,t)\|_{L^2_\omega(\Omega)} + \|\partial_t\phi\|_{L^2_\omega(Q)} \le e^{WT}\left(\|\nabla\rho\|_{L^2_\omega(\Omega)} + \|\chi\|_{L^2_\omega(Q)}\right) = D. \tag{8}$$

345

These bounds are proved in [3], [6] for $\omega = 1$. Their weighted counterpart results from multiplying (6) by $\omega^2 \Delta \phi$, integrating by parts, and making proper use of (7) to absorb an additional term via Gronwall. We thus realize that (6) does not exhibit a regularizing effect such as the heat equation since $\phi(\cdot, t) = \rho$ for all $0 \le t \le T$ if $b = \delta = 0$; hence $\|\Delta \phi\|_{L^2(Q)}$ is never bounded uniformly in δ. Duality combined with (5) as $\delta \downarrow 0$ thus yields

$$\chi = 0: \quad \|e_u(\cdot, T)\|_{H^{-1}_{-\omega}(\Omega)} \le \|u_0 - U^0\|_{H^{-1}_{-\omega}(\Omega)} + \sup_{\rho \in H^1_0(\Omega)} \frac{|\mathcal{R}(\phi)|}{\|\nabla \rho\|_{L^2(\Omega)}}, \tag{9}$$

$$\rho = 0: \quad \|e_{\beta(u)}\|_{L^2_{-\omega}(Q)} \le \|u_0 - U^0\|_{H^{-1}_{-\omega}(\Omega)} + \sup_{\chi \in L^2_\omega(Q)} \frac{|\mathcal{R}(\phi)|}{\|\chi\|_{L^2_\omega(Q)}}. \tag{10}$$

We stress that U need not be a finite element approximation of u. We use (9) and (10) below to study the time and full discretizations of (1).

3 Error Estimation for Time Discretization

We first illustrate the main ideas for an implicit time discretization of (1). Let $\tau_n = t^n - t^{n-1}$ be a variable time step and let U^n satisfy

$$\partial U^n - \Delta \beta(U^n) = f^n,$$

where $\partial U^n = (U^n - U^{n-1})/\tau_n$ and $f^n(\cdot)$ is an approximation of $f(\cdot, t)$ in $[t^{n-1}, t^n]$ (e.g., $f^n(\cdot) = f(\cdot, t^n)$ or $f^n(\cdot) = \tau_n^{-1} \int_{t^{n-1}}^{t^n} f(\cdot, s)\, ds = \bar{f}^n(\cdot)$). If $U(\cdot, t) = U^n$ for $t^{n-1} < t \le t^n$, then the residual $\mathcal{R}$ in (2) is given by

$$\mathcal{R} = \sum_{n=1}^{N} (\partial U^n + f - f^n) \chi_{[t^{n-1}, t^n]} - (U^n - U^{n-1}) \delta_{t^{n-1}},$$

where χ_I is the characteristic function of I and δ_s is a Dirac mass at $t = s$. Equivalently, we can split $\mathcal{R}$ into $\mathcal{R}_1$ and $\mathcal{R}_2$ as follows:

$$\mathcal{R}_1(\phi) = \sum_{n=1}^{N} \int_{t^{n-1}}^{t^n} \langle \partial U^n, \phi - \phi(\cdot, t^{n-1}) \rangle, \quad \mathcal{R}_2(\phi) = \sum_{n=1}^{N} \int_{t^{n-1}}^{t^n} \langle f - f^n, \phi \rangle. \tag{11}$$

If $f^n = \bar{f}^n$, we can exploit L^2 time orthogonality in conjunction with $\phi(\cdot, t) - \phi(\cdot, t^{n-1}) = \int_{t^{n-1}}^{t} \partial_s \phi(\cdot, s)\, ds$ and (8), to deduce

$$|\mathcal{R}_1(\phi)| \le \sum_{n=1}^{N} \int_{t^{n-1}}^{t^n} \|U^n - U^{n-1}\|_{L^2_{-\omega}(\Omega)} \|\partial_t \phi\|_{L^2_\omega(\Omega)} \le D\mathcal{E}_1,$$

$$|\mathcal{R}_2(\phi)| = \left| \sum_{n=1}^{N} \int_{t^{n-1}}^{t^n} \langle f - \bar{f}^n, \phi - \phi(\cdot, t^{n-1}) \rangle \right| \le D\mathcal{E}_2,$$

346

where

$$\mathcal{E}_1 = \Big(\sum_{n=1}^{N} \tau_n \|U^n - U^{n-1}\|_{L^2_{-\omega}(\Omega)}^2 \Big)^{1/2} \qquad \textit{time residual,}$$

$$\mathcal{E}_2 = \Big(\sum_{n=1}^{N} \tau_n^2 \int_{t^{n-1}}^{t^n} \|f - \bar{f}^n\|_{L^2_{-\omega}(\Omega)}^2 \Big)^{1/2} \qquad \textit{interpolation.}$$

If $\mathcal{E}_0 = \|u_0 - U^0\|_{H^{-1}_{-\omega}(\Omega)}$ denotes the *initial error*, on using (9) and (10), we immediately obtain

$$\|e_u(\cdot, T)\|_{H^{-1}_{-\omega}(\Omega)} + \|e_{\beta(u)}\|_{L^2_{-\omega}(Q)} \le e^{WT} (\mathcal{E}_0 + \mathcal{E}_1 + \mathcal{E}_2);$$

the discretization error thus accumulates in time in the L^2 norm. To make it computable, we can replace $\mathcal{E}_1$ by the upper bound

$$\mathcal{E}_1 \le T^{1/2} \max_{1 \le n \le N} \|U^n - U^{n-1}\|_{L^2_{-\omega}(\Omega)}.$$

Invoking H^1 space regularity of ϕ yields a weaker expression for $\mathcal{E}_2$ regardless of the choice of f^n [6]: $\mathcal{E}_2 = \sum_{n=1}^{N} \int_{t^{n-1}}^{t^n} \|f - f^n\|_{H^{-1}_{-\omega}(\Omega)}.$

4 Error Estimation for Full Discretization

Let $\mathcal{M}^n$ be a uniformly regular partition of Ω into simplices S with diameter h_S. Let $\mathbf{V}^n \subset H^1_0(\Omega)$ be the piecewise linear finite element space over $\mathcal{M}^n$ and I^n be the Lagrange interpolation operator.

The discrete problem can be written as follows. Let $U^0 \in \mathbf{V}^0$ be an approximation of u_0. Given $U^{n-1}, \Theta^{n-1} \in \mathbf{V}^{n-1}$, and having modified $\mathcal{M}^{n-1}$ and τ_{n-1} to get $\mathcal{M}^n$ and τ_n, we compute $U^n, \Theta^n \in \mathbf{V}^n$ according to $\Theta^n = I^n \beta(U^n)$ and

$$\langle \partial U^n, \varphi \rangle^n + \langle \nabla \Theta^n, \nabla \varphi \rangle = \langle f^n, \varphi \rangle^n \quad \forall \varphi \in \mathbf{V}^n. \tag{12}$$

Hereafter $\langle \cdot, \cdot \rangle^n$ is the vertex rule and $\partial U^n = (U^n - I^n U^{n-1})/\tau_n$ is the discrete time derivative. Mass lumping diagonalizes the mass matrix and together with the constitutive relation $\Theta^n = I^n \beta(U^n)$ being only enforced at the nodes, leads to a monotone problem easy to implement and solved via nonlinear SOR [5].

On combining (12) with (11), we easily arrive at

$$
\mathcal{R}(\phi) = \sum_{n=1}^{N} \int_{t^{n-1}}^{t^n} \langle R^n, \phi - \varphi \rangle - \sum_{n=1}^{N} \int_{t^{n-1}}^{t^n} \langle \nabla \Theta^n, \nabla(\phi - \varphi) \rangle
$$
$$
+ \sum_{n=1}^{N} \langle U^{n-1} - I^n U^{n-1}, \phi(\cdot, t^{n-1}) \rangle + \sum_{n=1}^{N} \int_{t^{n-1}}^{t^n} \langle \partial U^n, \phi - \phi(\cdot, t^{n-1}) \rangle
$$
$$
+ \sum_{n=1}^{N} \int_{t^{n-1}}^{t^n} \langle \nabla I^n \beta(U^n) - \nabla \beta(U^n), \nabla \phi \rangle
$$
$$
+ \sum_{n=1}^{N} \int_{t^{n-1}}^{t^n} \left(\langle R^n, \varphi \rangle - \langle R^n, \varphi \rangle^n \right) + \sum_{n=1}^{N} \int_{t^{n-1}}^{t^n} \langle f - I^n f^n, \phi \rangle
$$

for all $\varphi(\cdot, t) \in \mathbf{V}^n$, $t^{n-1} < t < t^n$; hereafter R^n stands for the interior residual $R^n = I^n f^n - \partial U^n$. The last three terms account for consistency in the constitutive relation, quadrature, and interpolation of the source term; we will refer to them collectively as *consistency terms*. The fourth term is similar to the time residual derived in §3, whereas the important new terms are the first three. They will be the focus of our next discussion.

In light of (8), we readily get

$$
\sum_{n=1}^{N} \langle U^{n-1} - I^n U^{n-1}, \phi(\cdot, t^{n-1}) \rangle \leq D \sum_{n=1}^{N} \| U^{n-1} - I^n U^{n-1} \|_{H_\omega^{-1}(\Omega)}.
$$

If we integrate by parts at the element level, and rearrange the contributions over each element edge, we obtain

$$
-\langle \nabla \Theta^n, \nabla(\phi - \varphi) \rangle = \tfrac{1}{2} \sum_{S \in \mathcal{M}^n} \langle\!\langle J_S^n, \phi - \varphi \rangle\!\rangle_S,
$$

where $\langle\!\langle \cdot, \cdot \rangle\!\rangle_S$ stands for the inner product in $L^2(\partial S)$ and $J_S^n = [\nabla \Theta^n \cdot \nu]_S$ is the jump of temperature flux across ∂S. With the convention that the unit normal ν points outward and $\nabla \Theta^n$ inside S is subtracted from $\nabla \Theta^n$ outside S, J_S^n is well defined. We next resort to approximation theory, and a nonoscillation property of ω at the element level, to select a piecewise linear φ satisfying

$$
\| \phi - \varphi \|_{L_\omega^2(\hat{S})} + h_S \| \nabla(\phi - \varphi) \|_{L_\omega^2(\hat{S})} \leq C h_S \| \nabla \phi \|_{L_\omega^2(\hat{S})},
$$

where $\hat{S}$ is the collection of all elements surrounding S; φ could be the Clément interpolant of ϕ. According to the regularity (8) of ϕ, we therefore deduce

$$
\sum_{n=1}^{N} \int_{t^{n-1}}^{t^n} \langle R^n, \phi - \varphi \rangle \leq C_1 D \sum_{n=1}^{N} \tau_n \Big(\sum_{S \in \mathcal{M}^n} h_S^2 \| R^n \|_{L_\omega^2(S)}^2 \Big)^{1/2},
$$
$$
-\sum_{n=1}^{N} \int_{t^{n-1}}^{t^n} \langle \nabla \Theta^n, \nabla(\phi - \varphi) \rangle \leq C_2 D \sum_{n=1}^{N} \tau_n \Big(\sum_{S \in \mathcal{M}^n} h_S \| J_S^n \|_{L_\omega^2(\partial S)}^2 \Big)^{1/2}.
$$

348

Collecting all previous estimates and using (9) and (10), we deduce the *a posteriori* error estimates of (3), namely,

$$\|e_u(\cdot,T)\|_{H^{-1}_{-\omega}(\Omega)} + \|e_{\beta(u)}\|_{L^2_{-\omega}(Q)} \le e^{WT}\sum_{i=0}^{4}\mathcal{E}_i + \text{consistency terms},\qquad(13)$$

where

$$\mathcal{E}_1 = C_1\sum_{n=1}^{N}\tau_n\Big(\sum_{S\in\mathcal{M}^n}h_S^2\|R^n\|_{L^2_{-\omega}(S)}^2\Big)^{1/2}\qquad \textit{interior residual,}$$

$$\mathcal{E}_2 = C_2\sum_{n=1}^{N}\tau_n\Big(\sum_{S\in\mathcal{M}^n}h_S\|J_S^n\|_{L^2_{-\omega}(\partial S)}^2\Big)^{1/2}\qquad \textit{jump residual,}$$

$$\mathcal{E}_3 = \sum_{n=1}^{N}\|U^{n-1} - I^nU^{n-1}\|_{H^{-1}_{-\omega}(\Omega)}\qquad \textit{coarsening,}$$

$$\mathcal{E}_4 = \Big(\sum_{n=1}^{N}\tau_n\|U^n - I^nU^{n-1}\|_{L^2_{-\omega}(\Omega)}^2\Big)^{1/2}\qquad \textit{time residual.}$$

5 Application

We study the evolution of a persistent corner singularity for a one-phase Stefan problem in two space dimensions. The example, still unpublished, is due to Athanasopoulos, Caffarelli, and Salsa. The key question is whether or not $\pi/2$ is the critical angle beyond which the interface immediately regularizes. To address this issue, the use of adaptive local refinements has been essential. Our simulations seem to indicate that the critical angle is actually larger than $\pi/2$.

Let $\Omega = (-0.1, 0.1)^2$, $T = 0.1$, and $g(t) = 2.1 - 10\,t$. If (ρ, ξ) denote polar coordinates, then the function

$$u^+(\rho, \xi, t) = 1 + \rho^{g(t)}\cos\big(\xi g(t)\big)\quad \text{if } 2g(t)|\xi| < \pi,\quad u^+ = 0 \quad \text{otherwise,}\qquad(14)$$

is a supersolution provided $2\,g^3(t)\rho^{g(t)-2} + \pi\,g'(t) \le 0$. Such a condition is not satisfied for all $t \le T$ and can only be enforced for $g(t) > 2$ in a shrinking domain Ω as $g(t) \downarrow 2$. It also provides some support to the above conjecture that $\xi_0 = \pi/4$, and thus the opening $\pi/2 = 2\xi_0$ could be critical. We only use u^+ to set up the Dirichlet condition on the parabolic boundary of Q in our simulations. We also use an angle shift of 0.2, and thus consider $\xi - 0.2$ in (14), to avoid grid orientation effects.

Given an error tolerance ε, a local equidistribution strategy is used to enforce the bound

$$\mathcal{E}(u_0, f, T, \Omega, \omega; h, U, \tau) \le \varepsilon.$$

349

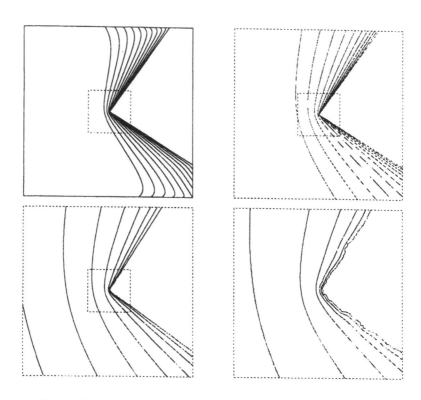

Figure 1: Zoom of interfaces with scaling factors 1, 4, 16, 64 for smallest tolerance $\varepsilon = 1$, at times $t = 0.01\,k$, $0 \leq k \leq 10$ (top) resp. $3 \leq k \leq 10$ (bottom).

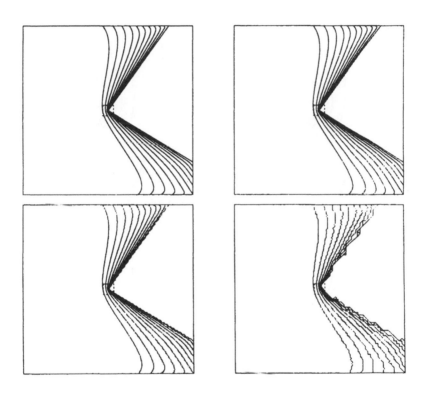

Figure 2: Interfaces for tolerances $\varepsilon = 1, 2, 4, 8$, at times $t = 0.01\,k$, $0 \le k \le 10$.

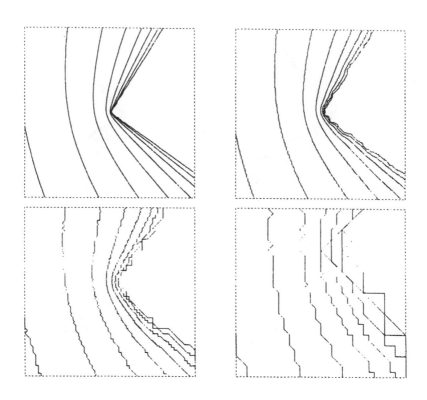

Figure 3: Zoom of interfaces with scaling factor 16 for tolerances $\varepsilon = 1, 2, 4, 8$, at times $t = 0.01\,k$, $3 \leq k \leq 10$.

352

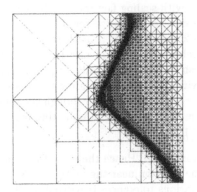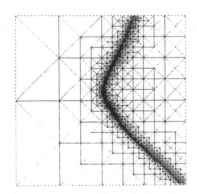

Figure 4: Mesh and zoom with scaling factor 16 for tolerance $\varepsilon = 4$, at time $t = 0.06$ (number of triangles $M = 15,273$).

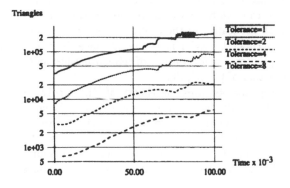

Figure 5: Number of triangles M vs. time for various tolerances; at $t = 0.05$, $M = 113,136$ for $\varepsilon = 1$, $M = 40,306$ for $\varepsilon = 2$, $M = 12,618$ for $\varepsilon = 4$, $M = 2,677$ for $\varepsilon = 8$.

353

In view of (3), this guarantees error control; we refer to [6], [7], [8], [9] for details. We run simulations for various tolerances $\varepsilon = 1, 2, 4, 8$ and fixed time-steps $\tau = 0.5, 1, 2, 5 \times 10^{-3}$.

Figure 1 displays the interfaces and zooms with scaling factors 1, 4, 16, 64 for the most accurate run with $\varepsilon = 1$. The angle seems to regularize for $0.04 \leq t \leq 0.05$, when it is already larger than $\pi/2$. Figures 2 and 3 depict a mesh study for zoom scaling factors 1 and 16, respectively. The relative location of interfaces, as well as time of regularization, are consistent with mesh refinement. Finally, Figure 4 contains a representative locally refined mesh, along with its zoom with scaling factor 16. It is clear that the interface is correctly captured by the algorithm even though the solution is very degenerate near the origin. This is due to the additional weighting factor

$$\omega(\rho) = \sqrt{\max(\rho^2, h_{\min}, \beta(U))},$$

with $h_{\min}$ given, which is essentially proportional to ρ. Hence the weight ω^{-2} in (13) compensates for the behavior $\beta(u) \approx \rho^2$ of the solution near the origin.

The number of triangles M vs. time for various tolerances is plotted in Figure 5: reducing the tolerance ε by a factor 2 entails an increase of M by a factor 4. The oscillations of M for $\varepsilon = 1$ are due to the upper limit $M_{\max}$ of triangles that the Fortran code is allowed to generate; when $M_{\max}$ is exceeded, the code automatically increases the tolerance ε. For $\varepsilon = 1$, we set $M_{\max} = 160,000$ and observe that oscillations likely occur after the corner smooths out, thereby not affecting our conclusions.

6 Conclusions

It may seem surprising at first glance that a problem governed by a parabolic partial differential equation can exhibit a stationary corner singularity. This is a hyperbolic effect consistent with the scaling at the interface. We could infer that the Stefan problem possesses a hyperbolic behavior near the interface. This is a structural property already used in [4], [5] for *a priori* design of refined meshes and a consequence of the *a posteriori* mesh design of [6], [7], [8].

The example of §5 corroborates the theory of [1], namely, that the following two factors may prevent smoothing: (a) the angle is not sufficiently large; (b) the heat fluxes of both phases vanish simultaneously. The theory predicts that failure of either (a) or (b) yields immediate smoothing, but it does not address the behavior for intermediate angles. This delicate question can only be explored numerically with a flexible and reliable adaptive technique: the critical angle appears to be larger than $\pi/2$.

The minimum value of mesh size is 1.2×10^{-5} for $\varepsilon = 1$ in §5. A comparable accuracy for a quasi-uniform mesh would require 10^8 elements to capture the singularity; adaptivity and local mesh refinement are thus essential in our study.

Acknowledgment: We would like to thank S. Salsa for bringing the example of §5 to our attention and for several illuminating discussions.

References

[1] I. ATHANASOPOULOS, L. CAFFARELLI, AND S. SALSA, *Degenerate phase transition problems of parabolic type. Smoothness of the front* (to appear).

[2] O.A. LADYZENSKAJA, V. SOLONNIKOV, AND N. URAL'CEVA, *Linear and Quasilinear Equations of Parabolic Type*, AMS, Providence, 1968.

[3] R.H. NOCHETTO, *Error estimates for multidimensional singular parabolic problems*, Japan J. Indust. Appl. Math., 4 (1987), pp. 111–138.

[4] R. H. NOCHETTO, M. PAOLINI, AND C. VERDI, *An adaptive finite element method for two-phase Stefan problems in two space dimensions. Part I: Stability and error estimates. Supplement*, Math. Comp., 57 (1991), pp. 73–108, S1–S11.

[5] ——, *An adaptive finite element method for two-phase Stefan problems in two space dimensions. Part II: Implementation and numerical experiments*, SIAM J. Sci. Stat. Comput., 12 (1991), pp. 1207–1244.

[6] R.H. NOCHETTO, A. SCHMIDT, AND C. VERDI, *A posteriori error estimation and adaptivity for degenerate parabolic problems*, Math. Comp. (to appear).

[7] ——, *Mesh and time step modification for degenerate parabolic problems*. In preparation.

[8] ——, *Adaptive algorithm and simulations for Stefan problems in two and three dimensions*. In preparation.

[9] ——, Adaptive solution of phase change problems over unstructured tetrahedral meshes, in *Grid Generation and Adaptive Algorithms*, M. Luskin *et al.* eds., IMA VMA (to appear).

R.H. Nochetto
Department of Mathematics
University of Maryland
College Park, MD 20742, USA

A. Schmidt
Institut für Angewandte Mathematik
Universität Freiburg
79104 Freiburg, Germany

C. Verdi
Dipartimento di Matematica
Università di Milano
20133 Milano, Italy